STUDIES IN PHYSICAL GEOGRAPHY

Ecology and Environmental Management

STUDIES IN PHYSICAL GEOGRAPHY
Edited by K. J. Gregory

Also published in the series

Man and Environmental Processes
Edited by K. J. GREGORY and D. E. WALLING

Geomorphological Processes
E. DERBYSHIRE, K. J. GREGORY and J. R. HAILS

Forthcoming volume

Applied Climatology
J. E. HOBBS

STUDIES IN
PHYSICAL GEOGRAPHY

Ecology and Environmental Management

A Geographical Perspective

CHRIS C. PARK
Lecturer in Geography, University of Lancaster

Routledge
Taylor & Francis Group
NEW YORK AND LONDON

First published 1980 by Westview Press

Published 2018 by Routledge
605 Third Avenue, New York, NY 10017
2 Park Square, Milton Park, Abingdon, Oxon OX14 4RN

Routledge is an imprint of the Taylor & Francis Group, an informa business

British Library Cataloguing in Publication Data

Park, Chris C
Ecology and environmental management.—
(Studies in physical geography; 3 ISSN 0142-6389).
1. Environmental protection
2. Ecology
I. Title II. Series
301.31 TD170.2

ISBN 13: 978-0-367-02216-7 (hbk)

Contents

Figures

Tables

Acknowledgements

The task of writing this book proved to be more stimulating and challenging than I could ever have anticipated at the outset. What could otherwise have become drudgery turned out quite the opposite, thanks largely to the encouragement, guidance and friendship of many people to whom I owe a great debt of thanks. Principal amongst these have been my two main mentors of recent years – Professors David Thomas and Ken Gregory. The former had the foresight to appoint me to his staff, and he has proven to be a stimulating source of inspiration and guidance over the past three years. Ken Gregory has been an endless source of advice, encouragement, friendship and inspiration for over six years; the original idea of writing this book was largely his, and he has lived with the manuscript as long as I have! Graham Sumner kindly agreed to read through some of the draft chapters of the manuscript, and they have benefitted from his comments. Dawson Publishing also proved to be a patient master to work for, and they have made the task of writing this book a very pleasant one. In a more general sense, my spirits were kept high during writing by numerous colleagues in the Department of Geography at Lampeter, and special thanks are owed to Paul and Viv Cloke for lightening the frequently tiresome task with cheerful distractions and perpetual encouragement. The Pantyfedwen Fund at St David's University College also provided help, in the form of financial assistance with the costly task of producing a readable manuscript. Any author owes his greatest debts of thanks to his close family, because it is they who share the agonies and the ecstacies of book-writing in the full. My two principal sources of inspiration, and my two main critics, have been my wife Brenda, who made many and wide-ranging sacrifices whilst I was working on the book, and Emma-Jane, who showed great restraint when repeatedly advised that her daddy could not play outside with her because 'he is working again'. Andrew Alexander arrived between Chapters 7 and 8, so he missed most of the action, but he did manage to slow down the work-rate much more successfully in the last two months than did both Brenda and Emma-Jane despite their efforts over the preceding year!

Preface

Against a background of increasing concern over the quality of the environment, the disruption of the earth's natural ecosystems, and the depletion of natural resources, there is a growing lobby of opinion that enlightened management of the environment is necessary so that future generations can enjoy a stable biosphere. During the latter stages of writing this book, a number of environmental issues have made 'headline news'. These include controversy over a proposed scheme of reclamation and drainage for agricultural use of a large proportion of wetland habitat around the Ribble Estuary, and controversy over a proposed cull of grey seals around the Orkney Islands which would decimate up to a third of the world population of grey seals. Such issues have assisted in awakening the interest of the layman in ecology and the environment.

But there have also been signs of growing interest in academic and scientific circles in environmental management. George Perkins Marsh wrote in 1864 that the subject of man's modification of the environment 'is a matter of suggestion and speculation only, not of established and positive conclusion'. Within the last decade, however, a wide range of applied books which review the basis of ecological conservation and resource management have appeared, such as those by Ehrenfeld (1970), Warren and Goldsmith (1974), Duffey, Morris, Sheail, Ward, Wells and Wells (1974), Duffey (1970, 1974), Edington and Edington (1977), Dasmann (1976), Duffey and Watt (1971), Usher (1973) and Van Dyne (1969). These tend, however, to be written from the standpoint of the ecologist. Most of the books currently available are not sufficiently comprehensive to find widespread interest in other disciplines, and so on the whole geographers have tended to remain largely unaware of ecological aspects of resource and environmental management. A number of recent biogeographical texts – such as those by Pears (1977), Cox, Hedley and Moore (1973), Tivy (1971), Robinson (1972) and Watts (1974) – have in some ways sought to rectify this situation, but these in the main concentrate on the description and evaluation of natural ecological systems, and they generally pay less attention to ecologically-orientated environmental management. A number of these also appear to be simply the biogeographer's equivalent of the books by ecologists. Simmons's *The Ecology of Natural Resources* (1974) appears to be the most positive contribution to date, yet this stresses the resource problem rather than the problem of ecologically-orientated planning.

Yet this lack of geographical books is in many ways unfortunate because many of the environmental managers of the future, and those faced with making important planning decisions, will probably be graduates from fields like geography and environmental science, who – by their very training – have studied in depth the man/environment relationship and its variations and controls in both space and time. As Reid (1969) points out, 'the science of ecology is developing very quickly, but there is a time lag between the discoveries that ecologists make, and the spread of information to people who need to be well informed.' The applied geographer must surely rank in the front line of those who need to be well informed in this context.

This book is thus designed to introduce the student with little or no previous ecological or biological training to a field which is expanding rapidly, and which is becoming increasingly important both as an academic subject in its own right, and as an applied science with clear relevance to society. It is aimed in particular at students of geography and environmental science, and it has three main aims:

(a) to introduce the basics of ecology, and the relationships between the biotic (living) and abiotic (non-living) components of the environment,
(b) to evaluate the need for planned use of the environment, and highlight some of the approaches to, and problems of, environmental management in the light of conflicting priorities and based on the principles of ecology,
(c) to produce an approach which is not solely or even primarily physical or human, but founded in one and written for the other.

If the reader finds the book half as interesting to read through as I found it stimulating to prepare, then my efforts will have been amply rewarded.

Chris C. Park,
Lampeter.
December 1978

For Brenda, Emma-Jane and Andrew

1
The Environment – Problems and Perspectives

> Geographers will surely not be happy in their work unless they can see some relation in its purpose to the current goals of human endeavour, and can relate its practice in some ways to the needs of the times (Linton, 1957).

Linton's plea for relevance of geographical endeavour anticipated the broadening of geographical horizons consequent upon the rise of interest in the past decade in the environment, its problems and its prospects. Geographers are becoming increasingly involved in all aspects of environmental management. Most of their attention has to date been focused on institutional, economic, perceptional and landscape aspects of environmental management, and ecological aspects have received less than their fair share of attention. The following chapters seek to clarify the ecological principles of relevance to environmental management, in so far as they are of concern and interest to the geographer. Before considering these principles, it is perhaps appropriate to reflect on the nature of the 'environmental crisis', and to explore some perspectives on the man/environment relationship. The nature of ecological and geographical interest in environmental problems also merits consideration in this chapter.

1.1 THE ENVIRONMENTAL CRISIS

There has been a marked growth of interest within the last decade in the quality of the environment, the disruption of the earth's natural ecosystems and the depletion of resources. Pollution, ecology and environment have been projected from the cloistered world of science into the forefront of public debate, and all aspects of man's use of his environment have been widely discussed with passionate interest. The speed and nature of environmental change (particularly man-induced change) in recent years have brought about a series of environmental problems of global magnitude – including population expansion, energy resources and utilization, the provision of food supplies, exploitation of raw materials, and environmental pollution. So drastic and quick have been the changes that Dasmann has perceived that 'the human race is like an ape with a hand-grenade. Nobody can say when he will pull the pin' (Dasmann, 1976).

There is no doubt that an emotional peak in public concern about the environment has been reached during the 1970s; 1971 was designated European Conservation Year and 1972 saw the important Stockholm conference on the environment. However, Abelson

(1971) has predicted that public interest in the environment will not be sustained for a long period of time because of reduced coverage of environmental issues by the press, a growing recognition that environmental improvement is very costly and a long-term investment (of time, money and resources), and an awakening to the fact that much information used as ammunition in the environmental debate might be erroneous. Downs (1972) has considered the changes in public interest through time in closer detail, and he has formulated a five-stage 'Issue Attention Cycle', diagnostic of most public issues including pollution and environmental quality. Stage one is a 'Pre-problem Stage', where the issue has not yet captured the public attention although certain experts and interest groups may be alarmed. Stage two is characterized by 'Alarmed Discovery and Euphoric Enthusiasm', when the issue is at the forefront of public attention, and the public respond by their enthusiasm to 'solve the problem' readily and with few long-term repercussions. In stage three comes a realization of the cost of significant progress, and an awareness that technological development is not always the best solution to problems. Stage four often witnesses a gradual decline of intense public interest, when more and more people realize how difficult and costly (to themselves as well as in general) the solution would be; and this is followed by stage five, a 'Post-problem Stage', when the issue moves into a 'prolonged limbo – a twilight realm of lesser attention or spasmodic recurrences of interest' (Downs, 1972). Within this framework Downs is of the opinion that public interest in environmental quality is now about mid-way through the cycle – suggesting that in the future attention will diminish as other issues of social, economic or ideological significance enter the limelight of public concern.

One symptom of this growing concern for environmental quality has been a recent explosion of interest in ecology and environment, although the history of interest in conservation can be traced at least as far back as the 1850s. Thoreau's widely acclaimed *Wolden, or Life in the Woods* (Thoreau, 1960) is an early expression of concern for all aspects of wild nature and wild environments; and Thoreau earnestly demanded that attention always be given first to the 'Laws of Nature' when there arises any conflict between nature and human society. Soon afterwards, in 1864, George Perkins Marsh published *Man and Nature; or Physical Geography as Modified by Human Action*, which was an examination of 'only the greater, more permanent, and more comprehensive mutations which man has produced, and is producing in earth, sea and sky; sometimes, indeed, with conscious purpose, but for the most part as unforseen though natural consequences of acts performed for narrower and more immediate ends' (Marsh, 1864, p. 19). Within the first half of the present century a number of important books have appeared, including Sherlock's (1922) *Man as a Geological Agent*, which evaluated how the soils and scenery of Britain have been transformed by human interference over a long period of time; and Tansley's (1949) *The British Isles and their Vegetation* which documented many changes in the natural countryside which can be attributed to man's influence. More recently Thomas's (1956) *Man's Role in Changing the Face of the Earth* provided a valuable compilation of man-induced landscape changes, and it stands as a landmark in the study of man/environment relationships. Within the last two decades there have appeared a number of geographical (e.g. Simmons, 1974) and ecological (e.g. Edington and Edington, 1977) books which focus on man's use and abuse of the environment.

What has recently been referred to as the 'Environmental Crisis' has commanded much attention, and it has been the subject of a number of recent books, some of which are listed in Table 1.1. Widely acclaimed 'state of the subject' reports have been offered in the *Blueprint for Survival* (Ecologist, 1972), the Massachusetts Institute of Technology *Study of Critical Environmental Problems* (MIT, 1970), and Ward and Dubos' *Only One Earth* (1972), commissioned by the United Nations Conference on the Human Environment held in Stockholm in 1972.

Table 1.1
SOME RECENT BOOKS ON THE ENVIRONMENTAL CRISIS

Author	Title	Date
Arvill	*Man and Environment – Crisis and the Strategy of choice*	1967
Berrill	*Inherit the Earth – the story of man and his changing planet*	1967
Calder	*The Environment Game*	1967
Nicholson	*The Environmental Revolution – a guide for the new masters of the world*	1969
Barr	*The Environmental Handbook – Action Guide for the U.K.*	1971
Ehrlich and Harriman	*How to be a Survivor – a plan to Save Spaceship Earth*	1971
Klotz	*Ecology Crisis – God's Creation and Man's Pollution*	1972
Sears	*Where there is life*	1972
Dasmann	*The Conservation Alternative*	1975
Harvey and Hallett	*Environment and Society – an introductory analysis*	1977

1.1a The nature of environmental problems

A compilation of all of the environmental changes which have occurred in recent years, and which have inspired identification of the 'Environmental Crisis' would be lengthy and beyond the scope of this book, so it will suffice here to identify general environmental changes and specifically ecological ones. The former would include various forms of pollution, the depletion of natural resources and an increasing reliance on energy-consuming and ecologically-damaging technologies (Clayton, 1971; Simmons and Simmons, 1973); whereas the latter would include the reduction and loss of ecological populations from toxic pesticides (see Chapter 7), the loss of habitat related to industrial, urban and agricultural expansion (see Chapter 6), and the loss of genetic variety related to mono-culture practices and habitat removal (see Chapter 4). Poelmans-Kirschen (1974) has summarized the most important issues central to the present environmental crisis:

accelerated growth of production potential within the last 25 years,

accelerated pace of scientific and technological discovery,

an exponential increase in population,

a standard of living high enough in some countries for a campaign for a better environment to be included amongst planning priorities,

a progress in those sciences (like ecology) which offer a 'macro-view' of phenomena, and

a basic and recurrent questioning of the goals of the 'consumer society'.

The *Blueprint for Survival* (Ecologist, 1972) has offered the thesis that indefinite growth cannot be sustained by finite resources, and this suggests a basic formula for environmental problems which could be of the form:

(increased population)+(increased per capita consumption)=environmental impact.

This form of relationship provides a valuable frame of reference *if* some means could be found of measuring the impact. The MIT *Study of Critical Environmental Problems* (1970) focused on the notion of an 'Ecological Demand' variable, defined as 'a summation of all man's demands on the environment, such as the extraction of resources and the return of wastes' (MIT, 1970). The most convenient measure of the demand appears to be *Gross Domestic Product* (GDP), which is the population size multiplied by a measure of the material standard of living of that population. The MIT study showed that global GDP levels are increasing on average at a rate of between 5 and 6 per cent per year (that is, they are doubling every 13·5 years on average), and so the environmental impact on the earth of finite size, with finite resources, is becoming increasingly acute.

Inevitably there are marked spatial variations in Ecological Demand, which relate to variations in population sizes and densities, and in per capita consumption. One study which has identified such spatial variations was reported by Ackerman (1959), who identified five types of *population/resource regions* on the basis of level of technology, population density and availability of resources:

TYPE 1 Technology-source areas of low population-potential/resource ratio (the 'United States Type').

TYPE 2 Technology-source areas of high population/resource ratio (the 'European Type').

TYPE 3 Technology-deficient areas of low population/resource ratio (the 'Brazil Type').

TYPE 4 Technology-deficient areas of high population/resource ratio (the 'China Type' of Ackerman, but up-dated by Zelinsky (1966) to be termed the 'Egyptian Type').

TYPE 5 Areas which are technology-deficient and possess few food-producing resources (the 'Arctic-Desert Type').

Equally inevitably there are marked temporal variations in Ecological Demand, and these stem from population changes (in size and distribution), from technological advances, and from changing social and economic values through time. The nature of environmental problems has changed radically in recent years, and Malone (1976) has rationalized the many changes into four basic themes. First, new hazards are to a far greater extent than before a side-effect of human activity. Second, the level of uncertainty in relating causes to effects is much greater than before, because of increasing awareness of the complexities of natural systems (see Chapters 2 and 3). Third, accumulating evidence is showing that the cumulative effects of small impacts over long periods of time may reach significant levels before these are detected or realized. Fourth, it is becoming increasingly apparent that the cumulative effects of some of the environmental impacts over space pose a threat of global dimensions.

There appear to be two fundamental problems concerning the present environmental crisis, and these relate to an emotionally-biased view of the crisis, and to the difficulties of distinguishing causes from effects in considering remedies.

Emotional concern

Concern over environmental problems is often emotionally biased rather than being based on balanced scientific appreciation of the problems involved. Thus Clayton perceives the environmental movement as 'a confused mixture of quasi-scientific concern and thorough-going sentimentality' (Clayton, 1971). Even environmentalists themselves stand accused of this, and Thomas (1975) has warned that the genuine concern of conservationists can at times become so over-sentimentalized that the tide of public opinion aroused may (through short-sightedness) work against the very cause it set out to support. The prophecies of ecological doom, allied with social and economic disaster, have done much to arouse emotional concern but little to foster penetrating and constructive analysis of environmental problems, and one of the simplest ways of arousing widespread concern has been to sketch possible future events by means of *scenarios* (as used, for example, by Ehrlich, 1968). The validity of the scenario approach may be questioned on several grounds, however (Hodson, 1972). One is that the approach may be counter-productive in terms of political or psychological reactions, in that people imagine that the problems are too large or too hard to try and solve, and hence solutions are not sought. Another is that many environmental problems are composed of a series of inter-related aspects, and these cannot be solved by one concerted effort but rather by a synthesis and co-ordination of individual solutions. This is extremely difficult to model effectively under scenario conditions. Finally, scenarios do not readily allow for adaption, or for the sorts of scientific discovery or technological improvement which could slow down or alter otherwise predictable sequences of events. The simplest short-term solutions to this problem of emotional pre-occupation lies in environmental education which is essential in creating environmentally-conscious people who can view environmental problems with a realistic perspective.

Distinction between cause and effect

The Biosphere (see Chapter 2) is a complex interacting system. The effects of a particular environmental modification may only become apparent after a lag time which can vary in length considerably, depending on the particular problem. Thus the effects of vegetation removal on water and sediment yield might be apparent within a season or two of removal, whereas the benefits of large-scale conservation schemes might only become apparent after a generation or two. Also the environmental effects might not occur at the same location as the initial impact. For example, water pollution emitted from a series of point sources (such as individual factories) can be carried downstream by streamflow, and an environmental impact of real concern might only occur with an accumulation of local small-scale pollution effects in a lake downstream, in the form of *eutrophication* (see Chapter 7). A number of factors produce these difficulties in isolating cause from effect, including the dynamic nature of the Biosphere and its major subsystems, the lengthy history of environmental modification in many areas, and a basic lack of research into the cumulative effects of many localized environmental impacts. Furthermore, many environmental impacts and problems are cumulative, and they can only really be appreciated at, or beyond, certain critical levels (such as the pollution problems of the Canadian Great Lakes) which might be difficult to identify until the threshold has been exceeded and large-scale solutions are required for what was previously an amalgamation of localized problems. The constraints imposed when individual disciplines examine different aspects of one environmental problem provides a further problem, which does relate in part to the problem of emotional concern. Cotgrove (1976) has expressed the fear that because of 'the bewildering disagreement over what are the most pressing environmental problems, what are their causes, and how they can best be tackled', environmental problems can often be used as levers to promote particular recipes for social action. This is really an extension of the idea that ecology and the ecological viewpoint are becoming 'social points of view' (see pages 33–4), and that conservation is an inherent part of the counter-culture which motivates certain sections of contemporary industrialized society (Dasmann, 1974).

1.1b Causes of environmental problems

The numerous symptoms of the current environmental crisis are matched by a range of viewpoints on the root causes of the environmental problems. These can be rationalized into four basic philosophies which stress man's changing religious view of his role in nature, recent population expansion, increased general affluence and economic growth, and changes in productive technology.

Religious view of man's role in nature

Some observers have sought an explanation for the exploitative attitude that has prompted much environmental modification in Western Europe and North America in the teachings of the Judeo-Christian religious tradition. This conceives of man as being superior to all other creatures, and it views everything else as being created for his use and enjoyment. White maintains that 'human ecology is deeply conditioned by beliefs about our nature and destiny – that is, by religion' (White, 1967), and Toynbee believes it possible to trace the background to the present environmental crisis back to a religious cause, 'and this cause is the rise of monotheism' (Toynbee, 1972). Toynbee cites the appropriate passage from the Book of Genesis, which reads 'Be fruitful and multiply, and replenish the earth and subdue it' (Genesis 1, 28), and he suggests that 'the injunction to *subdue*, which modern man has taken as his directive, is surely immoral, impracticable and disastrous' (Toynbee, 1972). Moncrief (1970), however, has suggested that White's simplistic 'religious' explanation is based more on fad than on fact, and he has pointed out that the environmental crisis appears to have a cultural basis in being directly related to forces of democracy, technology, urbanization, increasing individual wealth and an aggressive attitude towards nature.

Population expansion

Much of the recent debate centres around the conflicting views of whether a diminishing quality of the environment should be assigned to over-population, or to failures of the social system and/or technology. The *over-population* school favours the notion that increasing population size, and hence increasing population density, places increasingly greater demands on the planet's finite resources, and that there is thus a diminishing marginal return from population explosions such as that characteristic of the period after the Industrial Revolution in Europe. Many authorities assert that population expansion is a key element, if not *the* key element, in the man/resources/environment balance (Zelinsky, 1966; Ehrlich, 1968; Borgstrom, 1969), and accordingly the concept of an *optimum population* has evolved to stress the balance between population size, available resources and prevailing social structure at the national (Taylor, 1970) and global (Borgstrom, 1969) scales. Because of the assumed significance of population size and density, this school favours population control by a variety of means as a long-term remedy to the environmental crisis. The concept of population as the key limiting factor is not new, however. It can be traced back to the Malthusian philosophy that in an ideal situation an *equilibrium* population will be created and maintained in balance with the available resources and environmental potential (the maintenance being a function of population control by war, famine and disease). Whilst in its original format the Malthusian doctrine appears outdated, there are still a number of neo-Malthusian harbingers of doom advocating similar principles. Ehrlich (1968), for example, has prophesied the horror and social unrest which would be created by unchecked population growth in the long term, based on scenario simulations of possible international disasters which might arise if population continues to expand at recent rates. A modern counterpart to the concept of equilibrium or optimum population can be found in the concept of *carrying capacity* (see p. 97).

Increased affluence and economic growth

Considerable attention has also been directed towards the thesis that increased general affluence (that is increased material aspects of per capita consumption of goods and resources), related to economic growth, are key elements in the recent crisis (Figure 1.1).

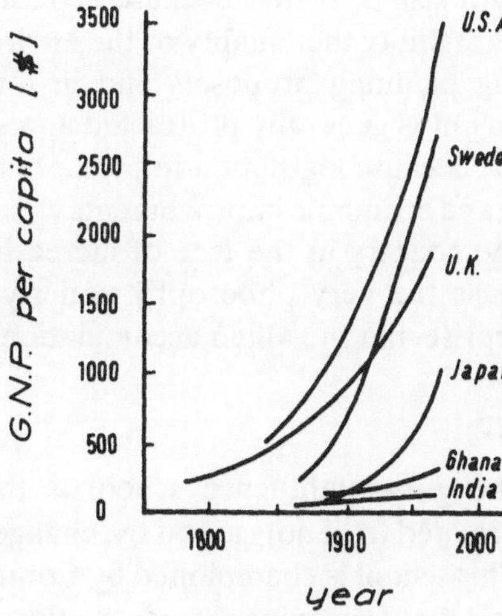

FIG. 1.1 INCREASING GENERAL AFFLUENCE DURING THE TWENTIETH CENTURY.
The graph shows variations in the rate of change of per capita Gross National Product, measured in American dollars, between various countries for the period 1800 to 1970. (After Kuznets, 1971.)

Schumacher (1974) has stressed the basic incompatibility of simultaneously maintaining both economic growth and the quality of the environment, pointing out that one must inevitably assume priority over the other in all formal planning strategies. The same philosophy underlies the *Limits to Growth* study by Meadows et al. (1972), which advocates a new perspective for considering the nature, magnitude and direction of future growth of economies at local, national and global scales. The *Blueprint for Survival* (Ecologist, 1972) ranks increased per capita consumption alongside increased population size as a cause of environmental problems, and it lists six ways in which individual governments encourage continued expansion in the tendency for economic growth to create the need for more economic growth. These are:

(1) the introduction of technological devices (to replace natural controls on the environment),

(2) industrial growth promotes population growth (there is thus a need for capital investment to maintain employment and productivity),

(3) economic growth is required to reduce unemployment,

(4) business enterprises tend to be self-perpetuating, in requiring surpluses for further investment for continued growth,

(5) a government's success is related to its ability to increase the 'standard of living' (that is, per capita gross national product),

(6) there is a general need to maintain and even expand the economy to ensure a healthy state of the stock market.

Packard (1960) maintains, furthermore, that industry contributes substantially to the growth process by planned obsolescence of the function, quality and desirability of consumer products, which means that demand for consumer goods is always increasing. The basic thesis of the growth school is that because economic growth is required for political, social and economic stability the 'quality of the environment' normally assumes lower priority in formulating planning proposals and in long-term planning because deterioration of the environment is generally protracted and socially less oblique than a deterioration in the economy; because legislation to promote the 'quality of the environment' at the expense of social and economic improvements would be politically unwise and socially unpopular in any one country in the face of increasing international economic inter-dependence; and because the very philosophy underlying western society is the prospect of freedom of enterprise and the allied accumulation of wealth.

Type of productive technology

In close association with the growth/affluence school is the philosophy that recent environmental problems are related to, if not caused by, changes in the type of productive technology in recent years. This school is championed by Commoner who, in *The Closing Circle*, elaborates on his thesis that 'sweeping transformations of productive technology since World War Two . . ., productive technologies with intense impacts on the environment, have displaced less destructive ones' (Commoner, 1972), and that the environmental crisis is thus the inevitable result of a counter-ecological pattern of productive growth. The environmental problems are held to be associated with productive processes concentrating on synthetic and non-biodegradeable materials (such as the increased use of plastics and detergents) rather than with radical changes in the overall output of goods or in material aspects of consumption (Figure 1.2). Commoner's conclusions have attracted some debate from those who highlight individual cases where the combined effects of population increase and greater affluence appear to be sufficient to account for most of the increases in the presence of specific pollutants. Holdren and Ehrlich (1972), for example,

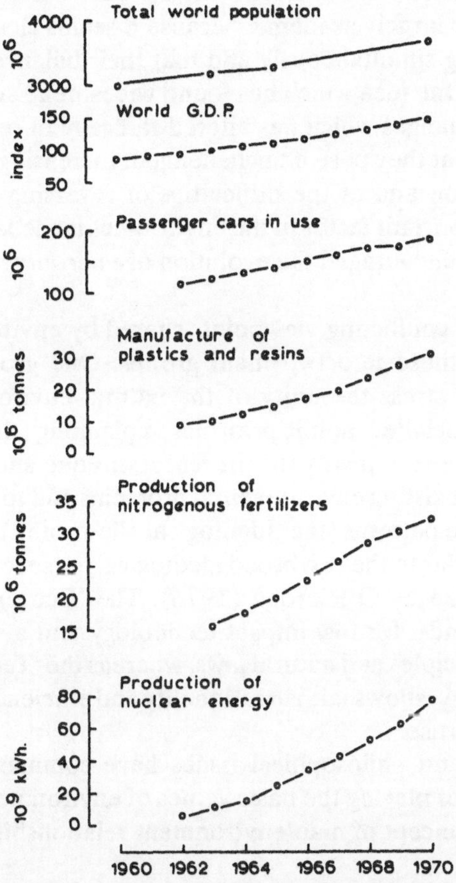

FIG. 1.2 CHANGING MATERIAL ASPECTS OF CONSUMPTION BETWEEN 1960 AND 1970.
Changes in a range of indicators of material wealth, based on data from the United Nations (1972).

maintain that this is the case for nitric oxide emitted from cars. The environmental case against modern productive technology is more clearly demonstrated in recent concern for 'Alternative Technology' (Dickson, 1974; Clarke, 1973) which questions the very nature of technology rather than just the use to which it is being put. The alternative technology movement itself stems in part from counter-culture activities in the United States during the 1960s (Dasmann, 1974), although the value and genuineness of the concept of low-impact technology has at times been questioned. Although experiments have been undertaken with a variety of technological developments, the various attempts share a series of common goals which include using minimum amounts of non-renewable resources, bringing about minimum environmental interference, the elimination of alienation amongst individual workers, and the establishment of regional and sub-regional units of self-sufficiency as opposed to state units of inter-dependence.

There are clearly a number of divergent (and at times conflicting) theories of what are the basic causal factors underlying the recent environmental problems of pollution and resource depletion. It seems fair to conclude that individual aspects of recent environmental changes can be associated with different factors. The search for one over-riding root cause would appear to be largely academic because it seems clear that each of the major causes could be operating simultaneously and that their balance will vary from place to place and through time. One idea which has found widespread support is that it is not the scale of environmental changes which has altered radically in recent decades, but rather our perception of the threat they pose if unchallenged. Increasing awareness of the ease of environmental degradation and of the difficulties of reversing trends and processes of degradation is thus an important factor in the environmental debate (Clayton, 1971). This increasing awareness has encouraged the evolution of *environmentalism* as a social movement.

There are a number of conflicting viewpoints shared by environmentalists, and Kruse (1974) has rationalized these into two main groups. One group favours a 'Limits to Growth' policy and they stress the unity of the natural environment and the need to integrate ecological and social/economic priorities in planning (the 'Utopian Viewpoint'); whereas the other group seek to justify the present status quo and they show great faith in human ingenuity to solve existing environmental problems and to integrate environmental values into existing culture patterns (the 'Ideological Viewpoint'). Although not identical, these two are broadly similar to the two broad ideological perspectives on the environment which have been identified by O'Riordan (1976). The 'Ecocentric Perspective' shows concern for values and ends, for low impact technology and a search for stability via a respect for ecological principles and natural laws, whereas the 'Technocentric Perspective' concentrates on technology, shows faith in rationality and efficiency and stresses the role of professionalism and expertise.

These environmental and philosophical issues have commanded much attention in recent years. One means of placing the basic values of environmentalism into perspective is with reference to the concept of man/environment relationships.

1.2 PERSPECTIVES ON THE MAN-ENVIRONMENT RELATIONSHIP

The relationship between man and his environment has traditionally been a focal point of geographical enquiry. In this context 'environment' refers to the sum total of conditions which surround man at a given point in space and time. Thus the environment of early man was dominated by natural factors such as vegetation, soils, climate and other animals, whereas modern man has increasingly surrounded himself with an environment of his own design and construction, which is geared to the provision of food, shelter and access. There are a number of ways in which the man/environment relationship can be perceived.

An *environmentally deterministic perspective* stresses that man is subordinate to, and hence largely controlled by, the natural environment. This approach was typified by geographical thinking early in the present century and was epitomized in the writings of Semple, who maintained that 'man is a product of the earth's surface' (Semple, 1911), and Huntingdon (1915), who considered in depth the influence of weather and climate on the course of human history and the evolution of society. Grossman (1977) has pointed out that the deterministic school before about 1920 was greatly influenced by three aspects of the contemporary scientific climate. These were a widespread adoption of Darwinian principles of adaption and survival, the use of deductive approaches to scientific enquiry, and a dogmatic acceptance of Newtonian concepts of cause-effect relationships.

A *teleological perspective*, in contrast, stresses that man is superior to nature, and he thus has the potential for complete control over all aspects of nature. This school is inevitably less popular than the deterministic school, but some recent observers have identified the roots of the present environmental crisis in the widespread and unquestioned adoption of the Judeo-Christian religious tradition (see pages 24–8).

More recently an *economic perspective* has emerged, and this stresses man's mastery over increasingly larger parts of the environment. This school favours continued economic and industrial expansion, and it sees in scientific research and industrial development the opportunity for increasing control over individual parts of the environment. Zelinsky (1966) has pointed out that to many people 'economic determinism' offers a more sound explanation for understanding the size and distribution of human populations than does the direct impact of the physical environment, but he also identifies two fallacious assumptions inherent in this line of thinking. These are that the number of inhabitants in a region is positively correlated with the level of economic development and activity; and that universal economic principles govern the interactions of people, resources and society.

Finally, the *ecological perspective* stresses that man is an integral part of nature, and thus his relationship with the natural environment should be symbiotic and not exploitative nor suppressive. Ward and Dubos (1972) maintain that man must accept responsibility for the stewardship of the earth, and apply appropriate environmental management strategies based on ecological principles. Although it receives widespread support at the present time this viewpoint is not completely new. Grossman (1977) has traced its roots to the 1920s when alternative viewpoints to the determinists were radically influenced by Barrow's concept of geography as human ecology, stressing mutual interaction between man and the environment.

Clearly each of these perspectives is of value in certain circumstances, and since environmental problems are generally multi-dimensional each perspective has a contribution to make to the man/environment debate. Whilst there is undoubtedly a current emphasis on ecological perspectives, it must not be overlooked that a complex series of factors condition human population densities and distributions. These include the direct impact of the physical environment, the workings of the economy, the cultural configuration of society, the impacts of social and physical disasters and the impact of specific social and political decisions (Zelinsky, 1966).

1.2a Symbiosis between man and environment

The relationship between man and his environment is basically two-way, in that man is affected by the environment and yet he also has the capacity to modify the environment.

The effects of the environment on man fall into three broad areas. First there are *biophysical limitations* on man. Man is subject, in a biological sense, to a restricted range of environmental conditions in that the human body can only function properly and survive within relatively fixed ranges of environmental conditions such as heat, light and space for movement unassisted (Barnett, 1968). With assistance (such as the synthetic environments offered to astronauts) these environmental ranges can be extended. Closely related to these biophysical constraints, there is increasing recognition of likely correlations between medical disorders and environmental factors, highlighted in well-defined spatial patterns in the incidence of non-infectious diseases. Cole (1971) has commented on the correlations between some forms of cancer and cardio-vascular complaints and the geochemistry of soils and waters; and Davies and Pinsent (1975) have evaluated the relationships between mineral levels (especially of trace elements) in soils and plants and diseases such as stomach cancer and multiple sclerosis. A second area is *perception of*, and *behavioural control* by, the environment. Many aspects of social and economic behaviour and stability are conditioned by people's perception of, and responses to, environmental hazards and stimuli such as flood, droughts and disasters (Burton and Kates, 1978); and in most, if not all, cases our perception is less than perfect. Perception of natural environmental factors can be radically influenced by frequency of occurrence and magnitude of event, and in this way environmental factors can significantly affect behavioural patterns and responses. Finally the *availability of resources* affects human activity and behaviour. Environmental factors can readily influence the availability and economic viability of basic ecological resources (such as food), and of other resources such as minerals, clean air and water supplies, suitable land resources and a high quality environment, which are of benefit to, if not essential for, the survival and well-being of man.

The effects of man on the environment are complex and numerous, and they can be either direct or indirect. *Direct impacts* are generally premeditated and planned; the impact is commonly felt relatively soon after the environmental modification; the effects are often long-term but reversible; and study of the impact is generally made simple because before-and-after studies can be integrated into the overall planning of the modification. Such changes include land-use changes, various constructional and excavational activities, the direct ecological impacts of agricultural practices, and the direct effects of weather modification programmes. *Indirect impacts*, on the other hand, are generally unplanned and often socially if not economically undesirable; the effects are often not manifested for some time after the original impact (the *time lag* will depend on such factors as the sensitivity of the system to change, the existence of threshold conditions, and interaction between different side-effects of the initial impact). Moreover, the effects are often long-term and cumulative, and in many cases irreversible due to complexity or a lengthy history of change; and the effects are commonly difficult to identify and evaluate, and so prediction of environmental impacts is often difficult, if not impossible. Examples of indirect impacts include the introduction of DDT and other toxic elements into the environment by spraying field crops, and the subsequent accumulation of the material into food chains over wide areas and affecting non-target species throughout food webs (see Chapter 3); the triggering off of long-term and perhaps long-range climatic modifications by particulate and gaseous atmospheric pollution; and the indirect local climatic effects associated with changed land surface configuration or material composition (see Chapter 2).

Man's capacity to modify the environment is clearly a dynamic one, varying in both time and space. The changes through time are related to changing sizes and densities of populations, changing cultural characteristics and associated changes in demand for (and perception of) environmental resources; coupled with changing environmental stability and resiliance conditioned by natural environmental change (such as that related to climatic change) and by the history of previous man-induced environmental modification.

Population change through time is clearly a key variable, and Dawson and Thomas (1975) identify three major phases in world population growth:

PHASE 1 A period of slowly increasing population, coincident with a stagnant or at best a slowly developing technology, running up to the Neolithic Revolution,

PHASE 2 A phase of moderate but at times irregular population growth after the Neolithic Revolution, in which population expansion proceeded at a rate permitted by man's knowledge of the earth's resources and by technological innovation, followed by

PHASE 3 A period of rapid population growth beginning with the Industrial Revolution in which total population numbers increased rapidly, and through time the rate of increase itself accelerated. Within this phase there has been a rapid, yet relatively contained expansion in world population in the period 1750 to about 1900, followed by a sharp acceleration in the rate of increase since about 1900 (the 'Population Explosion').

The manner in which man views himself in the time and space domain is a second key variable. Malone (1976) has outlined three major historical developments which have radically influenced man's views of himself, and of his role in nature. First there was the 'Age of Transcendence', a long period of about 1000 years during which man's interests spread beyond the material world for the first time and he began to explore the realm of human value preferences and ethical frameworks. The 'Copernican Revolution' during the sixteenth century brought about a crumbling of the concept of the anthropocentric (man-centred) universe and revealed man as simply the inhabitant of a small planet circulating about stars within a vast universe. Finally, the 'Darwinian Revolution' of the 1850s revealed that man is a highly complex product of a process that proceeded from molecules to cells to organisms to colonies and to tribes and nations over a period of several million years.

The spatial variations in man's capacity to modify the environment relate principally to the global distribution of population numbers and densities, and to spatial variations in cultural and technological practices (in part conditioned by population density and demand for resources); as well as to spatial variations in environmental characteristics and resource availability. Certain heavily populated areas, such as most of Western Europe and North America, have suffered massive environmental transformation at the hands of man through industrial development, urbanization, agriculture and land-use changes. Even remote areas such as the Arctic and Antarctic have undergone man-induced environmental change because, despite low population densities and hence relatively small pressures on available resources, these areas are often the most sensitive to interference, and they often suffer from the indirect environmental effects transmitted from distant more densely populated areas (such as the global effects of large-scale atmospheric pollution, and the mobility of soluble toxic chemicals via the water cycle (see Chapter 2)).

This inherently two-way interaction between man and environment has been conceived by Haggett (1972) as a *feedback relationship* (Figure 1.3) in which changes in population density place direct impacts on environmental quality; these in turn affect the population density by the feedback loops. Logically one could add a *threshold condition* to this conceptual model, whereby if the change in population density was less than the critical threshold level (related to available resources and prevailing social and economic factors)

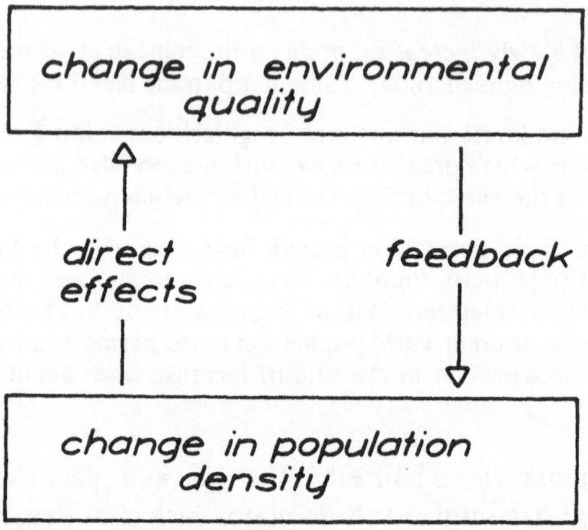

FIG. 1.3 FEEDBACK RELATIONSHIP BETWEEN POPULATION DENSITY AND ENVIRONMENTAL QUALITY. (After Haggett, 1972).

there would be no real change in environmental quality and the population could continue to expand unchecked by environmental factors; whereas if the change exceeded the threshold there would be an identifiable change in environmental quality. The former case could occur in regions of very low initial population densities such as many areas of Europe in pre-Neolithic times and many under-developed nations today. Haggett (1972) points out two complications behind the concept of this feedback relationship. One is that the feedback might only occur after a time lag of variable length, and the other is that the feedback might not occur in the same location.

One basic problem underlying interpretation of the man/environment relationship is that man's use of and reaction to the environment are very much influenced by prevailing social, cultural and economic climates. Hutchinson has observed that 'we not only need to understand the ecology of our environment much better than we do now, but in many cases we fail to make use of the knowledge we have because of prejudice, because of our values, and, most of all, because of our failure to look beyond today and tomorrow' (Hutchinson, 1970).

Inevitably most environmental problems and issues fall within the realms of both ecology and geography, and it is important to consider the approach of each discipline to environmental problems.

1.3 ECOLOGICAL AND GEOGRAPHICAL INTEREST IN ENVIRONMENTAL PROBLEMS

1.3a The development and contribution of ecology

The term '*oecology*' was first coined by Haeckel in 1869. He defined the subject as 'the entire science of the relations of the organisms to the surrounding exterior world, to which relations we can count in the broader sense all the conditions of existence. These are partly of organic, partly of inorganic nature' (quoted in Friederichs, 1958). The initial development of the subject was relatively slow, in part because of a suppression of interest during the late nineteenth century in aspects of the natural world other than those concerned directly with verifying or refuting Darwin's concept of the evolution of species (Fraser Darling, 1963). By 1904, however, Tansley was able to proclaim that 'ecology may now be considered almost a fashionable study' (Tansley, 1904). Its early popularity was in part related to the fact that in addition to being a respectable scientific discipline in its own right, it appealed also to the large numbers of laymen and enthusiasts who fostered the development of natural history during Victorian times. Indeed, in 1927 Elton announced that 'ecology is scientific natural history' (Elton, 1927); although the debate over the distinction between the two over-lapping areas of study has been continued to the present day (Disney, 1968, 1970). To most ecologists, however, their subject is much more precise and tangible than 'natural history'. Some view the subject in the same light as Friederichs, who viewed ecology as 'the science of the living beings as members of the whole of nature' (Friederichs, 1958), thus largely ignoring the effects of the living beings on the environment and of the environment on the living beings. But increasingly ecologists are tending to adopt the terms of reference suggested by Fraser Darling who defined ecology as 'the science of organisms in relation to their total environment, and the inter-relationships of organisms inter-specifically and between themselves' (Fraser Darling, 1963). The distinction between studying ecological relationships between species, and ecological relationships within the environment, is in part reflected in the basic division of ecology into *synecology* which deals with the study of plant communities in relation to their habitats, and *autecology* which is the study of the ecological relations of individual species.

Although ecology is by definition concerned with the plant and animal species of the natural environment (and more recently the man-made environment), most attention has been directed towards the study of plants and vegetation, in part because of the evolution of ecology from field studies of natural history in general and of vegetation in particular, and in part because static plants are simpler to study than highly mobile animals. This has fostered a lengthy and continuing debate on whether ecology is really an extension of botany. Major has attempted to resolve this problem by adopting a scale threshold between the two fields, maintaining that 'in general the areal scales of study in plant ecology and sociology, and plant geography, differ by a factor of at least 10 to 100' (Major, 1958).

The traditional view of ecology as the science of living things in relation to their environment has helped to place it in a valuable strategic position from which it can make important contributions to environmental management. Indeed, to many people ecology is almost synonymous with conservation and the environment. This is in part because, as

Friederichs points out, 'ecology has ceased to be a synthesised branch of biology . . . it has become a *viewpoint*' (Friederichs, 1958) which stresses the unity and balance of the environment. Because of its holistic nature Maddox has argued that 'ecology is a state of the mind' (Maddox, 1972); Dasmann (1974) has traced the rise of ecology as a unifying social movement; and Ash (1972) has advised the planner to give deep respect to the sentiments of the ecologist. Because of its increasing potential contribution to environmental problems, and hence to social and economic stability and balance, Simon and Gerondet have suggested that 'the need is for ecology to become not so much an abstract scientific discipline as the touchstone of human purpose' (Simon and Gerondet, 1970).

The roots of this recent change in emphasis, and diversification of viewpoints in ecology, can be traced in the evolution of the subject, which Pearsall (1964) has sub-divided into three phases prior to 1964. *Phase One* occurred before World War One, and it was dominated by the search for the fundamental units of vegetation and by the development of pioneer methods of vegetation classification. The surveying and mapping of vegetation distributions in Britain can be seen as a diffusion process starting with the semi-natural vegetation of northern Britain and progressing gradually southwards. A contemporary definition of ecology offered by Tansley epitomizes the ecological perspective of this period. Tansley saw ecology as 'those [plant and animal] relationships which depend directly upon differences of habitat amongst plants' (Tansley, 1904), although he was later (Tansley 1947) to date the development of modern ecology to this period, during which a more general scientific renaissance was to give birth to atomic physics, modern physical chemistry, biochemistry and genetics. *Phase Two* occurred between the two world wars, and it was to witness the development of more rigorous and scientific methodologies (such as the development of laboratory analysis, the beginnings of pollen analysis, the introduction of statistical analysis, and the advent of experimental ecology). This period also witnesses a major re-orientation of interest away from simple mapping of vegetation towards analysis of the relationships between habitat and vegetation characteristics, and a diversification of interest from natural terrestrial vegetation to freshwater and marine studies, increasing interest in animal ecology and in the dynamics of ecological stability and interaction (see Chapter 3). *Phase Three* is post-World War Two, and is characterized by a major conceptual re-orientation towards ecosystem studies, an increased experimental basis and a marked rise of interest in ecological processes. The study of ecological fluctuations through time, over a variety of time scales, has also commanded much attention during this period.

The major change in the present phase has been an increasing awareness of the role of ecology in relation to conservation, and this applied aspect of ecology became crystallized during the 1960s after the publication of Rachel Carson's *Silent Spring* in 1964. Despite subsequent criticism, this book played a large part in awakening the non-scientific world to the value of an ecological perspective on environmental problems. Lowe and Worboys (1976) have pointed out that the findings, concepts and approach of ecology have proved a rich source of ideas and analogies in the construction of environmentally-sound social planning programmes. In particular the concepts of niche theory, the ecosystem, energy flow, material recycling, tolerance limits and successional change (see Chapter 3) have been employed in a range of different planning contexts. Echoing this development, a number of ecological texts have recently been devoted specifically to the problems of conservation and resource management (for example Duffey and Watt, 1971; Ehrenfeld, 1970; Usher, 1973; Van Dyne, 1969).

These major changes in the focal points of ecology have been superimposed on a dramatic increase in the number and productivity of practising ecologists. A convenient means of identifying changes in the 'strength' of a subject through time is to examine the 'citation structure' of that subject, by an analysis of the dates of publication of books, monographs and scientific papers referred to in text books and 'state-of-the-art' reviews (Stoddart, 1967). Figure 1.4 shows the citation structure for *Introduction to Ecology* by Colinvaux (1973), *Biogeography* by Tivy (1971) and *Principles of Biogeography* by Watts (1974), totalling in all some 1763 references between 1850 and 1970. A dramatic rise in

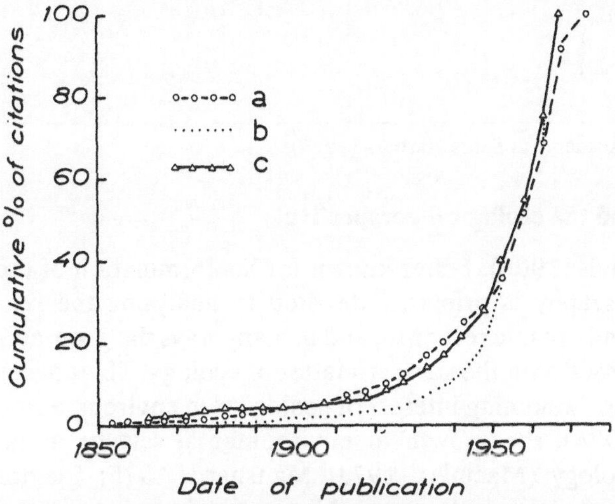

FIG. 1.4 CITATION STRUCTURE OF SOME BOOKS ON ECOLOGY AND BIOGEOGRAPHY.
Based on analysis of citations in Colinvaux's (1973) *Introduction to Ecology* (a), Tivy's (1971) *Biogeography* (b) and Watts' (1974) *Principles of Biogeography* (c). Data are plotted for five year intervals, and a total of 1763 articles and books are included (326 in (a), 555 in (b) and 882 in (c)).

the citations after about 1950 is very clear, and this highlights the recent explosion of interest and productivity in ecology and biogeography. Stoddart (1967) has observed that a two-phase growth of citations (1890 to about 1945; and post 1950) is common in science as a whole, and he related this to the steadily growing body of scientific knowledge coupled with a more tightly knit body of 'current literature' (which is cited mostly because it is current rather than because it has any real long-term significance or merit). The development of the subject has in part been assisted by the expansion of outlets for publication, most particularly in the form of scientific and general journals. The *Environmental Periodicals Bibliography* (1976), for example, lists some 270 environmentally-orientated journals from 16 countries, although American dominance in the field is clear from Table 1.2. Recent increased interest in ecology has consolidated membership of societies such as the British Ecological Society (founded in 1913), the Botanical Society of the British Isles (1836), the Linnean Society (1788) and the Zoological Society of London (1826), and it has also encouraged a diversification of career opportunities for ecologists into local and regional planning authorities, the Nature Conservancy Council and various other planning and conservation fields (Kelcey, 1976).

Table 1.2
COUNTRY OF PUBLICATION OF MAJOR ENVIRONMENTAL JOURNALS (1976)

Country of publication		Number of journals
United States of America		165
England		48
Canada		16
Australia		10
Netherlands		10
Switzerland		5
West Germany		3
Sweden	South Africa	2
New Zealand	Japan	2
India		2
Uganda	Denmark	1
New Mexico	France	1

Source: International Academy at Santa Barbara (1976)

1.3b Geography and the ecological perspective

William Morris Davis (1906), better known for his formulation of the Cycle of Erosion, observed that geography is primarily devoted to analysing the relationships between inorganic control and organic response, and in many ways the geographer has traditionally been closely concerned with the subject-matter of ecology. This concern has been crystallized of late with the awakening interest of ecologists in environmental problems (Warren and Goldsmith, 1974), the growth of environmental science as one bridge between geography and ecology (Macinko, 1973; Manshard, 1975; Clayton, 1976), and the renewed interest of the geographer in problems of resource evaluation and management (O'Riordan, 1971; Simmons, 1974; Coppock and Sewell, 1975). Applied geography seeks to evaluate the complex inter-relationships between man and environment and between nature and society (Hagerstrand, 1976), and the recent revolution of relevance within geography was anticipated a generation ago by Linton (1957).

More recently, Mead has identified two basic paradoxes in geographical enquiry, 'in that at a time of high general interest in geography, adjustments in attitude are demanded which must necessarily test the faith and resilience of its followers, and yet at the same time an increasing unity of approach to the subject is accompanied by an unparallelled diversification of interest within it' (Mead, 1969). Thus the internal structure of geography is in a constant state of flux, but divergent trends can be rationalized by dividing the realms of geographical activity into three broad modes of analysis (Figure 1.5). *Spatial Analysis* focuses on spatial variations in properties or series of properties of the environment. *Ecological Analysis* focuses on the inter-relationships between man/environment variables. *Regional Complex Analysis* employs a combination of the spatial and ecological approaches (Haggett, 1972).

An ecological perspective on geography has been demanded by Hewitt and Hare who concluded that 'the main needs of environmental geography today are a deeper fusion of ideas and results from the life sciences' (Hewitt and Hare, 1973); and Eyre has even suggested that 'a more ecological approach enhances the prestige of geographers within the academic world' (Eyre, 1964a). The ecological perspective is of value in several ways. Simmons (1966) has argued that the ecological approach aids assessment of cultural factors in land use, and it can be used to show how man manipulates ecological systems for his own ends. Whilst ecology is not the sole answer in environmental conservation, it does aid in integrating the social and biological sciences. Eyre (1964b), on the other hand, claims a conceptual value for ecology and he has demonstrated that by adopting an ecological viewpoint geographers can stand to rid themselves of 'naive determinism' and misrepresentation in both human and physical geography. Alternatively, Stoddart (1965) concluded that the main contribution of ecology to geography is in providing a methodology in the form of the ecosystem concept.

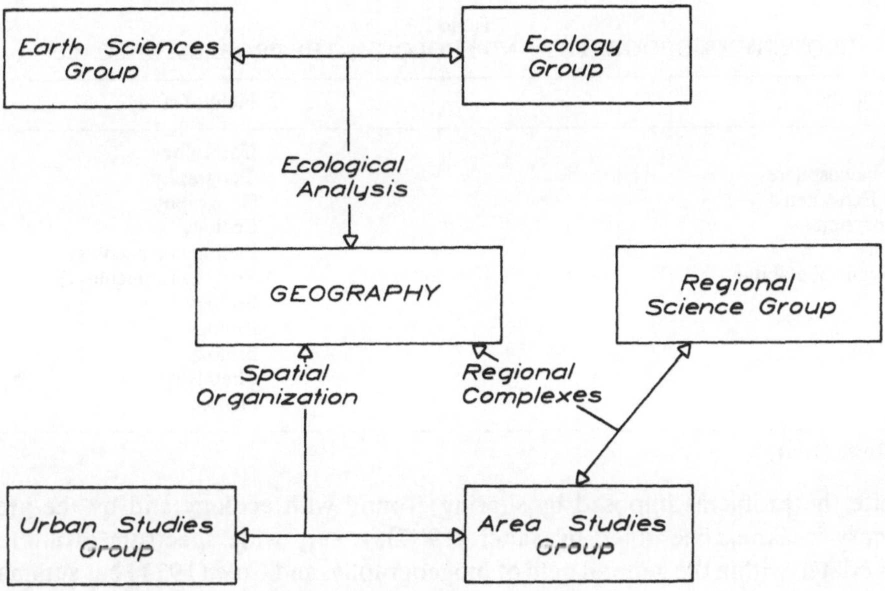

FIG. 1.5 MODES OF ANALYSIS IN GEOGRAPHICAL STUDIES.
(Generalized from Haggett, 1972).

Ecological studies within geography normally fall within the realms of biogeography which fulfils a vital role in bridging physical and human geography (Edwards, 1964), although there is still debate on whether or not biogeography is simply a part of ecology (Friederichs, 1958). Rowe (1961) has suggested that one convenient basis for distinguishing between ecology and biogeography is in terms of the scale of study adopted (Table 1.3) and more recently Fosberg (1976) has pointed out that ecology is generally concerned with the study of local ecosystems, whereas regional, continental or global ecosystems are better treated as a part of geography. Irrespective of scale there are a number of ways in which the study of ecosystems can be regarded as essentially biogeographical. If particular emphasis is placed on the spatial organization of and within ecosystems, then this is a traditionally geographical area of study. In many ecosystems it is the functional linkages between the living and non-living parts of the system which are of central concern, rather than simply the relationships between individual species of plants and animals (Cole, 1971), and here the geographer has a contribution to play, at least in Davis's (1906) opinion. Also, man as an integral part of the ecosystem is a common focal point of many ecosystem studies (Morgan and Moss, 1965). Simmons (1974) has suggested that attention within biogeography should centre on man's creation of new ecosystems, thus emphasizing a synthesis of ecology, cultural anthropology and economics. Modern biogeography is not without critics, however, and Sauer (1977) has outlined several of its major weaknesses in that few biogeographers are concerned in detail with biotic distributions, and most are still absorbed in field exploration at the expense of general theory. Furthermore, separate scholarly traditions within biogeography have not yet been integrated into a coherent body of thought, and Sauer argues that many schools of biogeography still accept deductive explanations without investigating the presumed processes involved.

Table 1.3
ECOSYSTEMS, GEOGRAPHY AND ECOLOGY – THE PROBLEM OF SCALE

Object of study	Field of study
Universe	Cosmology
Biosphere/Ecosphere	Geography
Regional Ecosystems	Geography
Local Ecosystems	Ecology
Ecosystems	Ecology (synecology)
Single Organism/Habitat	Ecology (autecology)
Organism	Biology
Organ	Biology
Cell	Biology
Molecule	Chemistry
Atom	Physics

Source: Rowe (1961)

Despite the problems imposed by sharing ground with ecology and by the areas of disciplinary weakness identified by Sauer (1977), a very wide spectrum of interests is accommodated within the general field of biogeography, and Cole (1971) has summarized the main studies into six major areas. First there are numerous *studies of plant and animal communities*, at a range of spatial scales, which concentrate on the physiology, composition and distribution of species. These are complemented by a range of studies in *historical plant and animal ecology* at varying time scales, which seek to reconstruct how the species distributions have changed through time. Thirdly, *soil studies* form an important area of biogeography, particularly those which concentrate on the development of soils and on the influence of cultural practices on soil development. A more recent development is the field of *geo-botanical and bio-geochemical studies* which aim to evaluate the influence of bedrock geology, soils, deposits and the like on plant distributions and on plant mineral content. Fifthly there are a large and growing number of *applied studies*, normally undertaken as integrated studies of vegetation, soils, geomorphology and land use, which are becoming increasingly important as the potential contribution of the geographer to resource evaluation and management is being realized. Finally, a major recent development is within *conservation and resource studies*, where the biogeographer benefits from a knowledge of both geography and ecology.

The two final themes are of particular relevance both to the recent increase in interest in biogeography and to this book. Taylor has stressed the value of biogeography to resource management, because 'the biogeographical approach automatically integrates the . . . physical assessment [of natural resources] with an . . . evaluation of socio-economic usage, in both potential and actual terms' (Taylor, 1974).

Although the citation structures for the two biogeography texts shown in Figure 1.4 are very similar to that of the *Introduction to Ecology*, this is more because they refer to essentially similar ecological research papers than because the history of the two allied fields is the same. Although Schimper published *Plant Geography on a Physiological Basis* in 1903, and Hardy's *The Geography of Flowering Plants* appeared in 1920, biogeography is widely judged to date from 1936 when Newbigin's *Plant and Animal Geography* first appeared. This provided a summary of the global distribution of plants and animals in relation to present day and former environmental controls, and it offered a valuable foundation-stone for biogeography to build on. The following decades, however, witnessed little real development of the subject, in part because of limited research opportunities and the shortage of suitable trained personnel (Cole, 1971) although the subject was advanced by plant ecologists concerned with the identification of plant communities and their controlling factors (for example 'Plant Geography' books were published by Raunkier, 1934; Klages, 1942; Cain, 1944; Wulff, 1950; Croizat, 1952; Polunin, 1960; Gleason and Cronquist, 1964; Good, 1964). Biogeography was further stimulated at the end of World War Two when shortages of food in Europe and the need to use previously undeveloped land in Africa and Australia produced the need to develop techniques for assessing land potential and for evaluating environmental factors (Cole, 1971). The major changes have been more recent, however, and they stem from the enhanced research opportunities offered from technological developments such as atomic absorption techniques for chemical analysis, and scanning electron microscopes for high magnification analysis, coupled with the development of sophisticated computer hardware and techniques, improved dating methods of organic and inorganic materials (Dimbleby, 1975), and developments in remote sensing for soil and vegetation studies (Paludan, 1976; Hesjedal, 1976). Biogeographical study of animal populations and their environmental relationships remains as yet relatively under-developed (Allee and Schmidt, 1951; Bennett, 1960), probably reflecting the allied lack of interest within ecology for animal populations; although plants and vegetation continue to attract widespread attention from biogeographers (Eyre, 1968; Kellman, 1975; Randall, 1978). The crystallization of biogeography as a coherent subject in its own right is evidenced in the number of biogeographical text books to appear in recent years (such as those by Dansereau, 1957; Watts, 1974; Cox, Hedley and Moore, 1973; Tivy, 1971; Seddon, 1971; Robinson, 1972; Pears, 1977).

Thus, through its concern with the ecological characteristics of the environment and its general concern with man/environment relationships, geography appears to be centrally placed to contribute effectively to the solution of many environmental problems. The main contributions of geography to conservation (Clayton, 1971) include the global viewpoint which geography commonly adopts, the capacity to reconcile ecological, social and economic claims on the environment and to identify the main substance of the claims in a rational manner, and the ability to encompass the views of a wide range of individual disciplines.

1.4 RATIONALE BEHIND THIS BOOK

It is clear that the most appropriate perspective to adopt on environmental problems is neither strictly ecological nor strictly geographical. Both have very valuable but distinctly different contributions to offer in environmental management, and the geographer must surely benefit from the cross-fertilization of ideas between the two disciplines. The treatment throughout this book is thus geared towards all geographers, not simply those interested in applied biogeography.

Both ecologists and geographers conventionally look at environmental problems on two scales – the global and the local. It seems fitting, therefore, to review the global-scale environmental system (the *Biosphere*) in Chapter 2, before focusing attention on more local problems and processes in ensuing chapters. The fundamental unit of ecological interest, and a convenient framework for elucidating ecological processes and implementing ecological resource management strategies, is the *Ecosystem* (Chapter 3). Ecosystems are dynamic systems, however, which vary in both time (Chapter 4) and space (Chapter 5); and consideration of both dimensions of change is essential in seeking to understand ecosystems, and in aiming to formulate realistic and effective ecosystem management and conservation schemes. Such schemes are required because of the pressures placed on ecosystem stability and on ecological resources by various forms of human activity (Chapter 6) and by pollution (Chapter 7). Ecological management (of species, habitats and entire ecosystems) is but one component of environmental management, however, and there are pressing needs to ensure that ecological aspects of land-use planning and development are evaluated realistically, and that appropriate consideration is given to the problems of conserving ecological resources at regional, national and international levels (Chapter 8).

2
The Biosphere – Global Ecological Systems Analysis

If geography is to play a full role in the study of environmental problems it is important that greater stress be given to the biosphere, which constitutes a vital resource base for man, and which has been altered more extensively by human activities than most other elements of the environment (Hill, 1975a).

Although there are a number of ways in which the unity of the physical environment can be conceived (Moss, 1974; Walton, 1968) it is convenient to regard the earth, its local environments and its living components as a system. The *systems approach* provides a unifying conceptual framework with a major emphasis on the flow of matter and energy within the system (Russwurm, 1974), and by evaluating the components and processes of the system the approach provides a good tool for preventing adverse environmental effects and for suggesting methods for restoring degraded environments (Perkins, 1975). This chapter is devoted to a general consideration of the biosphere, the global scale environmental system whose stability lies at the heart of the environmental crisis and which has, as Hill (1975a) points out, been extensively altered by human activities.

2.1 SYSTEMS AND THE ENVIRONMENT

Chorley and Kennedy define a system as 'a structured set of objects and/or attributes. These consist of components or variables that exhibit discernible relationships with one another, and operate together as a complex whole, according to some observed pattern' (Chorley and Kennedy, 1971). Systems can be classified on a *functional* basis into three broad types – *isolated systems*, which have boundaries closed to the import and export of mass and energy; *closed systems*, whose boundaries prevent the import and export of mass but not energy; and *open systems*, with open boundaries which allow free exchange of mass and energy with the system's surroundings.

By this classification the earth and its environments are clearly an open system, with energy exchange largely in the form of incoming and reflected radiation and mass exchange through meteorite impacts, rocket launches, and similar localized phenomena. However since the earth has a finite size and it has finite supplies of most resources it is perhaps more appropriate to view it as a closed system which is self-perpetuating (although in a strictly physical sense it is not closed to the solar system and its wider environment). This is the foundation of the 'Space-Ship Earth' concept first proposed by the United States Ambassador Adlai Stevenson in a speech on 9 July 1965 before the United Nations

Economic and Social Council in Geneva. He referred to the earth as 'a little space-ship in which we all travel together, dependent on its vulnerable supplies of air and soil', and the concept has since become the basis for much conservation thinking.

There are two major components in the earth's system – the living material (the organic or biotic component) and the non-living environment (the inorganic or abiotic component). Some authorities prefer to term the environment in which the organisms live and with which they interact the 'Ecosphere' (Cole, 1958), and to term the suite of organisms themselves the 'Biosphere' (Hutchinson, 1970). Dasmann maintains, however, that such a distinction 'involves separating the inseparable' (Dasmann, 1976). Accordingly he defines the biosphere as 'the thin layer of soil, rock, water and air that surrounds the planet earth, along with the living organisms for which it provides support, and which modify it in directions that either enhance or lessen its life-supporting capacity' (Dasmann, 1976). The biosphere can thus be envisaged as the basic global system, with biotic and abiotic components. This is valuable because, as Chorley and Kennedy (1971) point out, the real world is continuous, and sub-divisions of it merely represent subjective and often artificial portions of reality.

The real world can be visualized as a set of inter-locking systems, at various scales and of varying complexity, forming a *nested hierarchy* so that they are all to some extent mutually inter-dependent. The basis on which these sub-systems are identified and delimited is important. A simple sub-division could be based on three broad types of global sub-system:

natural systems; which are totally or wholly unaffected by human interference, such as the last vestiges of the Tropical Rain Forest,

modified systems; which have been affected to some extent by human interference, such as many local vegetation systems in lightly populated areas, and

control systems; where human interference by accident or by design provides a major regulator, such as most agricultural systems.

This form of sub-division reveals something about the significance of human impacts on the sub-system, but little about how the sub-systems operate, or how they inter-link to form the larger biosphere. Chorley and Kennedy (1971) suggest a *structural* classification of systems, which is more revealing and appropriate. They identify four basic structural types of system:

TYPE 1 *Morphological systems*; comprising a network of structural relationships between the form parts of systems,

TYPE 2 *Cascading systems*; defined by the pathway followed by throughputs of mass and energy,

TYPE 3 *Process-response systems*; a fusion of morphological and cascading systems so that the form components of the former react with and relate to the throughput or process components of the latter, and

TYPE 4 *Control systems*; which are process-response systems where the key components are controlled by some intelligence (generally man).

Within this framework most earth systems are process-response systems if they are relatively natural and control systems if they are modified by human interference.

2.2 THE BIOSPHERE AS A SYSTEM

The limits to the biosphere can be defined in the form of an envelope around the earth's surface, wherein plant and animal life can exist without protective devices. The envelope comprises a shallow layer of air, water, soil and rock generally less than about 30 km in thickness. The upper limits are conditioned by lack of oxygen, shortage of moisture, increasing cold and decreasing pressure with increasing altitude in the atmosphere, and although bacterial activity has been detected by NASA at over 15 km, most life forms favour lower levels in the atmosphere with more favourable environmental conditions. The lower limits, whether in terms of soil depth or ocean depth, are conditioned by lack of oxygen, lack of light and increasing pressure with increasing depth, and although forms of organic life have been detected at depths of over 9 km in deep ocean trenches, over land the biosphere generally extends downwards as deep as the deepest tree roots or the depth of burrowing of earthworms and similar creatures (Robinson, 1971). Within the biosphere so defined can be identified the organic components (plants, animals including man, and micro-organisms) and the inorganic environmental components. Both of these are intimately related to a series of large-scale cyclic mechanisms which involve the transfer of energy, water, chemical elements and sediment throughout the biosphere. The relationships are two-way in that the cyclic mechanisms influence the organic and inorganic components, and these in turn influence the cyclic mechanisms. In the natural state, therefore, the biosphere can attain a state of equilibrium which is self-sustaining and ecologically efficient. With environmental modification this state of equilibrium can be disrupted either partially or wholly, and this can lead to widespread and recurrent environmental and ecological damage. Because many parts of the biosphere operate as complex inter-dependent process-response systems, initial environmental impacts may become magnified through negative feedback effects (Chorley and Kennedy, 1971) and if critical environmental or ecological thresholds are exceeded then large-scale disequilibrium can rapidly follow. This state of balance of the biosphere is thus the key to environmental management, and its disruption is the cause of the recent environmental crisis (Whitakker and Likens, 1975).

Because of the delicate yet critical nature of this state of balance Dasmann (1973) has stressed the urgency of protecting natural areas and communities and wild species because of the need to increase understanding of the functions of the biosphere and of the operation of balance mechanisms. Perkins has further demanded that 'the development of large and diverse areas of unspoiled land should be prohibited until we are capable of understanding the capacity of that ecosystem to withstand modification' (Perkins, 1975). The series of large-scale cyclic mechanisms which provide the corner-stone of biosphere stability and which thus effectively control the capacity of different ecosystems to withstand modification can be considered in general terms in this chapter. The fundamental biotic processes which influence these cycles, which control ecosystem stability and dynamics, and which are of central concern to environmental management, are evaluated in Chapter 3.

2.2a The energy system

The natural system

There are three basic sources of energy in the biosphere – these are gravity, internal forces within the earth, and solar radiation. The latter is the most important source because of its abundance, it can be converted by green plants (by the process of *photosynthesis*) into energy forms which can be used by plants, animals and man, and it provides the main source of power for other major systems – in particular the water cycle and atmospheric circulation (Clapham, 1973). The sun provides a relatively steady and continuous emission of energy through thermo-nuclear fusion and this energy travels through the solar system in the form of electro-magnetic radiation. This solar radiation is composed of energy of a range of wavelengths (Figure 2.1a) from the short wavelengths of gamma rays (below about 0·00002 μ or microns), through X rays (0·0002–0·002 μ) and ultraviolet (0·002–0·3 μ) to the visible range of the spectrum (0·4 to about 0·7 μ), and into the long wavelengths of the infrared (about 0·8–200 μ) and radio waves (over 200 μ). Because of their physiological structure plants and animals normally respond only to the visible light parts of the spectrum, which generally accounts for approximately a quarter of the total radiation from the sun. The key to understanding the energy system lies in the two fundamental laws of thermodynamics, which state:

LAW 1 In a system of constant mass, energy cannot be created or destroyed, but it can be transformed. Thus it can be converted from electrical energy, for example, into mechanical energy.

LAW 2 Energy is dissipated as heat energy (or thermal waste) as work is done. Work is done when one form of energy is transformed into another.

The amount of solar energy which reaches the earth's outer atmosphere (the *solar constant*) is relatively fixed through time, and it is equal to about 1·94 langleys per minute (± 5 per cent). Not all of this energy reaches the earth's surface and becomes available for green plants (Figure 2.1b). In the order of 51 per cent of the solar radiation eventually reaches the earth's surface either directly (26 per cent) or by earthward re-radiation in a diffuse form after being absorbed either by particulate matter within the atmosphere (11 per cent) or by the cloud cover (14 per cent). The other 49 per cent of the solar radiation is effectively 'lost' to the earth's surface either by being absorbed into the atmosphere by collision between particles (14 per cent) or by being reflected upwards from the atmosphere (7 per cent), the upper surface of clouds (24 per cent) or from the earth's surface itself (4 per cent). These losses from reflection are of vital importance to the energy balance of the earth's surface and thus to the plant and animal world; and the amount of reflection (*albedo*) varies quite considerably between different materials (Table 2.1). By altering the earth surface material such as by urban development it is thus possible to modify the energy balance over the localized area.

Although in general it averages 51 per cent of the solar constant, the actual amount of solar radiation delivered to the earth's surface (the *insolation*) varies from season to season and from year to year with small random changes in the solar output and changes in transient atmospheric conditions (Sellers, 1965). The insolation also varies spatially

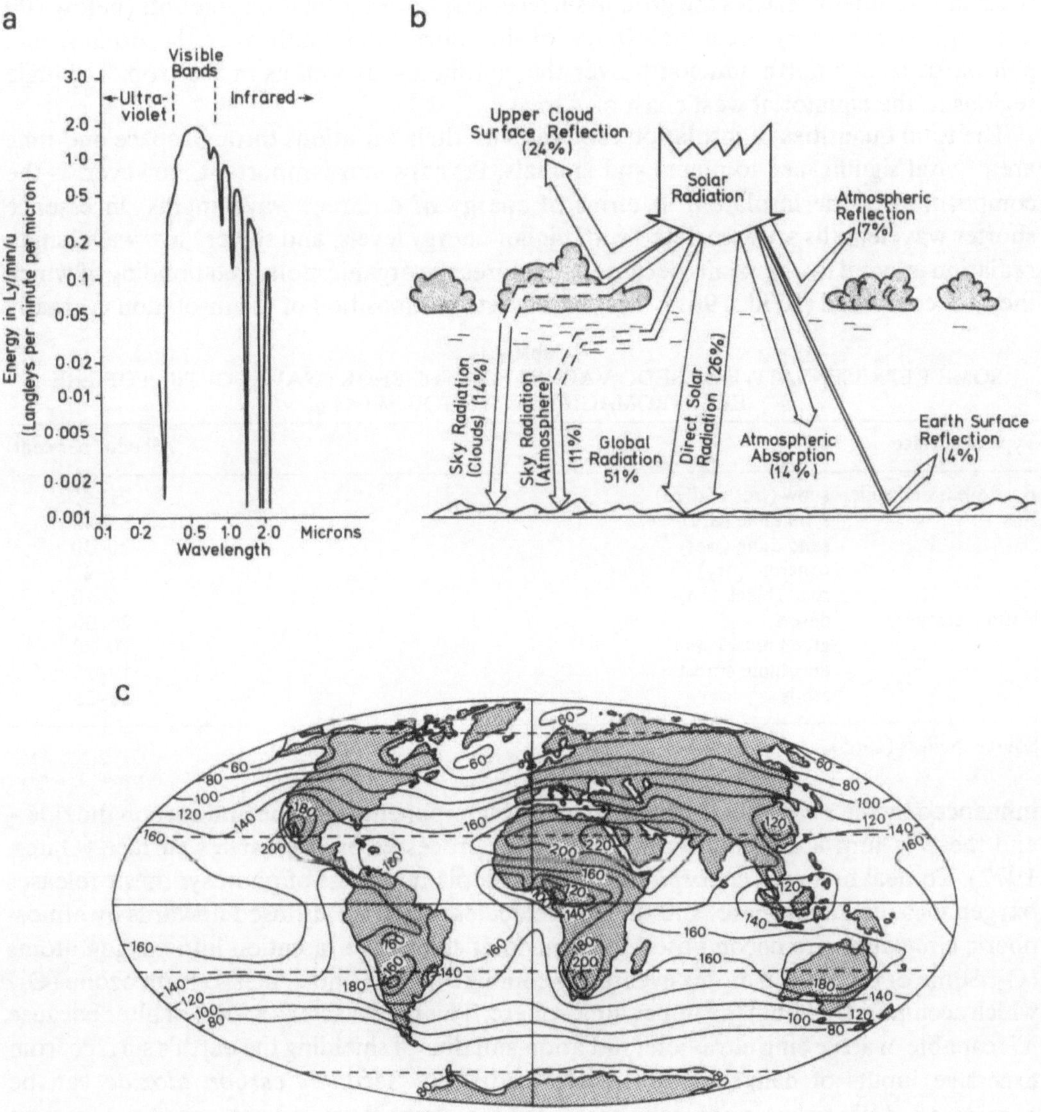

FIG. 2.1 THE ENERGY SYSTEM.
(a) Distribution of incoming solar radiation by wavelength composition (microns). (After Rumney, 1970.)
(b) The distribution of solar radiation between various components of the Energy System. (After Rumney, 1970.)
(c) Spatial variations in incoming solar radiation across the earth's surface; units are kilolangleys/yr. (After Sellers, 1965.)

(Figure 2.1c). The highest insolation amounts (in excess of 200 kilolangleys per year) are received over the world's major deserts such as the Sahara and Libyan deserts of North Africa and the Arabian deserts of Saudi Arabia, where up to 80 per cent of the solar constant eventually reaches the ground surface. The lowest insolation amounts (below 100 kilolangleys per year) occur polewards of 40° north and south over the oceans, and polewards of 50° north and south over the continents, as well as in the tropical jungle regions in the equatorial west coast of Africa.

The total quantities of insolation received and their variations through space and time are of vital significance to plants and animals. Perhaps more important, however, is the composition of the insolation in terms of energy of different wavelengths. In essence shorter wavelengths are associated with higher energy levels, and so very low wavelength radiation is lethal to organisms because it fractures the organic molecular bonding of which they are composed (Reid, 1969). The wavelength composition of the insolation is greatly

Table 2.1
SOME REPRESENTATIVE ALBEDO VALUES FOR THE SHORTWAVE PORTION OF THE ELECTROMAGNETIC SPECTRUM (<4 μ)

Type of surface		Albedo (per cent)
Bare areas and soils –	snow (fresh fallen)	71–95
	sand dune (dry)	35–45
	sand dune (wet)	20–30
	concrete (dry)	17–27
	road (black top)	5–10
Natural surfaces	desert	25–30
	green meadows	10–20
	deciduous forest	10–20
	crops	15–25

Source: Sellers (1965)

influenced by the presence of two atmospheric components – ozone and carbon dioxide – and these in turn are influenced by ecological processes on the earth's surface (Gates, 1972). To deal first with *atmospheric ozone*, the plant process of photosynthesis releases oxygen into the atmosphere, and the O_2 molecules which are diffused upwards by atmospheric circulation are decomposed by ultraviolet shortwave radiation into oxygen atoms (O). Some of the oxygen atoms eventually combine with O_2 molecules to form ozone (O_3) which accumulates within the upper atmosphere. This ozone acts as a form of filter because it is capable of absorbing ultraviolet radiation and thus of shielding the earth's surface from excessive inputs of dangerous ultraviolet radiation. Secondly *carbon dioxide* can be considered. Although it is naturally present in the atmosphere in small quantities, carbon dioxide is also a basic component in the process of photosynthesis, being taken in by the leaves of vegetation and used within the plants in the assimilation of carbon to carbohydrates. Carbon dioxide strongly absorbs radiation in certain longwave infrared portions of the spectrum; the absorption bands of CO_2 in the earth's atmosphere capture some of the infrared re-radiated from the earth's surface and they re-radiate it in two directions. Some is re-radiated back into space via the atmosphere but some is re-radiated back to the earth's surface. This latter helps to maintain a relatively warm flow of radiation between the atmosphere and the ground, and it thus helps to stabilize temporal variations in the earth's surface temperature, which is most important for plants and animals. Clouds and water vapour in the sky also serve to absorb and emit infrared radiation and contribute to this heating effect (the 'greenhouse effect').

Man-induced changes

There has been a growing concern in recent years that human modification of the environment can have drastic and lasting impacts on all phases of the energy system, and detailed accounts are offered by Landsberg (1970), Gates (1972), Singer (1975) and the Massachusetts Institute of Technology (1971). The two main modifications are to induce changes in the *quantity* and the *quality* of the energy available at the earth's surface. Although affecting the solar constant would appear at the moment to be beyond human capability it is possible to influence the quantity of energy reaching the earth's surface by affecting one or more of the reflection components in Figure 2.1b. Thus reflection from cloud surfaces can be affected by altering the sizes, locations or types of cloud surface involved (for example by cloud seeding or by large-scale weather modification schemes), or by changing the cloud's albedo by altering the composition of the cloud cover (for example by increasing dust particles in the atmosphere by the emission of particulate pollution). The energy lost to the atmosphere itself by collision between particles and that reflected from the particles can be affected by altering the overall quantity and spatial variations and size composition of dust particles (for example by induced wind erosion or by particulate pollution). The earth surface reflection component can be influenced by altering the nature of the ground surface and hence changing the albedo (Table 2.1) (for example by land-use changes and urban development). Landsberg (1970) has suggested that the average change in global radiation caused by urbanization is in the order of a 15 to 20 per cent reduction. The human influence on energy balances can also be more direct, such as by the production of energy in industrial operations (Table 2.2). Available data suggests that urban and industrial energy may exceed the annual net production by photosynthesis by a factor of more than a hundred.

The spectral composition of the radiation reaching the earth's surface can also be dramatically affected by human modifications, particularly within the ecologically critical ultraviolet and infrared portions of the spectrum. The shortwave ultraviolet radiation which is extremely harmful to living matter is in part absorbed by stratospheric ozone layers. If these are affected this can trigger off dramatic and possible irreversible ecological changes. There is accumulating evidence that the ozone content of the atmosphere can be affected by particulate and chemical pollution, and also mounting evidence that sustained increases in global pollution throughout the 1960s have been mirrored by a general, if rather variable, increase in atmospheric total ozone levels (Massachusetts Institute of Technology, 1971).

As pointed out earlier, the infrared radiation plays an important role in heating the earth's atmosphere by the 'greenhouse effect'. Much of the energy in certain of the long wavelengths which is reflected from the earth's surface, the atmosphere and clouds is absorbed by oxygen, ozone, carbon dioxide and water vapour in the atmosphere. Some of this is then retained as heat and the rest is re-radiated at longer wavelengths, and so the balance between the factors which absorb the infrared radiation is vital to the heat balance of the earth's surface and thus to ecological stability and survival. The carbon dioxide content of the atmosphere can be markedly influenced by the burning of fossil fuels, by transport technologies and by atmospheric particulate pollution, and there is accumulating evidence (Machta, 1971) that within and since the 1960s the general carbon dioxide levels in the atmosphere have increased substantially. Some authorities have associated this increasing carbon dioxide level with an intensification of the 'greenhouse effect', and with a corresponding increase in global temperatures in recent years.

Ecological significance

The state of balance of the energy system can be of critical importance to ecological stability and to general environmental equilibrium. Ecologically the energy system is important because solar radiation powers photosynthesis (see Chapter 3); the spectral composition of the radiation is important to the survival and stability of living organisms; the energy system can influence local and large-scale climate which is physiologically important; and vegetation growth is limited by the concentrations of atmospheric carbon dioxide. The energy system is important also because energy provides the driving force for the water cycle; energy supply affects vegetation type, density and stability and these can in turn affect flows of sediment and chemicals at the local scale; and energy affects climate which can, in turn, radically affect the other major sub-systems of the biosphere.

Table 2.2
SOURCES AND MAGNITUDES OF CERTAIN ENERGY RELEASES

Description of energy release	Magnitude (W/m^2)
Net solar radiation at earth's surface (dependent on latitude)	approx. 100
Urban industrial area estimate	12
1970 energy production distributed evenly over all continents	0·054
1970 energy production distributed evenly over the whole globe	0·0157
Annual net photosynthetic production of the continental vegetation cover	0·13
Global average heat flux from the earth's interior	0·062

Source: Massachusetts Institute of Technology (1971)

2.2b The water cycle

The natural system

Like energy, moisture can be present in the environment in more than one form but, unlike energy, water follows an essentially cyclic pathway through the environment. Through most of the cycle the moisture is in the form of water, although this can be transformed into water vapour by evaporation or by plant and/or animal transpiration. Moisture can also be transformed from water vapour by condensation around nuclei to produce clouds and perhaps eventually rain. The cycle is powered by both solar energy and gravity, and its major components are illustrated in Figure 2.2a.

The large-scale global cycle involves a relatively simple mechanism of the evaporation of water from the oceans to form atmospheric water vapour; the transfer of this across the oceans and over land masses by atmospheric circulation and local scale air/land interaction; the release of this water vapour as precipitation over land and sea; and the eventual transfer of this water which has fallen on land back to the sea by a range of hydrological processes of which the most important (in terms of volumes involved and speeds of movement) is surface runoff in streams.

The fate of the precipitation which falls over land areas is determined largely by the ground surface conditions. Some precipitation falls directly into stream channels, lakes and other water bodies (*direct precipitation*) and this can be carried directly back to the oceans. That which falls over vegetated ground interacts with the vegetation in a number of

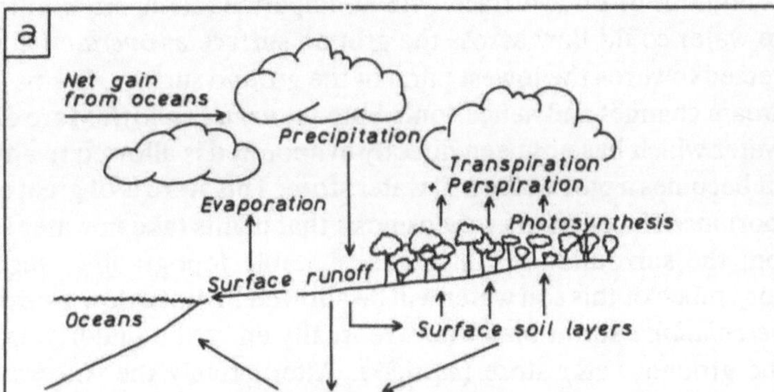

FIG. 2.2 THE WATER CYCLE.
(a) General components of the water cycle over land. (After Clapham, 1973.)
(b) Approximate values of the flow and storage components in the global water cycle. (Based on data from Ward, 1977.)

ways. A certain variable proportion will land on the leaves and branches of the vegetation (*interception*), and this can be evaporated back into the atmosphere as water vapour. If the weight of collected water exceeds the supporting capacity of the leaves, they will bend downwards and the water will be allowed to continue on its groundward path. Water can also be dislodged from the leaves by the direct impact of other water droplets. Another variable amount will trickle down the stems of the vegetation (*stemflow*) and although this reduces the speed of travel the bulk of the water will eventually reach the ground surface below the vegetation. The remainder of the precipitation will fall unimpeded by the vegetation and reach the ground surface beneath (*throughfall*). This fraction is clearly related to the type, density and structure of the vegetation because, for example, relatively more precipitation will fall unimpeded through scattered woodland than through dense forest. The total amount reaching the ground surface from these three basic routes (direct precipitation + stemflow + throughfall) can have several fates. On a sloping ground with saturated soil conditions or on a surface with an impermeable horizon in the underlying soil much of the water could flow across the ground surface as *overland flow*, its path of travel being directed towards the lowest parts of the ground surface. In this way the water can reach the stream channel and hence contribute to *surface runoff*. More often than not, however, that water which has not been directly evaporated is allowed to enter the soil by *infiltration* and it becomes a part of the soil water store. This store is of great ecological and agricultural importance since it is via root osmosis that plants take in water (and nutrients in solution) from the surrounding soil. Under suitable topographic, vegetational and climatic conditions much of this soil water will be allowed to move downwards through the soil horizons (*percolation*) and it may thus eventually enter the underlying bedrock and contribute to the ground water store (*aquifer*). Alternatively the soil water will move downslope either through the pore spaces of the soil structure (*interflow*) or by flowing along impermeable bands within the soil horizons (*throughflow*). The throughflow, interflow, percolation water and ground water can be released into streams via springs, natural soil pipes and channel bed and bank seepage, and although the first three pathways can rapidly produce streamflow, the ground water store normally has a very long residence time and water can be effectively locked away from the remainder of the cycle for thousands of years (although it is still cycling slowly).

Although these hydrological processes maintain the cyclic flow of water through the environment, in the order of 95 per cent of the total moisture in the biosphere is not available to the cycle except over extremely long periods of geological time because it is 'stored' by being chemically bound in the rocks of the earth's crust. Of the remainder some 97 per cent is stored in the oceans, 2 per cent is stored in polar ice caps and as permanent glaciers, and the rest is available as fresh water, atmospheric moisture, ground water, soil water and water in storage in plants and animals. Figure 2.2b shows approximate values for the storage and movement of water in these components of the cycle, and the significance of the ocean areas is clearly evident. Although the interaction between the cycle and vegetation provides a small part of the overall cycle in terms of storage and movement of water, this sector is ecologically very important to ensure ecological diversity and stability; it affects the sediment and chemical cycles through landscape stability and erosion; and it affects the energy system because water is the basic limiting factor without which plants and animals would not exist.

The overall balance of the cycle and the equilibrium between its components both vary considerably in time and space. Variations through time arise because of the effects of climatic change over a variety of time scales (Lamb, 1972) and because of the changing impacts of man on the water cycle through land-use changes, urban and industrial development and water resource development schemes. Variations through space arise because of the spatial patterns of climatic factors and the related atmospheric circulation patterns (Table 2.3), and these are complimented at the local scale by minor variations in topography, natural vegetation and land use which can all modify the balance of the cycle components.

Table 2.3
AVERAGE WATER BALANCE OF LAND AREAS, BY CONTINENT

Continent	Precipitation (cm/yr)	Evaporation (cm/yr)	Runoff (cm/yr)
Africa	69	43	26
Asia	60	31	29
Australia	47	42	5
Europe	64	39	25
North America	66	32	34
South America	163	70	93
Total land areas	73	42	31

Source: Budyko (1971)

Man-induced changes

There are a number of means by which human interference can modify the balance, speed of flow and component parts of the natural water cycle (Overman, 1968; Walton, 1970; Gregory and Walling, 1973). Several aspects of the atmospheric components of the cycle can be affected by accident or design. The spatial movement of the water vapour across oceans and land areas, for example, could perhaps be influenced by changes in circulation patterns brought about by weather modification programmes, or by man-made energy emissions and induced changes in the atmospheric ozone and water vapour levels. The precipitation component is related in part to the formation of cloud droplets and rain drops, and changes in atmospheric dust can be brought about by particulate pollution and accelerated wind erosion. The precipitation component can be directly controlled under suitable conditions by cloud seeding with silver iodide crystals. The evaporation component is related to the surface composition of the local environment and the local energy budget, and so changes in the surface conditions (such as through urban development or reservoir construction) can induce changes in evaporation losses at the local scale. The stream runoff component can be directly affected by increasing the effective lengths or densities of the channel network (such as by the installation of drainage ditches, field drains and artificial channels) or by decreasing them (such as by direct building operations and channel removal or infilling). The size and hydraulic efficiency of the channels themselves can be altered by channel clearance, re-alignment and diversion and by direct engineering works. The infiltration component can be influenced by changing the surface conditions of the materials, such as by stabilizing stream channel beds and banks with concrete or revetments; by local land-use changes from continuous natural vegetation cover to agricultural cover; by irrigation schemes and field drainage installations. Ground water can be affected by pumping and abstraction works which will induce down-draw of the water-table, and by increased inputs related to reservoir development and upstream drainage basin changes. The vegetation components can be directly influenced by vegetation removal, afforestation, cropping and grazing; and since the vegetation cover influences the local water cycle and – via its effects on landscape stability and erosion – it also influences the sediment and chemicals cycles (which in turn affect the local water cycle) this is one area in which man's impacts can trigger off a series of wide-ranging and perhaps long-term environmental repercussions.

The most radical hydrological effects are naturally those related to the simultaneous and perhaps inadvertent modification of more than one of the components of the water cycle. These can be very important locally, and through replication of effects over wide areas the cumulative effects on the regional and global scales can be quite marked. To these often accidental effects must be added the planned use of the water cycle through water resource management, which can have radical effects on the spatial and temporal availability of water of suitable quality and quantity to be of value to the biosphere. The data listed in Table 2.4 show that in 1965 a global total of 2300 cubic kilometres of water was used in irrigation schemes, although only 600 of these were returned directly to the cycle – the rest were used in plant growth, evaporation and soil infiltration, and so the cycle was 'short-circuited'.

Table 2.4
ESTIMATED GLOBAL WATER NEEDS AND WATER RESOURCES

(km^3/yr)	1965		2000	
	withdrawal	return	withdrawal	return
Needs:				
Municipal water supply	98	56	950	760
Irrigation	2300	600	4250	400
Industry	200	160	3000	2400
Energy	250	235	4500	4230
	2848	1051	12700	7790
Resources:				
Precipitation on continents	108000			
Runoff	37000			
Evapotranspiration	71000			
Evapotranspiration from agricultural areas	3560			

Source: Lvovich (1969)

Ecological significance

These ways in which the water cycle can be affected and modified by human actions are important because the water cycle is of vital importance to ecological stability and diversity in a number of ways. Water is a basic requirement for plant and animal growth, because it provides the hydrogen required for the formation of carbohydrates (which are essential for tissue development). Water is also the medium through which chemical elements and nutrients are available to plants via root osmosis, and the transfer of chemical elements within plants takes place in solution. Furthermore, water is required for many chemical reactions which are essential for physiological growth and fitness within plants and animals. Water also has indirect effects on the local ecology because it influences the local micro-climate and local landform stability. The water cycle is also intimately linked with the other major systems of the biosphere. As the main means by which chemical transfers can take place the water cycle is of fundamental significance to the chemical cycles. Water also affects landforming processes through detachment, entrainment and transport of sediment, and in this way the water cycle is closely related to both the sediment and chemical cycles. The cycle is also related to the energy system via its close dependence on local land-surface characteristics.

2.2c The chemical elements cycles

The natural system

The availability and transfer of a wide range of chemical elements in the environment are key factors influencing the stability and continued survival of the biosphere, and the sub-system for these is composed of a series of inter-locking and delicately balanced *biogeochemical* cycles. The most important elements which cycle in this manner are carbon, oxygen, hydrogen, nitrogen, phosphorus and sulphur. The flow of these elements through the environment is cyclic because of the finite supply and the relatively constant form of the individual elements. Individual cycles can be identified for each of the major and minor chemical elements (Table 2.5) but they all have in common a basic two-part structure involving an *inorganic component* (comprising the abiotic parts of the environment, with sedimentary and atmospheric phases) and an *organic component* (comprising the plants and animals – living and dead – and their physical and chemical interactions). Figure 2.3 summarizes the major parts of a general terrestrial biogeochemical cycle, although in this general form the model would be equally applicable to a marine cycle.

Table 2.5
MAJOR AND MINOR CHEMICAL ELEMENTS IN THE ENVIRONMENT

(a) *Major elements:* plants require large amounts of			
	(i) oxygen	(ii) carbon	(iii) hydrogen
(b) *Minor elements:* plants require relatively large amounts of			
	(i) nitrogen	(ii) phosphorus	(iii) potassium
	(iv) calcium	(v) magnesium	(vi) sulphur
(c) *Trace elements*: plants also require small amounts of over 100 elements, including			
	(i) iron	(ii) manganese	(iii) cobalt

Living vegetation takes in chemical elements in solution form via *root osmosis* from the soil and it converts the elements into forms which can be readily used in plant growth by *biochemical processes* (which include photosynthesis). The elements can be released from the plants in a number of ways. When individual plant organs are separated from the plants (as for example by leaf fall from trees during spring), *decomposition* of the organs by bacterial activity in the soil and humus releases the organic compounds generally in a soluble inorganic form which can then be stored in the soil or another biogeochemical reservoir. When whole plants die the same form of bacterial activity releases elements to the reservoir. Alternatively, if the vegetation is burned either naturally by lightning or accidentally or on purpose at the hand of man (such as controlled heather moorland burning or accidental forest fires) then a portion of the elements will be released to the atmosphere on *combustion*, and the remainder as ash can be decomposed by the bacteria in the soil and sub-stratum. Elements are released to *grazing animals* when they eat and digest living vegetation, and although the organic phase of the cycle will be considered in greater detail in Chapter 3 it is important to note the significant role played by both plants and animals in the chemical cycles. The grazing animals can release elements either on

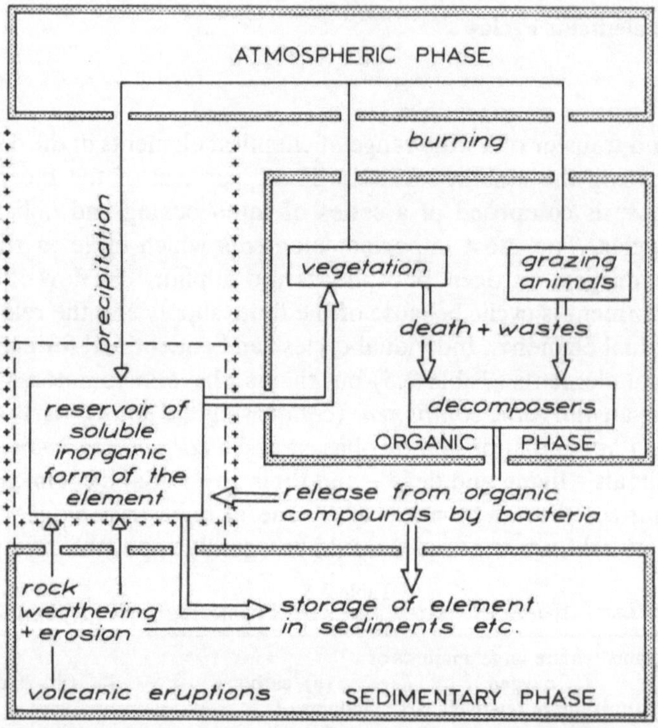

FIG. 2.3 GENERALIZED FORM OF TERRESTRIAL BIOGEOCHEMICAL CYCLE. SOLID ARROWS DENOTE MAJOR PATHWAYS. (After Clapham, 1973.)

death by *decomposition*, or by *combustion*; but they also produce supplies of the main chemical elements throughout their life-cycle by the production of *waste material* and *faeces* which can be decomposed by bacterial activity. The elements which have been released from the organic phase of the cycle are thus in one of two forms: first the combustion material which has been issued to the atmosphere, and second the decomposition material which has been issued to bacteria in the soil and sub-stratum. The former can be 'washed out' by precipitation and then contribute to the general reservoir of the soluble inorganic forms of that particular element where this is stored in the *sedimentary phase* of the cycle, or it can simply contribute to the reservoir if this is held within the *atmospheric phase*. The latter can be released by bacterial activity and it can either contribute directly to the reservoir of the element, or by interaction with the water and sediment cycles it can become encorporated into the sedimentary phase of the cycle and be stored in sediments, soils and sedimentary rocks for a variable but generally long period of time. This sedimentary storage can later be released to the general reservoir by normal processes of rock and soil weathering and erosion, which can also release chemical elements from various igneous sources within the sedimentary phase.

The importance of the inorganic phase is clear from this generalized cycle (Figure 2.3), because it provides a major reservoir for almost all of the nutrient elements, although the flow through the inorganic phase is generally much slower than that through the organic phase. The importance of the two inorganic components – the sedimentary and atmospheric phases – is also clear. Although the latter is an important part of only some of the chemical cycles and the former is basic to all, these components do make the biogeochemical cycles intimately linked with the stability and progress of the energy system and the sedimentary cycle; and the water cycle provides the flow pathways which link the three together. The strength of this linking is in many ways conditioned by the solubility of the individual chemical elements. With relatively soluble substances such as nitrogen the amounts of the elements present at different phases of the cycle are related to the water budget. With insoluble substances such as phosphorus, however, there is relatively little removal of the element from the soil (except by soil erosion) and so the amount present is largely related to the sediment cycle. The relative importance of the atmospheric and sedimentary phases differs between individual elements (Clapham, 1973). Sulphur, for example, follows a cycle in which both phases are present and important. In the phosphorus cycle the atmospheric phase is of little significance and the sedimentary phase assumes much greater importance (Figure 2.4a). The nitrogen cycle, on the other hand, has an atmospheric phase which is much more important than the sedimentary phase because of the large reservoir of atmospheric nitrogen, the direct contributions to this from the organic phase via denitrification by decomposers, and the capability of plants to take in atmospheric nitrogen through leaf surfaces by the process of *biological fixation* (Figure 2.4b).

Man-induced changes

There are a number of ways in which human modification of the biosphere can influence the balance and stability of the biogeochemical cycles, particularly in terms of the initial inputs into the cycle, extraction from the cycle and changes in the speed and direction of operation of the processes within the cycle. For example initial inputs in the atmospheric phase can be affected by increasing the quantities of elements already present in the natural state (such as carbon from air pollution associated with car exhaust and industrial gaseous emissions) and by introducing into the atmosphere 'alien' elements not normally found there (such as through the production of gases by combustion of some chemicals). Similarly inputs in the organic phase can be increased by application of fertilizers and other 'natural' elements in increased quantities, as well as by introducing toxic elements such as DDT for which stable biogeochemical cycles have not yet fully evolved. Inputs into the sedimentary phase can also be affected by intentional action, such as the dumping of industrial chemicals.

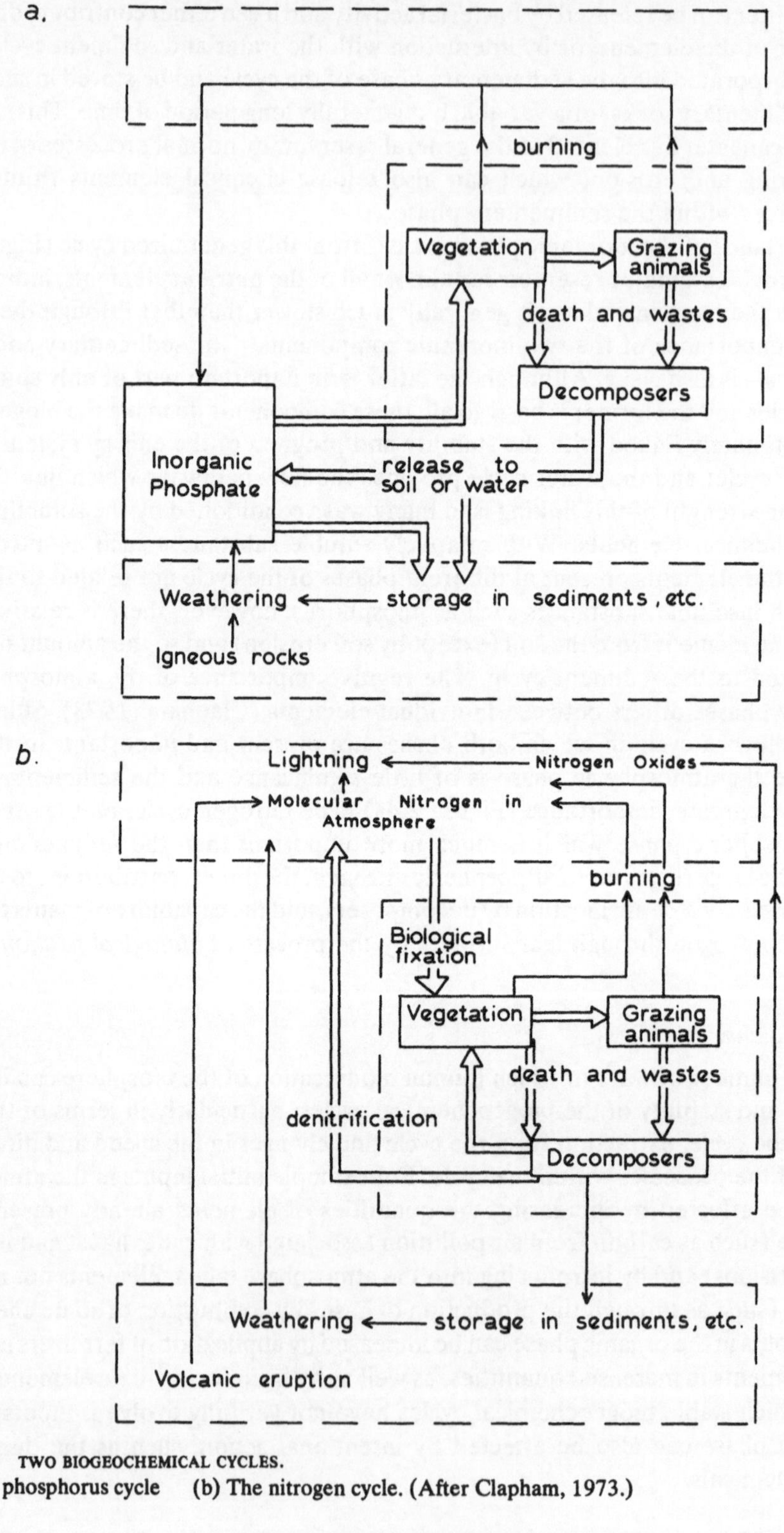

FIG. 2.4 TWO BIOGEOCHEMICAL CYCLES.
(a) The phosphorus cycle (b) The nitrogen cycle. (After Clapham, 1973.)

The ecological effects of introducing 'alien' elements into the various phases of the cycles are particularly important because not only do the new elements themselves often prove to be toxic to plants and animals in the food chain which they enter (see Chapter 3), but also the new elements can undergo chemical reaction and interaction with the existing 'natural' chemicals and this can disturb a previous state of balance in the natural cycles. The extraction of chemical elements can also produce man-induced changes either by extraction from the sedimentary phase (such as phosphorus and nitrate extraction) or from the organic phase (such as with cropping of vegetation, whole tree harvesting, and the removal of grazing animals from an ecosystem).

These man-induced changes in inputs and outputs in the biogeochemical cycles will often have lasting ecological effects only at the relatively local scale, because the overall supplies of the individual elements are relatively fixed, and because the form in which the chemical elements are used cannot readily be changed so drastically that they cannot be re-used in the organic phase of the cycles (the basis of *re-cycling* in the natural environment). The changes thus represent spatial transfers which can lead to spatial imbalance in biosphere stability at the local scale, rather than wholesale modification of the global cycle.

The third major area of human modification lies in affecting the speed of operation of the cyclic processes. Burning, whether intentional (such as heather moorland firing) or accidental (such as forest fires started by campers), can effectively and efficiently destroy all of the vegetation and most of the relatively immobile grazing animals and decomposers from the organic phase of a local cycle, and at the same time increase the atmospheric store of certain chemical elements at the local scale. The atmospheric phase can also be influenced by planned weather modification schemes, in that cloud seeding, for example, can influence precipitation input to the soluble reservoir of certain elements. The encorporation into, and release from, storage in the sedimentary phase of certain elements can also be influenced by land management practices, and accelerated erosion can disturb local geochemical balances. Agricultural activities such as crop rotation, field drainage, high density stock grazing and factory farming; water resource management schemes such as reservoir construction, large-scale water transfers and ground-water abstraction; and industrial activities which issue chemical pollutants into streams, lakes, the oceans and the atmosphere can all radically affect total supplies and the speed and pathways of cycling of certain elements. Disturbing the balance of the biogeochemical cycles can also have extensive environmental effects. McCarl and his fellow workers (1975) have demonstrated, for example, that the addition of 1 ton (imperial) of 'man-made' nitrogen into the world's nitrogen cycle can lead to the inadvertent production of some 800 tons of flow in the cycle overall.

Ecological significance

Like the energy system and the water cycle the biogeochemical cycles are of vital importance to plants and animals. Chemical elements provide the basic substances for tissue and organ development in both plants and animals, and the availability of nutrients in delicately balanced quantities is critical to ecological stability. The species composition of plants and animals in an area are both closely related to availability of chemicals, and a change in the balance can lead to a corresponding and possibly irreversible ecological change. Shortages or excesses of natural elements can be complemented by the introduction of alien elements which might be highly toxic and thus affect the entire food web of certain environments (such as reported findings of DDT, released far away, in Arctic penguins). The biogeochemical cycles are also closely related to the other biosphere sub-systems via their effects on plants and animals. In particular the density and characteristics of the vegetation cover, which are both strongly dependent on chemical balance and availability, can markedly influence the sediment and water cycles as well as the localized energy balance.

2.2d The sediment cycle

The natural system

The final major sub-system of the biosphere is that concerning the movement of clastic sediment through the earth's environment. As earth scientists since James Hutton (1795) have repeatedly demonstrated, the surface of the earth is constantly being weathered, eroded and sculptured by a suite of land-forming processes (of wind, water, ice and wave action) which produce varying quantities of sediment (Table 2.6). Large-scale and long-term geological processes such as *geosyncline development* and *tectonic uplift* serve to create a cyclic movement of this sediment over the very long time scale (Figure 2.5a) based

Table 2.6
ESTIMATED TRANSFER OF MINERAL MATTER FROM CONTINENTS TO THE OCEANS BY VARIOUS PROCESSES

Transfer	Amount of sediment (10^6 tonnes/yr)
(a) *Eroded from continents*:	
by streams	9·3
by wind	0·06–0·36
by glaciers	0·1
TOTAL	9·46–9·76
(b) *Deposited in oceans*:	
shallow waters (below 3 km depth)	5–10
deep waters (above 3 km depth)	1·2
TOTAL	6·2–11·2

Source: Judson (1968)

on dynamic interactions between the earth's mantle and crust, the water cycle and the atmosphere. Allen (1975) has described this as the 'geocycle', although it is the slowest and least complete of the biosphere sub-systems and it depends on a state of balance between the levelling effects of erosion and the constructive effects of tectonic uplift which might only be apparent over extremely long periods of time. The ultimate depositional environment for all sediment is the sea bed, and sediment accumulates in geosynclines where it is buried and eventually compacted by the weight of overlying sediment. Instability of the earth's crust can be triggered off by excessive compaction, and this can lead to uplift by tectonic forces to yield 'new' upland areas of metamorphic rocks which are then exposed to the agents of erosion and weathering. Alternatively the burial and compaction of the geosynclinal sediments may continue until the sediments are altered by pressure heating, deformation and fusion at depth, to become igneous magma which can then be released to the earth's surface by volcanic and other igneous activity.

Whilst this large-scale, long-term cycle is of fundamental importance to the long-term stability and form of the biosphere, it is the weathering, erosion and deposition components which are of more immediate concern to plants and animals in the biosphere and to the other sub-systems of energy, water and chemicals. In this context erosion and transport by running water are clearly of much greater significance than that related to wind and glaciers (Table 2.6). There is evidence of distinct spatial patterns of weathering, erosion and deposition of sediment (Stoddart, 1969), and a number of studies have sought to identify the factors controlling rates of erosion (Gregory and Walling, 1973). Principal amongst these appear to be rock type, climate, vegetation and drainage basin characteristics such as slope, drainage density and relief. Fournier (1960), for example, related world suspended sediment yields to various variables describing relief and climate; and Langbein and Schumm (1958) related suspended sediment yield to runoff for small drainage basins in the United States and found that peak yields occurred in semi-arid areas, with lower rates in areas of more and less effective precipitation. This relationship was explained by the combined operation of relationships between two variables and climate. First the erosive influence of precipitation increases with its amount. Secondly, and opposing this, the protective effects of vegetation increase with increasing precipitation. A state of balance thus appears to exist between the erosional influence of water and the protection offered by vegetation, which yields peak erosion rates in semi-arid areas.

Many of the important local controls of erosion are summarized in the form of a 'Universal Soil Loss Equation' (Smith and Wischmeier, 1962), which is of the general form:

$$A = R.K.L.S.C.P$$

where A is the average annual sediment loss (kg/ha); R is a rainfall factor; K is a 'soil erodibility' factor; $L.S$ is a slope length and steepness factor; C is a cropping and soil management factor; and P is a measure of conservation practices.

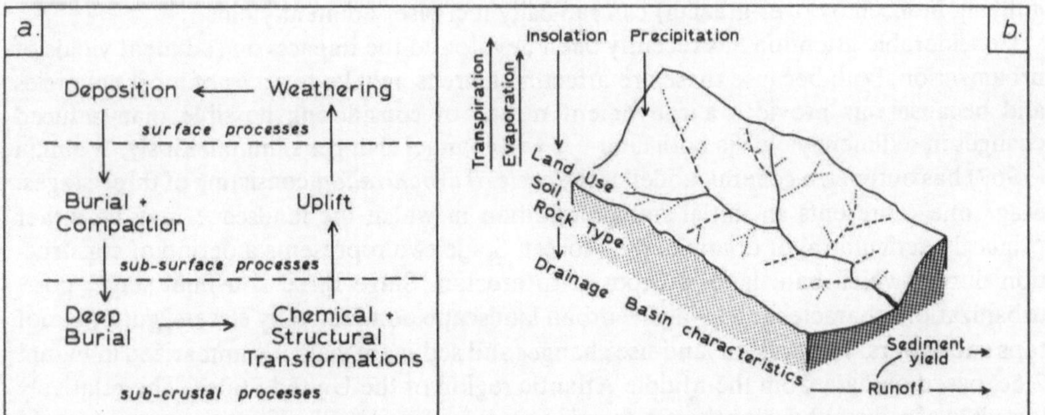

FIG. 2.5 THE SEDIMENT CYCLE.
(a) Major elements in the cycle of erosion, sedimentation and uplift. (After Haggett, 1972.)
(b) Major components and flows in the drainage basin system. (After Gregory and Walling, 1973.)

A convenient means of summarizing the flow of sediment on the land surface is in terms of *drainage basin processes*, since the drainage basin is widely regarded as the fundamental geomorphic unit (Chorley, 1969) because of its integrity and its systems properties. Figure 2.5b illustrates the major geomorphological components of the drainage basin, in terms of the influence of drainage basin characteristics in transforming inputs into outputs of water and sediment yield. Thus any changes in either the inputs or the drainage basin characteristics should produce changes in the water and sediment outputs.

Man-induced changes

There are a number of ways in which man-induced environmental changes can affect sediment yield and thus modify local sediment balances. Inputs can be affected by weather modification and by changes in the local energy budget which affect evaporation and transpiration losses. It is through modifying drainage basin characteristics, however, that the impacts of man and his activities are generally more pronounced; and changes in land use are the most widespread form of interference. Both the type and the density of the vegetation cover affect the protection of the land surface to erosion (Figure 2.6a) by preventing direct soil splash, by reducing downslope soil creep and movement, by stabilizing surface conditions (through root binding) and by reducing the amounts of water available through transpiration and evaporation losses. Thus, for example, forest offers greater protection than grassland or desert scrub; and a dense forest offers greater protection than a dispersed forest cover. Any direct or indirect changes in the vegetation cover will thus lead to changes in sediment yield. Figure 2.6b illustrates the variations in sediment yield from individual drainage basins in northern Mississippi under different types of land use and changing amounts of precipitation, and there are clear differences in yield between cultivated drainage basins and mature plantations. Changes in sediment yield can also be induced by man-induced changes in channel network and channel form characteristics. A contraction in the channel network (such as by building over a former channel) can lead to reduced sediment yield, whereas an expansion (such as through gullying induced by over-grazing) can radically increase sediment yields.

Considerable attention has recently been devoted to the impacts on sediment yields of urbanization, both because these are affecting increasingly large areas of most countries and because this provides a convenient means of considering possible man-induced changes in sediment yield via both land-use and channel changes simultaneously. Wolman (1967) has outlined a general model of the *cycle of urbanization* consisting of three stages. Stage one represents an initial stable condition in which the landscape may be either primarily agricultural or dominated by forest. Stage two represents a period of construction during which bare land is exposed to erosion. Stage three is a final stage, post-urbanization, characterized by a new urban landscape dominated by streets, gutters, roof tops and sewers. The cycle of land-use changes and sediment yield is summarized in Figure 2.6c, based on data from the Middle Atlantic region of the United States. The relatively low yield of sediment during the pre-farming era is replaced by significant increases in yield during farming, and yield then declines in the period prior to construction, when much of the farmland may be left as grass or even allowed to return by natural succession (Chapter 4) to brush or forest whilst awaiting development. The onset of clearance for construction is accompanied by a rapid rise in sediment yields as large areas of bare ground are exposed to natural erosion. When the entire drainage basin, or substantial parts of it, have been developed, sediment yields in the post-construction period might be expected to decline to values as low as those experienced before the farming era because of the lack of exposed channel for erosion and the common lack of 'natural' channel beds and banks. Other examples of sequences of sediment yield changes related to land-use and drainage basin changes are offered by Gregory and Walling (1973), Judson (1968) and Cooke and Doornkamp (1974). These include the effects on sediment yield of *over-grazing* and *reservoir construction*.

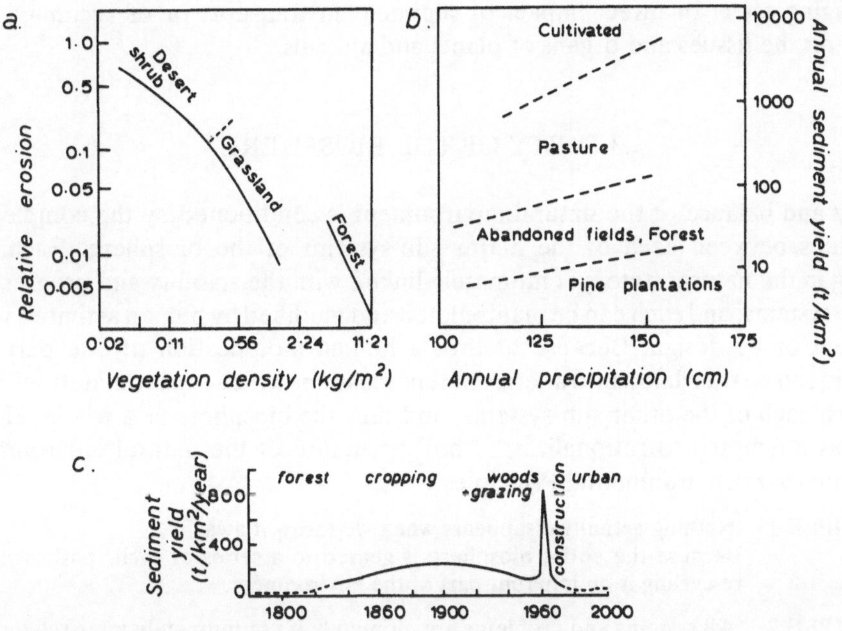

FIG. 2.6 SOME RELATIONSHIPS BETWEEN LAND USE AND SEDIMENT YIELD.
(a) Relationships between relative erosion and vegetation type and density. (After Langbein and Schumm, 1958.)
(b) Relationship between land use and sediment yield in northern Mississippi. (After Ursic and Dendy, 1965.)
(c) Suggested pattern of variations in sediment yield through a cycle of urbanization. (After Wolman, 1967.)

Man-induced changes in the sediment cycle can also be of a direct nature through reclamation of land from the sea and from the direct movement of superficial material (through the construction of embankments and through mining and quarrying operations, for example).

Ecological significance

It is clear that the sediment cycle is intimately related to each of the other major sub-systems. The water cycle provides water for the detachment, entrainment and transport of sediment; it provides the ocean basins and local topographic variations for sediment accumulation; and it also greatly influences vegetation distribution and density which in turn can affect sediment yield. The chemical cycle is also related via its effects on vegetation stability and density; and the rock weathering part of the sediment cycle provides an important source of chemicals for the various biogeochemical cycles. The sediment cycle at the local scale can also exert considerable influence over the nature of micro-relief and the surface characteristics of the earth's surface; and at the global scale it affects the distribution of land and sea areas. Each of these might have an effect on the energy system. Although it is largely via its connections with the other major sub-systems that the sediment cycle exerts an indirect influence over the biosphere and its organic components, it can exert direct influence via the stability of the erosional/depositional environment which can markedly affect ecosystem stability in both time and space, and via the damaging effect of direct impact of sediment in transport or of accumulations of sediment on the tissues and organs of plants and animals.

2.3 UNITY OF THE BIOSPHERE

The unity and balance of the natural environment is conditioned by the complex inter-relationships between each of the major sub-systems of the biosphere. Each is self-sustaining in the natural state, yet intimately linked with the stability and progress of the other sub-systems; and each can be manipulated and modified by human activities whether by accident or by design. Because of this, a human modification of one part of one sub-system can be translated into a series of repercussions which can ultimately influence if not disturb each of the other sub-systems, and thus the biosphere as a whole. Holliman (1974) has attempted to rationalize the holistic nature of the natural environment by formulating four 'Environmental Principles':

PRINCIPLE 1 'Nothing actually disappears when we throw it away'
Because the entire biosphere is geared to a series of cyclic pathways and so re-cycling is an inherent part of the environment.

PRINCIPLE 2 'All systems and problems are ultimately if not intimately inter-related'
Holliman adds that 'it does not make sense to squabble over which crisis is most urgent; we cannot afford the luxury of solving problems one by one. That is both obsolete and ecologically unsound anyway.'

PRINCIPLE 3 'We live on a planet whose resources are finite'

PRINCIPLE 4 'Nature has spent literally millions of years refining a stable ecosystem'
This is evidenced in the complexity and stability of the biosphere in its 'natural' state.

Because of this integrity of the biosphere increasing attention has been given in recent years to ensuring that resources of the biosphere are used wisely and that conservation measures are applied where appropriate (UNESCO, 1970). It was realized in the 1960s, however, that scientific knowledge of the ecological productivity of the biosphere and of its capacity for sustained maximum production is as yet limited; and so the *International Biological Program* (IBP) was launched in 1964 to apply standardized methods of study on the global scale to produce comparable results on productivity and productive capacity of the biosphere. The program was designed to yield data on spatial variations in primary productivity and in production processes (such as photosynthesis and biological nitrogen production) in terrestrial, freshwater and marine communities. Although essentially an ecological study there are a number of ways in which the geographer has been able to offer his services to the program (Newbould, 1964) by establishing background environmental information for the selected study sites and by providing measurements of environmental variations (such as in radiation intensity) at or near to the study sites.

More recently in 1970 UNESCO launched a *Man and the Biosphere Programme* (MAB) based on a series of fourteen detailed projects (Table 2.7) concerning man's interaction with the environment. The basic philosophy of the programme is to use standardized approaches to systems analysis and modelling within each of the projects; and a number of the projects (numbers 1, 2, 3, 4, 5, 6, 7 and 11) have specific geographical bases whilst the remainder relate primarily to human intervention in the biosphere. One outcome of the programme has been the proposal to set up a world-wide network of protected areas (*Biosphere Reserves*) to be used in the programme of monitoring research and training on ecosystem changes, and to be used in the conservation of species and of genetic diversity.

Table 2.7
'MAN AND THE BIOSPHERE' PROJECTS

Project number	Theme
1	Ecological effects of increasing human activities on tropical and sub-tropical forest ecosystems
2	Ecological effects of different land uses and management practices on temperate and Mediterranean forest landscapes
3	Impact of human activities and land-use practices on grazing lands – savanna and grassland (from temperate to arid areas)
4	Impact of human activities on the dynamics of arid and semi-arid zone ecosystems, with particular attention to the effects of irrigation
5	Ecological effects of human activities on the value and resources of lakes, marshes, rivers, deltas, estuaries and coastal zones
6	Impact of coastal activities on mountain and tundra ecosystems
7	Ecology and rational use of island ecosystems
8	Conservation of natural areas and of the genetic material they contain
9	Ecological assessment of pest management and fertilizer use on terrestrial and aquatic ecosystems
10	Effects on man and his environment of major engineering works
11	Ecological aspects of urban systems, with particular emphasis on energy utilization
12	Interactions between environmental transformations and the adaptive, demographic and genetic structure of human populations
13	Perception of environmental quality
14	Research on environmental pollution and its effects on the biosphere

Clearly the organic components of the biosphere are of central concern to the overall stability of the biosphere itself, and there is a clear need to consider in greater detail the ecological processes and relationships which are important in this way. Chapter 3 is devoted to such matters.

3
The Ecosystem as the Fundamental Ecological Unit

Chaos and confusion are not to be introduced into the order of nature because certain things appear to our partial views as being in some disorder (Hutton, 1795).

Stability in the biosphere is influenced in no small way by the organic parts of the overall system, and consequently much attention has been centred around ecological aspects of environmental stability. Many ecologists have adopted the ecosystem as a fundamental unit of study, and a flexible framework for analysis; and the ecosystem concept is of great value to all aspects of environmental management. This chapter will consequently focus principally on the ecosystem concept itself, on ecosystem form and function, and on productivity and equilibrium in ecosystems. It is important also to reflect on the problems and prospects inherent in ecosystem analysis, and on the value of the approach to environmental management.

3.1 ORDER IN NATURE

The organic part of the environment is characterized by considerable diversity. For example the wildlife in Britain includes over 1500 species of flowering plants, 200 species of birds which breed here and a further 200 species of irregular visitors, 60 species of mammals, 30 types of whales and seals which visit British seas, 30 species of freshwater and 50 species of marine fish (with 20 that live in both), 6 species of reptiles, 6 species of amphibians, and numerous invertebrates (Arvill, 1967).

Table 3.1
SIMILARITIES AND DIFFERENCES BETWEEN PLANTS AND ANIMALS

Similarities	Differences	
	Plants	Animals
Both are composed of cells, with nuclei, chromosomes, enzymes, etc.	Most have stiff cell walls	Most have flexible cell walls
Both need nutrition for growth and health	Usually make food by photosynthesis	Take food (energy is obtained from oxidation of glucose)
Both digest food, excrete wastes, grow and reproduce	Sessile organisms (do not move)	Motile organisms (free to move about)

Source: Robinson (1972)

Within the animate parts of the environment the basic distinction is between plants and animals. Whilst higher forms of vegetable life are readily distinguished from animal life both in general appearance and in physiology (Table 3.1), lower forms of organisms are often difficult to categorize. Thus, for example, certain single-celled organisms (such as *Euglena*) behave like plants in as much as they can convert solar radiation into usable energy forms by the process of photosynthesis, yet if kept for long periods in the dark or if provided with organic matter they live as animals. The distinction between *plant* and *animal* is thus not without problems.

A more convenient and meaningful basis on which to sub-divide living creatures is by *feeding habit*. The simplest division is between those organisms which can produce their own food from simple inorganic materials – the *autotrophs* (meaning, literally, 'self-feeders'); and those which cannot and which thus have to depend on eating autotrophs –

Table 3.2
SIMPLE CLASSIFICATION OF MAIN TYPES OF ORGANISMS

Autotrophs:	contain chlorophyll-like substances, make organic matter from inorganic ingredients, need energy (light and chemical)
(a) Phototrophs:	derive energy from light by photosynthesis; examples include mosses, ferns, conifers, flowering plants, algae and photosynthetic bacteria
(b) Chemotrophs:	derive energy from inorganic substances by oxidation; examples include certain bacteria and blue-green algae
Heterotrophs:	feed on the organic matter produced by autotrophs. This is available in three forms – living plants and animals; semi-decomposed plants and animals; and organic compounds in solution
(a) Saphrophytes:	feed on soluble organic compounds from dead plants and animals. Some absorb components that are already dissolved, others (such as some fungi and many bacteria) break down undissolved foods by secreting digestive enzymes onto them
(b) Parasites:	organisms that at some time or times in their lives make a connection with the living tissues of another species on which they rely for food. Food may be in the form of soluble or insoluble compounds. There are many animal parasites, many parasitic fungi and a few parasitic plants (such as mistletoe)
(c) Holozoic organisms:	eat by mouth, and can absorb large particles of undissolved food and soluble compounds. All higher animals are holozoic

Source: Reid (1969)

the *heterotrophs* ('other feeders'). The former are often referred to as *primary producers* and the latter as *consumers*. The autotrophs can be further sub-divided according to the source of energy necessary for converting inorganic into organic forms; and the heterotrophs can be divided according to the way in which the organism actually takes in the organic matter from the autotrophs (Table 3.2).

3.1a Food chains and food webs

The distinction between producers and consumers affords a basis for a hierarchical structure of organisms. The green plant is the most important component. Plants manufacture their own foods from carbon dioxide which is taken in through leaf walls from the atmosphere, and from inorganic salts (like phosphorus and nitrates) and water which are taken in from the soil via plant roots, by the process of *osmosis*. The manufacturing process involves the conversion of water and carbon dioxide into starch and sugar in the presence of sunlight. The process is termed *photosynthesis* ('building by light'). A key factor in the process of photosynthesis is the presence of chlorophyll, a green energy-trapping pigment within the leaves of the plant. Thus all organisms which contain chlorophyll are technically classified as 'plants' regardless of whether or not they have roots in the case of land plants, or swim around in water as free cells as with green algae, or in cell clusters. Animals do not contain chlorophyll and thus they cannot photosynthesize and produce their own food. They are thus fully dependent on plants and on other animals for food. Hence they are consumers. Animals thus lie above plants in the hierarchy – they eat them! For example

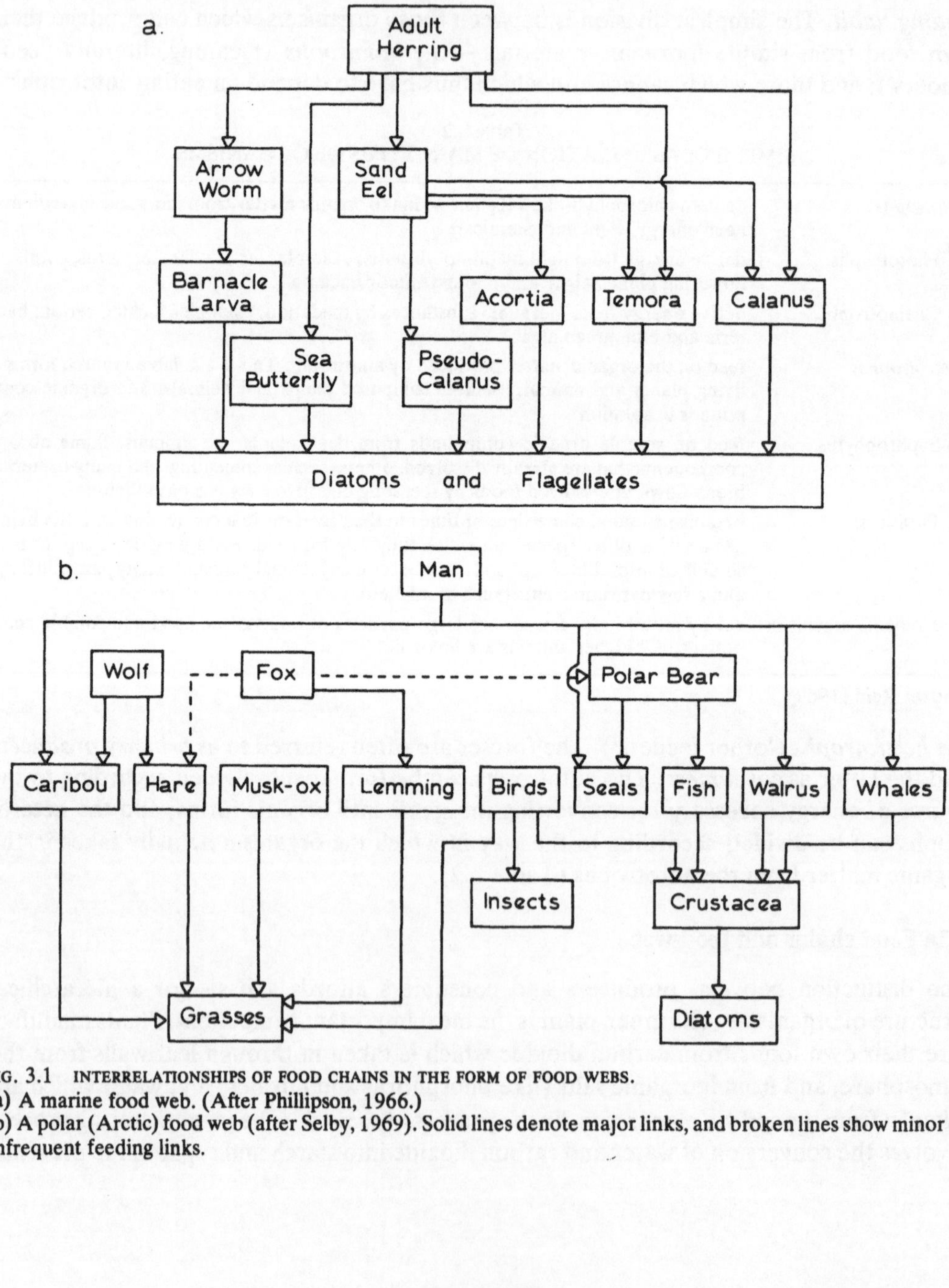

FIG. 3.1 INTERRELATIONSHIPS OF FOOD CHAINS IN THE FORM OF FOOD WEBS.
(a) A marine food web. (After Phillipson, 1966.)
(b) A polar (Arctic) food web (after Selby, 1969). Solid lines denote major links, and broken lines show minor or infrequent feeding links.

sheep and cattle eat growing grass, goats and giraffes eat both leaves and young shoots, and insect larvae eat stems, shoots and leaves (Reid, 1969). Level three in the hierarchy will be occupied by animals which eat other animals, that is, *carnivorous animals*. It is often possible to envisage a fourth level which might be occupied by man. Man has the convenience of being able to use the plant energy direct from the first level (as food, for fuel and allied uses), and also of being able to manage and use the animals from levels two and three for milk, meat and animal products. All organisms can thus be seen as part of a chain of eating habits – a *food chain*. A simple marine food chain could be a diatom (autotroph), eaten by a crustacean (herbivorous consumer), eaten by a herring (carnivorous consumer), possible eventually eaten by man.

Although the food chain is a convenient way of summarizing feeding relationships between organisms, the relationships are rarely simple and linear. Cousens points out that 'the food chain, a valid and useful concept, is an over-simplification for many purposes, for most animals vary their diet and in turn are preyed upon by a variety of carnivores' (Cousens, 1974). The simple food chain is generally very closely inter-connected with a number of other overlapping chains to form a *food web* (Figure 3.1). Although a large number of species may be present in a lateral sense within the web – that is, it is a wide food web, such as in Figure 3.1b – the webs are rarely longer than four vertical links because of the progressive loss of energy at each of the levels (see pages 75–82).

Although the concept of food webs is more realistic than that of food chains, in some ways the concept is still sounder in theory than in practice. The operational problem of 'where to draw the boundaries between food webs' is a real one, and because many animal species can alter their feeding habits and their food sources in times of food shortage, then there might be a large area of overlap between adjacent food webs. The boundaries should logically be drawn on the basis of the frequency of eating relationships between species; but this raises the question 'if one species (A) eats another (B) only once in its life span, does the food web for A overlap with and include all of the components of B's food web?' To some extent food web boundaries are drawn to maximize within-web unity and between-web variability. The inter-locking nature of the food web is illustrated in Smith's (1977) study of a small sub-Antarctic island (Marion Island) dominated by a tundra type of vegetation, where three food webs – a terrestrial one dominated by introduced cats as the top carnivore; a marine littoral one; and a terrestrial freshwater one – were interlocked throughout the whole 290 km^2 of the island.

3.1b The ecosystem concept

The hierarchical structure of food webs is not confined to organic components of the environment, nor is it confined to living plants and animals. The green plants rely heavily on the surrounding atmosphere for carbon dioxide and on the underlying soil for water and nutrients, and so the physical environment is an integral part of the system. Furthermore, plants die and their structures are broken down by decomposer species to release energy and chemicals within the plant organs and cells; animals produce waste material and they eventually die, and these both produce organic material which can be broken down; man produces wastes and he also dies and is 'broken down' slowly after burial or instantly on cremation. In this way *decomposition* is a fundamental part of the ecological system.

The various parts of this hierarchical structure are linked to each other functionally

(Figure 3.2) to produce a dynamic system of inter-relationships between organisms, and between organisms and environment. Tansley (1935) proposed the term *ecosystem* (a contraction of ecological system) for this dynamic structure, and he visualized the ecosystem as being composed of two major parts. These are first the *biome*, the whole complex of organisms (both plants and animals) naturally living together as a sociological unit; and secondly the *habitat* or physical environment. In Tansley's view 'all parts of such an ecosystem – organic and inorganic, biome and habitat – may be regarded as interacting factors which, in a mature ecosystem, are in approximate equilibrium; it is through their interactions that the whole system is maintained' (Tansley, 1935). Lindeman has stressed that the term applies to 'any system composed of physical-chemical-biological processes, within a space-time unit of any magnitude' (Lindeman, 1942). The ecosystem is a conve-

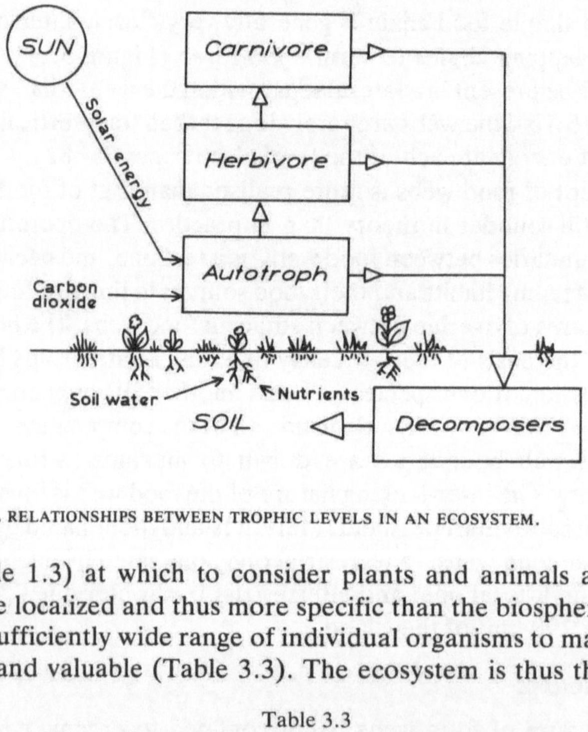

FIG. 3.2 FUNCTIONAL RELATIONSHIPS BETWEEN TROPHIC LEVELS IN AN ECOSYSTEM.

nient scale (Table 1.3) at which to consider plants and animals and their interaction because it is more localized and thus more specific than the biosphere in its entirety, and yet it includes a sufficiently wide range of individual organisms to make regional generalizations feasible and valuable (Table 3.3). The ecosystem is thus the sum of all natural

Table 3.3
FIVE LEVELS OF ECOLOGICAL ORGANIZATION

Level	Unit	Definition
1	The organism	an individual plant or animal
2	The population	a group of individuals of one species
3	The community	the sum of the different populations of species within a given area
4	The ecosystem	the sum of the communities and the 'non-living' environment in an area
5	The biosphere	the sum of all ecosystems

organisms and substances within an area, and it can be viewed as a basic example of an open system in physical geography (see Chapter 2). Thus, for example, the woodland ecosystem has a series of major inputs and outputs. The former include energy, rain water, soil minerals and soil water; and the latter include the standing crop of timber and logs (in a managed woodland), the animals within the system which migrate, water losses and heat loss (Figure 3.3).

Ecologists have considered many aspects of ecosystems since Tansley's formulation of the concept, and Evans (1956) has isolated three particularly important areas of enquiry. These are the quantities of matter and energy that pass through ecosystems and their rates of movement; the kinds of organisms which are present in different ecosystems; and the roles occupied by the organisms in the structure and organization of the ecosystem.

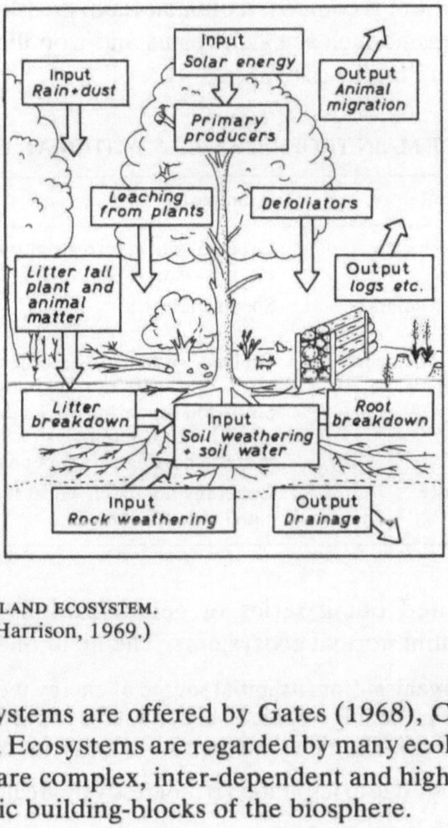

FIG. 3.3 DYNAMICS OF A WOODLAND ECOSYSTEM.
(After Ovington, 1972, from Harrison, 1969.)

General reviews of ecosystems are offered by Gates (1968), Clapham (1973), Harrison (1969) and Pears (1977). Ecosystems are regarded by many ecologists to be the basic units of ecology because they are complex, inter-dependent and highly organized systems, and because they are the basic building-blocks of the biosphere.

3.2 ECOSYSTEM FORM AND FUNCTION

The notion of a dynamic integrated system is a valuable one, and it is necessary to consider both the components and the flows within the systems.

3.2a Trophic levels

The main physical components within an ecosystem are represented in the various levels within the feeding hierarchy. These are termed *trophic levels*, after the classic work of Lindeman (1942). Lindeman pointed out that living organisms can be grouped into a series of more or less discrete trophic levels with each level depending on the preceding one for its energy supply. Thus ultimately the energy requirements of the ecosystem have to be met by solar radiation trapped by the autotrophic green plants, and converted into chemical energy in the plants themselves. Each trophic level within the ecosystem (denoted Λ in Lindeman's terminology) can be given a number, denoting its feeding level – in a vertical sense – relative to the autotrophs at trophic level one (Λ_1). Thus Λ_1 are the primary producers, Λ_2 are the primary consumers, and so on. The decomposer part of the system can be in trophic levels 2, 3 or 4 depending on the source of food, because it receives dead material from trophic levels 1, 2 and 3 and waste material from levels 2 and 3. The decomposer part of the system is composed of bacteria in grassland, for example; of fungi in woodland; of soil organisms such as earthworms and woodlice; and of aquatic species such as worms, bacteria and molluscs (Table 3.4).

Table 3.4
SUMMARY OF MAIN TROPHIC LEVELS WITHIN AN ECOSYSTEM

Trophic level	Feeding habit	Example
1	Primary producers (autotrophs)	Green plants in terrestrial systems, plankton in marine systems
2	Primary consumers (grazers)	Sheep and cattle
3	Secondary consumers (carnivores)	On land – lion eats wildebeest hawk eats fieldmouse Soil – soil organisms which decompose herbivore remains Water – herring eats copepod
4	Top carnivore	Generally not eaten whilst it is alive; e.g. lion in savanna

Lindeman (1942) pointed out a series of generalizations about the relationships between trophic levels within normal ecosystems. The main ones are:

(a) the more remote an organism from its initial source of energy the less probable that it will be dependent only on the preceding trophic level as a source of energy (that is, species at Λ_3 and above tend to be 'generalists' rather than 'specialists' in terms of feeding habit),

(b) the relative loss of energy due to respiration is progressively greater for higher trophic levels;

(c) species at progressively higher trophic levels appear to be progressively more efficient in using their available food supply, because increased activity by predators increases their chances of encountering suitable prey species, and in general predators are less specific than their prey in food preferences,

(d) higher trophic levels tend to be less discrete than lower ones, because of points (a) and (c),

(e) food chains tend to be reasonably short. Four vertical links is a common maximum, because of points (b) and (d).

An example of trophic studies is offered in the study of trophic relationships in a small woodland stream (Linesville Creek, Pennsylvania) by Cummins, Coffman and Roff (1966). Samples of bed sediment and stream water were taken from the stream and all organisms were removed from the samples in the laboratory. The gut content of each of the larger species was analysed (to see, quite literally, 'who eats what'), and this enabled the trophic structure of the creek to be established. The decomposer part of the system was clearly of considerable importance both because of the relatively large weight of decomposer species (30 per cent of the weight of organic matter in the creek) and the large number of individual decomposer organisms (80 per cent of the total number of individuals within the creek), as summarized in Table 3.5.

Table 3.5
TROPHIC STRUCTURE OF LINESVILLE CREEK, PENNSYLVANIA

Trophic level	Number of individuals (no/m²)	Dry weight biomass (gm/m²)
1 Primary producers:	4·6 (±1·2)×10¹⁰	387 (±97)
2 Primary consumers:	981·5 (19·0)*	1·313 (31·8)*
3 Primary consumers: (decomposer)	4187·0 (80·0)*	1·228 (29·7)*
4 Secondary consumers:	9·0 (0·1)*	1·593 (40·9)*
TOTAL CONSUMERS:	5177·5 (100)*	4·134 (100)*

Note: * relative percentage of total consumer level
Source: Cummins, Coffman and Roff (1966)

Although the trophic level concept opened up new horizons for analysing and modelling ecosystems there are a number of problems inherent in the concept itself, and in Lindeman's initial formulation (Lindeman, 1942). As Odum points out, 'the concept of trophic levels is not intended primarily for categorizing species. Energy flow through the community is stepwise in fashion due to the second law of thermodynamics [see page 44], but a given population of a species may be, and very often is, involved in more than one step of trophic level' (Odum, 1968). Slobodkin (1962) has also expressed concern at Lindeman's implication that there is always a fixed number of links in the passage from autotrophs to any given species, regardless of the route chosen. For some species the trophic position changes through the life-cycle of the individual – thus the feeding relationships of the tadpole are different to those of the frog – and many species can change their feeding habits depending on the availability of food supplies. Because many consumers do not fit well into the trophic scheme suggested by Lindeman (1942), Kercher and Shugart (1975) have suggested an 'effective trophic position' based on the distance from the food source to any member of the food web (like trophic position as identified by Lindeman, but derived from the notion of *ecological efficiency* (see pages 80–1)). The effective trophic position is thus a numerical index of energy intake per unit of time rather that an ordinal index based on number of food web links. There are a variety of associations between trophic levels and food web structure since both are based on feeding relationships, but as Slobodkin points out 'the two concepts are *not* identical; the first (trophic level) being a simplifying assumption whilst the second (food web) is purely descriptive' (Slobodkin, 1962).

3.2b Ecological pyramids

Clearly the concept of trophic levels provides scope for describing and analysing ecosystems of various types and sizes, involving different species but based on effectively similar structural relationships. Elton noted that 'the animals at the base of a food chain are relatively abundant, whilst those at the end are relatively few in number; and there is a progressive decrease in between the two extremes' (Elton, 1927). This suggested to Elton a 'Pyramid of Numbers' (Figure 3.4a). The general form of pyramid of numbers is characteristic of food webs where primary producers are small in size and thus they need to be very numerous to support the smaller numbers of larger consumers that feed on them. A sheep, for example, can and needs to consume a very large number of blades of grass. In

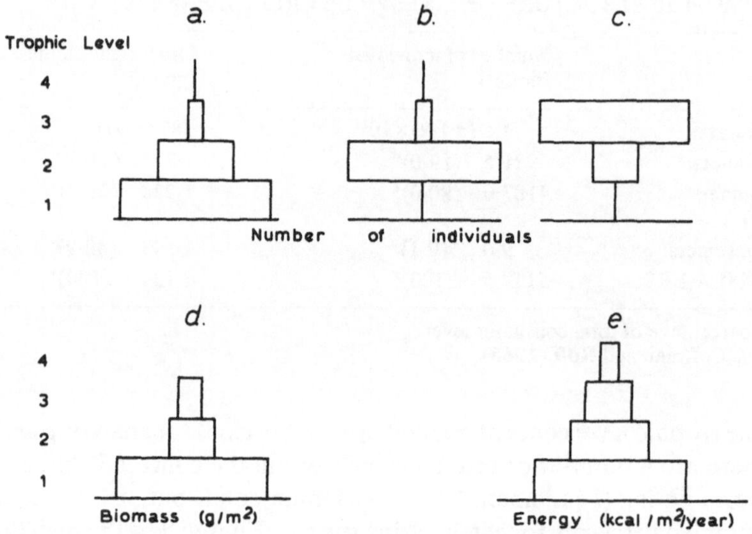

FIG. 3.4 GENERALIZED FORMS OF ECOLOGICAL PYRAMIDS.
(a) The pyramid of numbers for normal trophic relationships. The examples of a tree (b) and inverted pyramid characteristic of parasites (c) are also shown for comparison. (d) shows the normal form of the pyramid of biomass, and (e) the pyramid of energy.

some situations, however, the primary producers are large individuals like trees, and in such cases each individual plant can support a large number of herbivores (Figure 3.4b). Alternatively the plant/parasite food web typically has an inverted pyramid of numbers (Figure 3.4c) because a large number of parasites (Λ_2) and hyper-parasites (Λ_3) can be supported by one primary producer (Λ_1).

The value of the pyramid of numbers is that it allows comparisons to be drawn between the structure and food web diversity of different ecosystems. Thus a marine system can be compared with a woodland or riverine ecosystem on a meaningful basis. The comparison will be distorted by scale, however, because the pyramid of numbers takes no account of the size of the individual plant or animal. Thus a blade of grass does not compare favourably with an oak tree, yet both are at trophic level one; nor does an eagle compare favourably with a whale, yet both are at trophic levels three or four. For this reason ecologists generally prefer to compare systems on the basis of *biomass* (the total weight of organic matter) of each trophic level, rather than the number of individuals present. The 'Pyramid of Biomass' provides an ecologically more meaningful indication of food web structure (Figure 3.4d) and the normal form is a pyramid wide at the base (that is a large biomass at Λ_1) and thin at the apex. The sum total of biomass at each trophic level within the ecosystem represents the *standing crop* of that ecosystem at that point in time. This is the total weight of organic matter within the system. The standing crop and pyramid of biomass only represent ecosystem 'bulk' at a single point in time, however, and they offer no insight into either the total amount of organic matter produced through time or the rate of production. Ecological productivity is of basic importance to ecosystem structure and function, and so Elton (1927) suggested a 'Pyramid of Energy' to express the total amount of energy used at each trophic level per unit of area per unit of time (generally expressed in kcal/m^2/yr) (Figure 3.4e). The pyramid of energy allows comparisons to be drawn between the productivities of different types of ecosystems, such as the desert and the tropical rain forest systems.

There are a number of weaknesses behind the ecological pyramid approach. One is that the Eltonian pyramids depend heavily on the fact that there is typically a correlation between body size and trophic level. This is not always so, however, and inverted pyramids can arise either as temporary distortions of the steady state such as that due to the presence of an abundant immigrant species which is uncharacteristic of the ecosystem (Evans and Lanham, 1960), or as a consequence of very heavy predation and rapid growth in the lower levels of food chains (Odum and Odum, 1955). A second area of weakness is the problem of over-generalization inherent in the pyramid approach. Slobodkin (1962) raises two important questions on this matter. He asks 'is there a characteristic ratio between standing crops of species at different locations in a food chain?', and 'is there a maximum or characteristic number of possible links within a food chain?' Despite these forms of problems, ecological pyramids do offer a valuable and simple means of demonstrating trophic aspects of ecosystem structure.

3.2c Ecological niche

In a stable ecosystem each species will have established itself in a given location within the feeding web. That is, it has access to a given source of energy. Green plants derive their energy directly from the sun; animals derive energy indirectly through eating and absorbing energy bound into complex organic molecules within green plants. In this way ecosystems tend to have relatively stable populations of species and relatively stable food web networks.

The concept of an *ecological niche* was first suggested by Grinnell (1917) and developed extensively by Elton (1927). In any given kind of ecosystem a given function may be performed by a range of species. Simmons (1966) quotes the example of major grassland ecosystems which all have large herbivorous animals with essentially similar roles. In North America this role is played by the bison, in Africa by antelopes and zebras, and in Australia by the kangaroo. At the same time, in different ecosystems there are effectively similar roles to play. Owls in woodlands thus play the same role (predatory consumer) as hawks in open country.

Whittaker, Levin and Root (1973) have pointed out that the term *niche* is used in one or other of three senses. Many ecologists take it to mean the role played by a species within a community (it is a *functional concept*). It can also be used to refer to the distributional relationships of a given species to a range of environments or communities (a *habitat concept*). In other cases niche is used as a combination of these two (combining intra-community and inter-community factors). The authors stress that the term should apply exclusively to the functional role played within the community, although they do point out that a large number of factors affect the niche structure of natural ecosystems (Table 3.6).

Table 3.6
SUMMARY OF MAIN FACTORS WHICH AFFECT ECOLOGICAL NICHES

Inter-community (HABITAT) variables:	factors such as elevation, slope exposure, soil moisture, soil fertility as affected by parent materials, etc., and community gradients consequent upon these factors
Intra-community (NICHE) variables:	factors such as height above ground, relationship to intra-community pattern, seasonal time, diurnal time, prey size, and ratio of animal to plant food
Population response variables:	factors such as density, coverage, frequency of utilization, reproductive success and fitness

Source: Whittaker, Levin and Root (1973)

There are two important characteristics of ecological niches which make the concept of fundamental concern to environmental management. One is that the greater the diversity of niches within an ecosystem the greater the diversity of energy flow pathways and hence the more stable the ecosystem. Fluctuations of species populations within the system will be less extensive than those in systems with less diverse niches because predators have a wider choice of prey. The second is that the removal of a species (or a group of allied species) often leaves a niche empty and this can soon lead to a reduced energy flow through the entire system, which can radically affect the structure and stability of the ecosystem as a whole. Species can be removed from ecosystems in a variety of ways such as through pollution (see Chapter 7), over-exploitation, loss of habitat or induced local environmental change (see Chapter 6). The maintenance of a diverse range of species and of habitats is thus of basic concern to conservation.

The concept of ecological niches is also important because most trophic levels within ecosystems have one or more *dominant species*. McNaughton and Wolf (1970) define dominance as 'the appropriation of the potential niche space of certain subordinate species by another dominant species', and they point out that since many species have overlapping niche spaces then dominant species are not mutually exclusive. The same authors also considered the relationship between *dominance* and the *width of the ecological niche*, and they found that species which are most dominant within a community tend to have the broadest niches. This led them to suggest a 'Competitive Exclusion Principle' which states that species with identical niches cannot co-exist. This can be explained by assuming that dominant species are specialists which have evolved adaptations to a single dimension of environmental variation which is most likely to be limiting in a given environmental situation. McNaughton and Wolf (1970) also summarized the niche characteristics of a normal ecosystem (Table 3.7).

A state of dynamic equilibrium can exist within ecosystems and this balance is controlled largely through the flow pathways of energy and chemical elements through the system.

Table 3.7
SUMMARY OF NICHE CHARACTERISTICS IN AN AVERAGE ECOSYSTEM

Species numbers:	there are relatively few species with high abundance and a large number of species with low abundance
Niche widths:	average niche width within a community decreases as the number of species increases
Dominant species:	dominant species tend to have broader niches than less dominant or subordinate ones
Niche size:	species are added to the system by compression of existing niches
Site factors:	community dominance is at a minimum on the most equitable sites

Source: McNaughton and Wolf (1970)

3.2d Energy in ecosystems

Energy derived ultimately from solar radiation passes through the hierarchy and becomes an output from the system through respiration from successive trophic levels (Figure 3.5a). Chemicals, on the other hand, derive ultimately from the soil and bedrock in terrestrial ecosystems and from water in aquatic systems, and they cycle through the ecosystem via the decomposers (Figure 3.5b).

Gates (1968) has presented a very detailed discussion of energy flow mechanisms and their controls in natural ecosystems. The movement of energy through the ecosystem is closely governed by the two laws of thermodynamics (see pages 44–5). The initial source of energy for ecosystems is the sun, but only about 43 per cent of the solar radiation which reaches the earth's surface is of suitable wavelengths to be useful in the process of photosynthesis (Bassham, 1977). Solar energy enters the green plants at trophic level one through leaf walls, and in the presence of chlorophyll in the leaves of the plant the energy is transformed into a useable form by photosynthesis. Some of the energy converted in this way is lost by *plant respiration*, and the remainder is used in plant growth and it becomes stored in a chemical form in plant tissues. When the plants are eaten by primary consumers the energy stored within the plants enters trophic level two from which some is respired by the animals and the rest is used for growth and movement. This transfer process continues along the food chain. When the plants and animals die, if their bodies fall to the ground the decomposer organisms inherit the remanent energy which can then be released back to the atmosphere by respiration or used in the growth and movement of the decomposers. Data collected by Ovington (1961) confirms that a large fraction (often a third to a quarter) of energy fixed in forests annually is contributed to the forest floor as litter fall (mostly in the form of leaves), and that over a half of the net energy production during the development of a 55-year-old pine forest was released by decomposition.

The amount of solar energy trapped at any trophic level within an ecosystem is a good index of the amount of activity within that part of the system. Energy transfer through the trophic structure has two important characteristics. The first is that only a small but variable fraction of the energy available at any one trophic level in the system is passed on to the next level – the rest is lost as respiration. Bassham (1977) has offered an estimate of the maximum magnitude of energy efficiency for land plants as about 6·6 per cent of the total solar radiation. This calculation is based on four estimates:

(a) the theoretical maximum efficiency for the photosynthetic reduction of carbon dioxide to starch and glucose is 28·6 per cent,

(b) the fraction of total solar radiation which is photosynthetically active is 43 per cent,

(c) the fraction of total energy received by the plant which it can actually absorb is about 80 per cent,

(d) the fraction left after respiration by plants is often about 67 per cent.

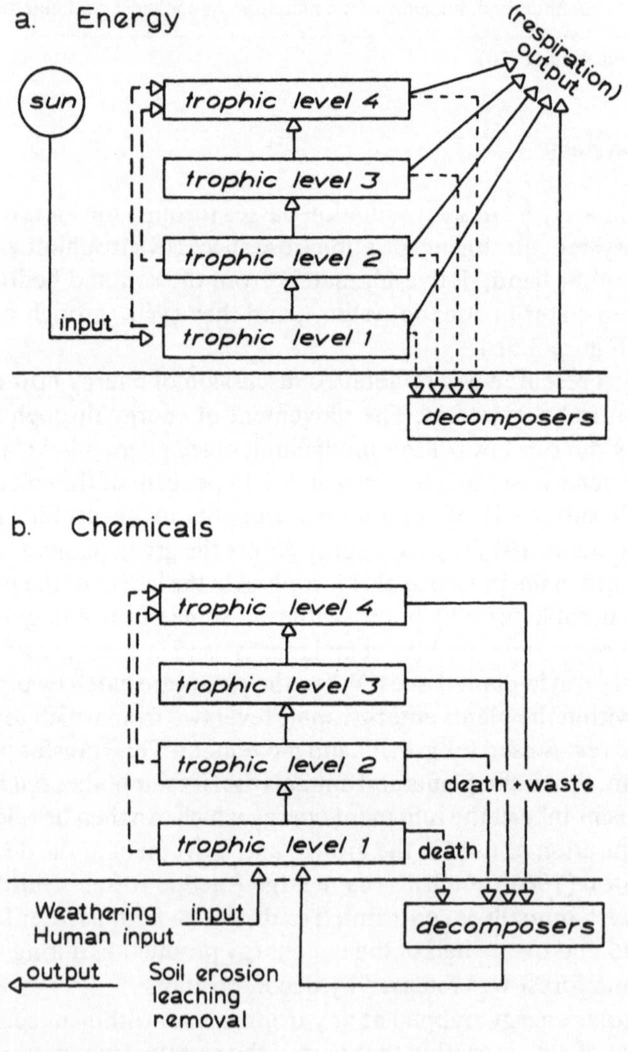

FIG. 3.5 FLOW MECHANISMS WITHIN THE ECOSYSTEM.
(a) Generalized pattern of energy flow.
(b) Generalized pattern of nutrient cycling.
In each figure the solid lines show major pathways, and the broken lines show minor pathways.

The estimated maximum plant efficiency is thus (0·286×0·43×0·80×0·67) which is 0·066 (6·6 per cent).

The second characteristic is that because the animals at trophic levels two and above are mobile they respire proportionally much more than the plants at trophic level one, and so the relative loss of energy increases at successively higher trophic levels. Hence there are rarely more than about four vertical links in a normal food web. Elton (1927) realized that the upper limit to the number of links in a food chain is a function of energy loss and that it is normally three or four.

Although the general form of energy transfers within natural ecosystems follows the pattern shown in Figure 3.5a, different ecosystems tend to direct energy along specific pathways and in this way the dominant flow paths may differ quite substantially between different types of system. Hughes (1974) has highlighted the basic differences between a

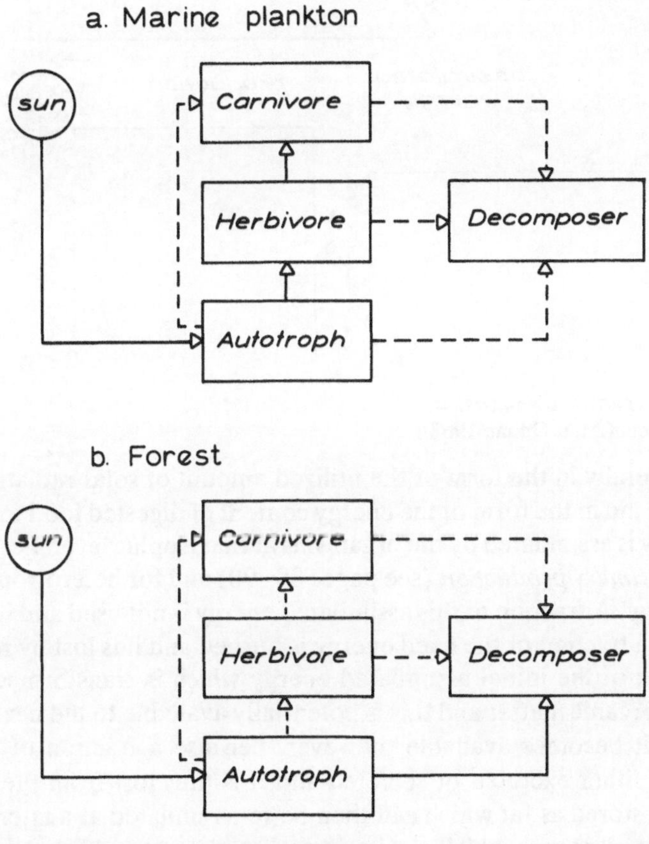

FIG. 3.6 ENERGY FLOW IN MARINE AND FOREST ECOSYSTEMS.
Major flows (solid lines) are distinguished from minor flows (broken lines). (After Hughes, 1974.)

marine food web where the structure of the ecosystem directs energy mainly through the consumer component, and the energy flow from plants and animals to the decomposers is relatively small because planktonic plants are small and can be replaced rapidly; and a forest food web where the ecosystem structure directs energy flow mainly through the decomposer component because the primary producers are too bulky and indigestible to be eaten in large quantities by herbivores (Figure 3.6).

It is important to be able to observe and quantify the direction and scale of energy flows at each trophic level within the ecosystem, and to this end Odum (1968) has devised an 'Energy Flow Diagram' as a blueprint (Figure 3.7). The diagram starts with an initial total

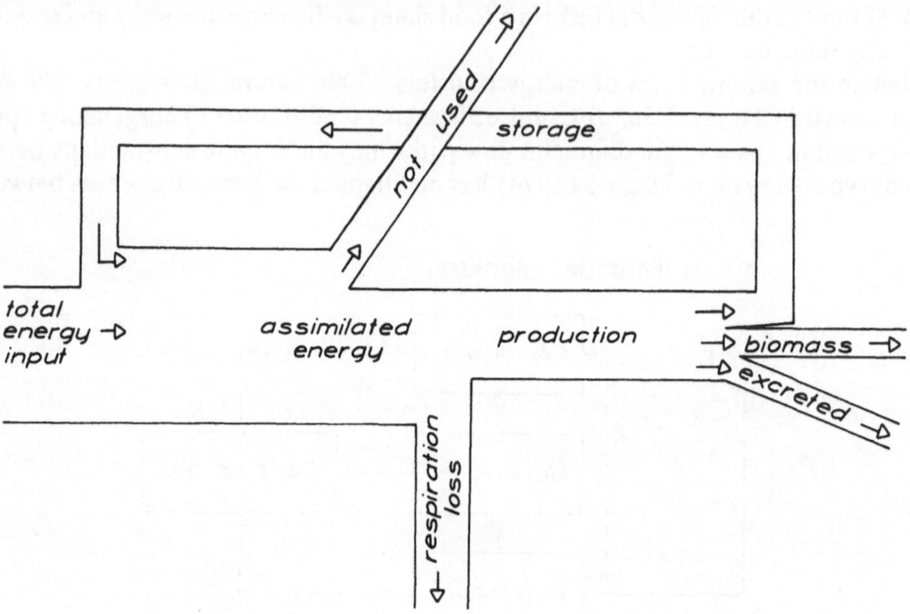

FIG. 3.7 THE ODUM *ENERGY FLOW DIAGRAM*.
See text for explanation. (After Odum, 1968.)

energy input, generally in the form of the utilized amount of solar radiation for plants at trophic level one, and in the form of the energy content of digested food for higher trophic levels. This energy is assimilated by the organisms at that trophic level. For autotrophs this represents *gross primary production* (see pages 89–90) and for heterotrophs it represents metabolized energy. A fraction of this assimilated energy is not used and it is lost from the budget. Similarly a fraction of the fixed energy is burned and it is lost by respiration. This leaves the fraction of the initial assimilated energy which is transformed into a new or different form of organic matter and this is potentially available to the next higher trophic level. Not all of it becomes available, however, because a fraction of the assimilated organic matter is either excreted or secreted and it is thus lost from the balance, and a further fraction is stored as fat which can then be re-assimilated at a later date. Thus the amount of *biomass* or growth which can be passed to higher trophic levels is a fraction of the initial assimilated energy. An example of the energy flow diagram, based on a study of energy flow in Root Spring in Massachusetts over the period 1953 to 1954 by Teal (1957), appears in Figure 3.8. There are two main inputs into the system – primary production by algae within the stream (21·7 per cent of total input) and debris carried along from upstream (77·7 per cent) in the form of leaves, fruit and branches of terrestrial vegetation, with a small supplementary input from organisms (especially caddis larvae) which migrate from upstream (0·6 per cent). There is a minor loss from the system in the form of emigration of organisms (1·1 per cent of total loss), and two major ones. These are energy which is encorporated into stream deposits (28·3 per cent) and a large loss by respiration and friction in the form of heat (70·6 per cent). The energy flow diagram is valuable because it provides the basis for evaluating energy budgets within ecosystems; it offers a

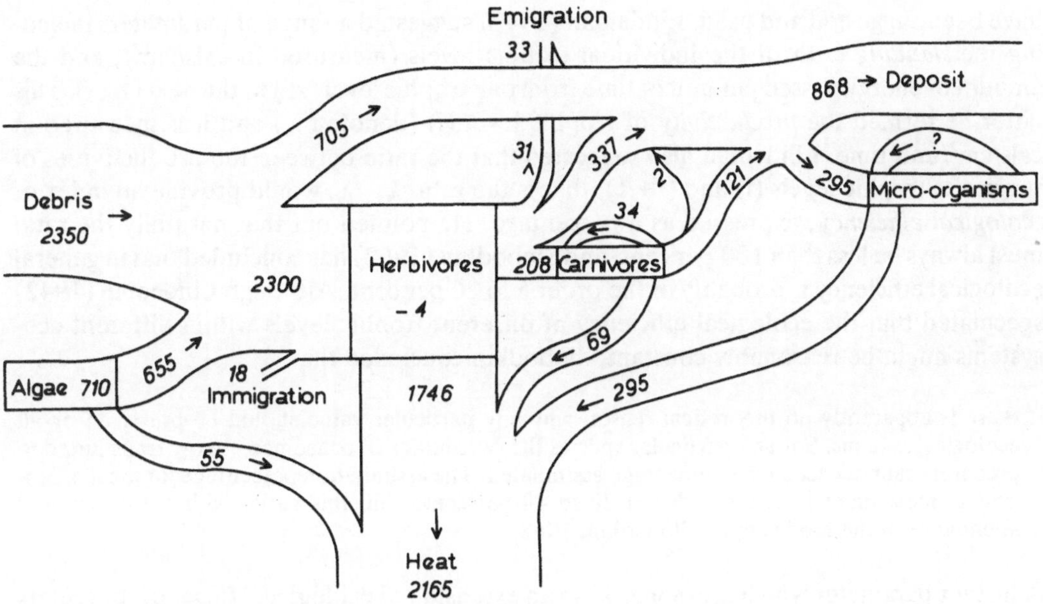

FIG. 3.8 *ENERGY FLOW DIAGRAM* FOR ROOT SPRINGS, MASSACHUSETTS (1953–4)
Based on the Odum Energy Flow Diagram (FIG. 3.7); data are in kcal/m^2/yr. Numbers in boxes indicate changes in standing crop; arrows indicate directions of flow. (After Teal, 1957.)

useful conceptual framework for energy studies; and it summarizes a complex web of flow pathways through ecosystems which is simple to follow and allows comparisons to be drawn between different ecosystems.

There are often quite pressing difficulties in measuring energy flows in ecosystems (Brock, 1967). Although it is possible to measure the energy content of light input it is difficult to determine what fraction of the light is absorbed in a photosynthetically active form. Determination of energy dissipation within ecosystems is also difficult, although attempts can be made with the use of calorimeters. Energy flow is often calculated indirectly rather than measured directly, via measurements such as the turnover of biomass. Because it is generally difficult to study energy in natural ecosystems (because of their large size and multi-variate nature), it is often convenient to simulate natural ecosystems under laboratory conditions by adopting a *microcosm approach* (Beyers, 1963) where environmental conditions can be manipulated at will and where experimental replication can be used to test field observations and theoretical predictions. Apart from these operational problems of measurement there are difficulties in establishing good estimates of energy budgets for many ecosystems because of the storage and differential release of energy from stores. Ovington (1961) showed the massive energy release from decomposers in a pine forest ecosystem; and data collected by Olson (1963) shows wide variations in the period during which an ecosystem as a whole may show a positive net storage (production minus loss) of energy in the form of dead organic matter unincorporated into mineral soils.

Despite the problems of establishing meaningful estimates or measurements of energy flows in ecosystems a number of parameters describing energy relations within ecosystems have been suggested and used. Lindeman (1942) suggested a range of parameters including the *standing crop* of the individual trophic levels (measured in cals/cm^2); and the amount of energy passed per unit of time from one trophic level (Λ_i) to the next (Λ_{i+1}). This latter he termed the *productivity* of trophic level (i) (denoted λ_i) and it is measured in cals/cm^2/unit time. Lindeman also suggested that the ratio between the productivities of species at trophic levels (i) and (i + 1), that is the ratio λ_{i+1}/λ_i, would provide an index of *ecological efficiency*, expressed as a percentage. He pointed out that naturally the ratio must always be less than 100 per cent, and Slobodkin (1962) has concluded that in general ecological efficiency is probably of the order 5 to 20 per cent. Although Lindeman (1942) speculated that the ecological efficiency of different trophic levels within different ecosystems might be reasonably constant, Slobodkin concluded that

> there is apparently no theoretical reason why any particular value should be preferred by all ecological systems. For any particular species the percentage of consumed energy transmitted to predators cannot exceed the percentage assimilated. The assimilation percentage for most organisms at most times is of the order of 20 to 40 per cent – but this varies with the nature and abundance of the food supply (Slobodkin, 1962).

A further parameter which is in some ways an extension of ecological efficiency, but refers specifically to energy conversion in autotrophs, is the *photosynthetic efficiency*. This is the ratio between energy fixation by autotrophs (measured as the energy content of the organic matter produced) and incident radiation at wavelengths suitable for photosynthesis. Wassinck (1959) suggests an overall photosynthetic efficiency for field crops in Holland of between 1 and 2 per cent, but he adds that efficiencies in the order of 7 to 9 per cent are possible for short periods of time. Ovington (1961) estimated the photosynthetic efficiency of a pine plantation at Thetford, England, to be about 2·5 per cent, and Beyers (1963) quotes efficiency values of experimental laboratory aquatic microcosms of between 2·7 and 4·0 per cent, with a mean value of 3·1.

Many studies of energy flow concentrate on establishing energy budgets for single trophic levels or for single species because this is one way of characterizing population metabolism and of evaluating the potential of that trophic level or species for ecological exploitation. O'Connor (1964) has suggested a generalized form of energy budget for the single species (Figure 3.9a) based on a breakdown between assimilated and 'wasted' energy. An example, based on the data of Engleman (1961), is shown in Figure 3.9b. Because the parameters in Engleman's study were measured independently from one another the energy balance shown does not strictly sum correctly, but the independent measurements are more valuable than estimates based on balancing the budget with only a few actual measurements because they allow the accuracy of the data to be evaluated. Several studies of energy flow at single trophic levels are summarized in Figure 3.10, and these allow comparisons to be drawn between the ecological efficiencies and exploitation potential of different ecosystems. The data for trophic level one, for example, highlight the large energy losses in photosynthesis in the managed spruce forest (Figure 3.10b), the British grassland (Figure 3.10c) and the marine phytoplankton (Figure 3.10d); the marked variations in standing crop levels between the ecosystems; the relatively small

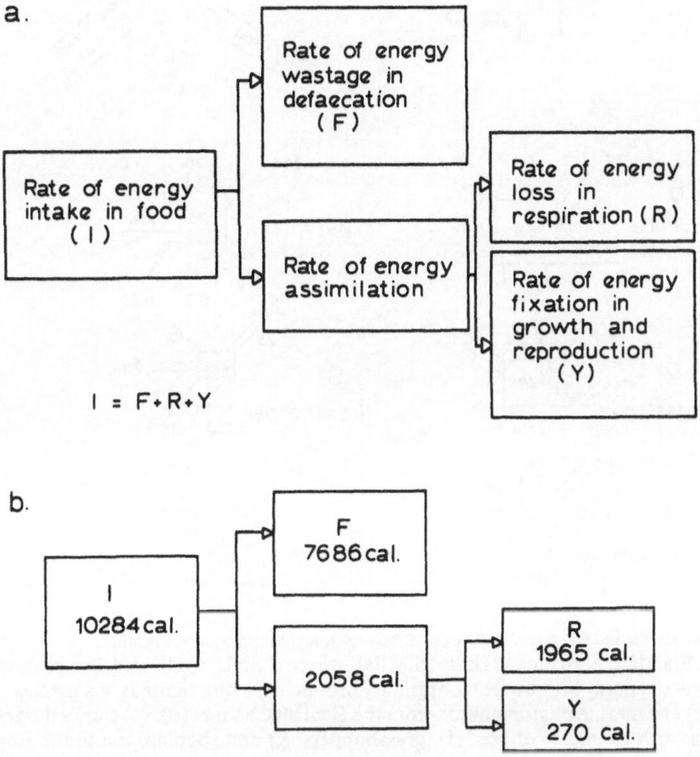

FIG. 3.9 ECOLOGICAL ENERGY FLOW AND POPULATION METABOLISM.
(a) Generalised form of energy budget. (After O'Connor, 1964.)
(b) An example from a mite population in an old-field community, based on data from Engleman (1961).

contribution of dead organic matter to decomposers in the marine phytoplankton; and the relatively high respiration losses and nėt primary productivity in the spruce forest compared with the grassland. Similar figures for trophic level two in various ecosystems also highlight basic ecological differences. The standing crop of beef (Figure 3.10f), for example, is much larger than that of the grasshoppers (Figure 3.10g) or the marine zooplankton (Figure 3.10h); the beef both loses more in respiration and produces more for carnivores than do the grasshopper or zooplankton systems; and the zooplankton system has extremely high energy losses in the form of urine and faeces.

Much research remains to be done before energy flow mechanisms and budgets within ecosystems are sufficiently well understood to allow effective and ecologically sound management, and it is one of the prime tasks of the *Man and the Biosphere* programme to

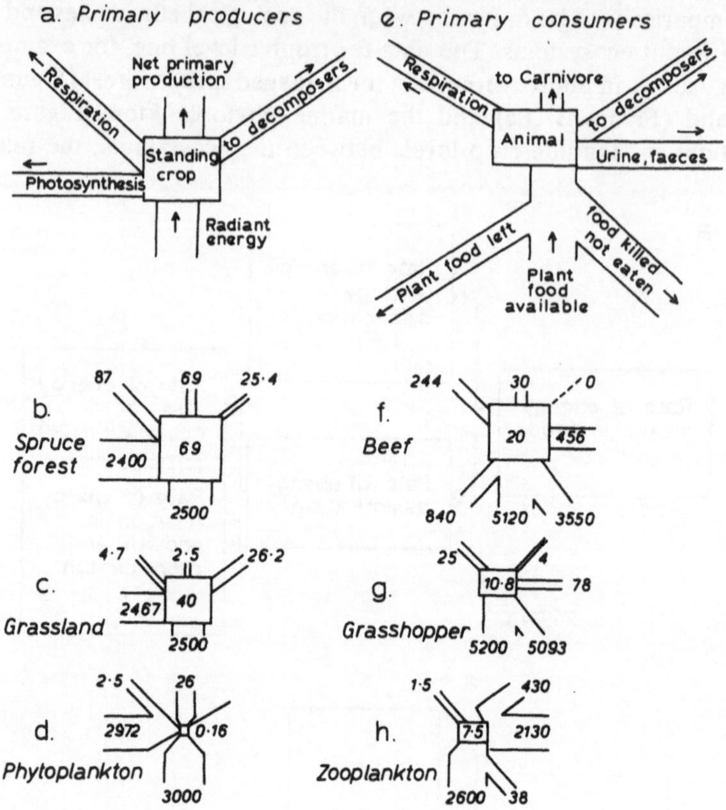

FIG. 3.10 SUMMARY OF ENERGY FLOW CHARACTERISTICS IN A RANGE OF ECOSYSTEMS.
The units used are Standard Nutritional Units (SNU's), where 1 SNU = $10^9 \times 4.184$ joules/ha/yr.
(a) Generalized form of energy flow budget for primary producers, with examples of a managed spruce forest (b), British grassland (c) and marine phytoplankton from the English Channel (d). (e) shows the generalized form for primary consumers, with examples of beef (f), grasshoppers (g) and zooplankton in the English Channel (h). (After Reid, 1969.)

study ecosystems in a variety of environments and to collect data on dynamic aspects of ecosystem structure and operation (Table 2.7).

3.2e Nutrients in ecosystems

Unlike energy, the flow pathways of chemical elements within ecosystems are essentially cyclic (see Chapter 2). There are a large number of chemicals which cycle through ecosystems including major elements such as nitrogen (N), phosphorus (P), potassium (K), calcium (C), magnesium (Mg), sulphur (S) and iron (Fe) which are required in relatively large quantities, and a large number of trace elements required in smaller quantities but nonetheless essential. Each chemical element is of importance in different ways (Table 3.8). The basic source for many of the chemicals is from physical and chemical weathering of bedrock and superficial deposits, although human inputs in the form of fertilizer applications and chemical pollution can be important in some areas (see Chapter 7). The chemicals are withdrawn from the underlying material by autotrophic green plants by the process of *root osmosis*, in solution in the water. Once taken in by the plants the elements are subsequently passed up through the trophic structure along the feeding web. The dead organisms from each trophic level, and waste material from trophic level two upwards, are worked upon by decomposer organisms which break down the material and release the chemical elements to the cycle again. In this way the cyclic mechanism is maintained and the decomposers clearly play an important role. The equilibrium of the

Table 3.8
ECOLOGICAL SIGNIFICANCE OF SPECIFIC NUTRIENTS

Element	Role in Biosphere
Oxygen	very active chemically; it combines with most elements in nature and the process of oxidation speeds metabolic processes. Oxygen forms an average of 70 per cent of the atoms in living matter, and it is a basic building block of carbohydrates, fats and proteins
Hydrogen	an important constituent of living matter; on average it constitutes about 10·5 per cent of the atoms
Carbon	another important constituent (18 per cent on average) of living matter
Nitrogen	an important constituent of protein molecules
Sulphur	an important, though small, constituent of some proteins
Phosphorus	a constituent of protein molecules, including ATP, the chemical energy-storing compound which is important in photosynthesis. Normally the element which limits the growth of ecosystems

Source: Bradshaw (1977)

cycle can be disrupted by soil erosion, the leaching of elements deep into soil profiles beyond the reach of plant roots, water flow into streams and extraction for resource use (such as guano). In these cases the outputs of minerals from the ecosystem exceed the inputs and this leads to degeneration of the soil structure. It hence induces a breakdown of ecosystem equilibrium and stability. Such changes might be triggered off in a number of ways, including the over-grazing of grassland ecosystems which can lead to soil erosion and destruction of soil structure, and thus promote ecosystem deterioration.

Gersmehl (1976) proposed a relatively simple model of ecosystem mineral cycles in general (Figure 3.11) based on three principles of nutrient cycling. One is that the total amount of mineral nutrients in an ecosystem depends on the rate of movement of nutrients into and out of the ecosystem. The second is that the amounts of nutrients within the living biomass and litter and soil components of the ecosystem are a function of the transfer rates of those nutrients between the components. The final principle is that with time an ecosystem tends towards an equilibrium condition in which the quantity of nutrients within the system and within each compartment remains the same. Although nutrient cycles have been studied in a variety of ecosystems, and generalized forms of nutrient budgets and flow mechanisms have been established for a number of chemical elements (for example Figure 2.4), detailed measurements within specific ecosystems over relatively long periods of time are required before accurate modelling and prediction of likely consequences of modifications to nutrient cycles can become a viable reality. Indeed Meentmeyer and Elton point out that 'the lack of representative data [on spatial variations in nutrient stores and rates of transfer] has hindered the incorporation of nutrient cycling within biogeography' (Meentmeyer and Elton, 1977). Although the *International Biological Program* has recently collected valuable nutrient cycling data for many specific ecosystems, information on global-scale spatial variations and on large-scale global equilibrium simply does not exist at the present time. Meentmeyer and Elton (1977) have thus considered methods for estimating rates of nutrient cycling throughout the world, and they suggest that the most feasible method of estimation is to base calculations on climatic inputs of energy and

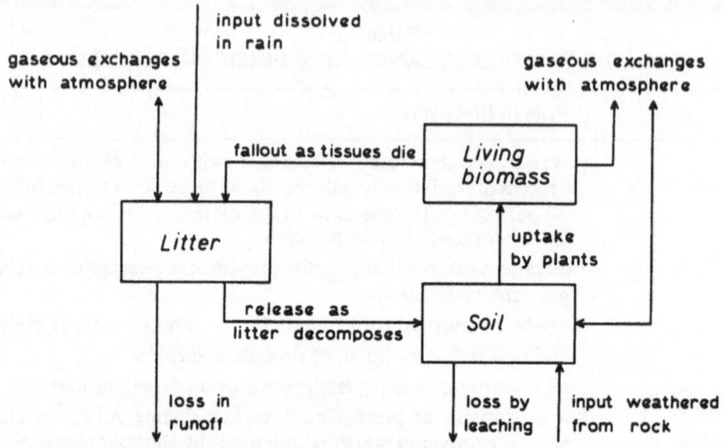

FIG. 3.11 GENERALIZED MODEL OF NUTRIENT CYCLING IN ECOSYSTEMS.
(After Gersmehl, 1976).

moisture into regional ecosystems (data on these are available on the global scale), and to use actual evapotranspiration estimates to predict *net primary productivity*. Net primary productivity could then be used as an index of nutrient cycling characteristics on a world scale.

Ecologists have tended to devote most of their attention within nutrient cycling to three main areas of enquiry – the evaluation of nutrient budgets within specific (and generally experimental) ecosystems; the impact of ecosystem management and ecological exploitation of stability of nutrient cycles; and the effects of pollution on nutrient cycles.

One example of analysis of nutrient budgets is the *Hubbard Brook Ecosystem Study*, which has produced much of the currently available information on woodland ecosystems. The study is located in the White Mountains of New Hampshire, USA, where a series of adjacent watersheds are being studied as ecosystems. The study has been under way since the early 1960s, and it has four main aims. These are to analyse the size and compositional characteristics of the forest trees; to determine the productivity and biomass of the forest; to study and model the forest dynamics; and to establish the basis for studying nutrient cycling within the forest community. An ecosystem approach is thus being used to quantify nutrient budgets and cycles within the northern hardwood forest (Bormann and Likens, 1967). The forest is drained by a series of perennial streams and the major losses of nutrients from the ecosystem are in the form of particulate matter removed in surface drainage waters and in solution in surface and sub-surface drainage waters (Bormann, Likens and Eaton, 1969) (Figure 3.12b). A basic part of the nutrient cycle is the movement of elements from the forest canopy to the soil (Figure 3.12a), and two main processes for this have been isolated. The first is the transfer of nutrients as leaves and various parts of the trees and other plants fall to the ground as litter, which is then leached by percolating water and decomposed by organisms (Gosz, Likens and Bormann, 1972). The second process is the transfer of nutrients from plants to soil as precipitation passes through the forest canopy. This adds nutrients directly to the available nutrient pool within the soil without the intervention of any process of decomposition on the forest floor (Eaton,

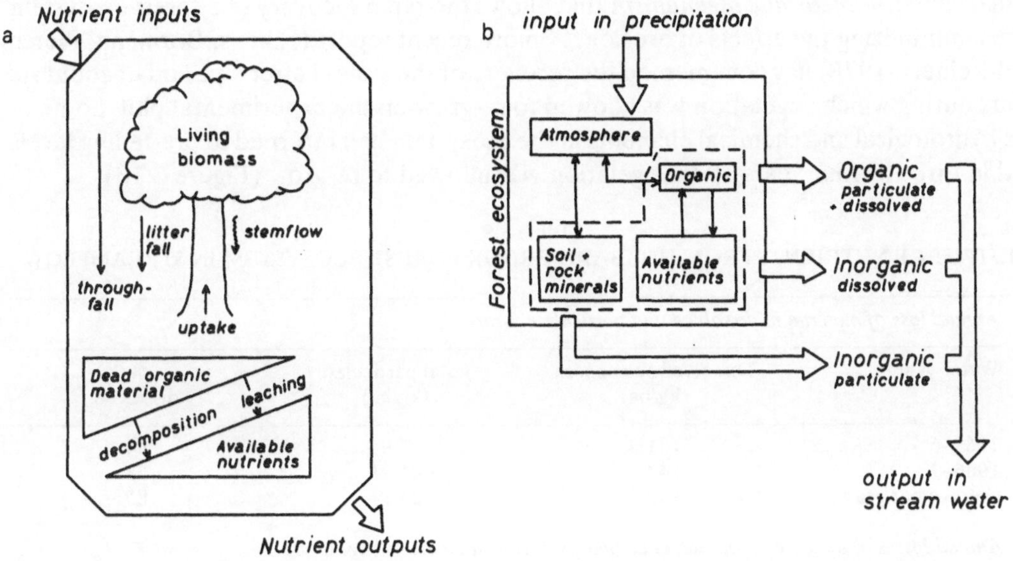

FIG. 3.12 MAIN PATHWAYS OF NUTRIENT CYCLING WITHIN THE HUBBARD BROOK ECOSYSTEM STUDY.
(a) The forest ecosystem sub-cycle, showing processes of movement of nutrients through and from the forest canopy to the soil, and nutrient outputs from the forest and soil. (b) Removal of particulate and solution material from the forest ecosystem in surface and sub-surface drainage waters. (After Bormann, Likens and Eaton, 1969.)

Likens and Bormann, 1973). Monitoring of nutrient cycling over a period of years within the ecosystem allowed Bormann, Likens and Eaton (1969) to evaluate the overall losses of nutrient material from the area in the form of dissolved substances and particulate matter. Clearly most of the losses (Table 3.9a) occurred in solution and certain elements were lost in greater amounts than others. Thus significant amounts of calcium, magnesium and sodium were lost from the system (Table 3.9b). Overall, however, nutrient losses were quite low in relation to the total nutrient budget, and this led Bormann, Likens and Eaton (1969) to conclude that the Hubbard Brook ecosystem is very stable overall.

As an experiment designed to test this apparent stability some 15.6 ha of the forest ecosystem was completely felled during the winter of 1965–6, but the system was altered as little as possible otherwise – none of the forest products were removed from the site, and soil disturbance was minimized. Regrowth of vegetation on the area during the summer of 1966 was inhibited by the aerial application of herbicides, and then the area was left to regenerate itself by natural secondary succession (see Chapter 4). Over this period of vegetation clearance the nutrient budget and equilibrium changes were monitored in detail (Bormann, Likens, Siccama, Pierce and Eaton, 1974). The export trends for both dissolved and particulate matter after deforestation were found to be non-synchronous.

The first response was a mobilization of nutrients from storage in the soils and from leakage into stream water, and so the dissolved nutrient losses were experienced without any appreciable time-lag. A sharp rise in particulate matter output was not observed until nearly two years after deforestation; by which time the export of dissolved nutrients was in decline, probably because of a reduction in the amounts of readily available nutrients in the system due to depletion. Bormann and his colleagues (1974) interpret these findings as evidence of a *homeostatic mechanism* that allows the rapid recovery of a forest ecosystem whilst minimizing the effects of erosion. A more recent report (Likens, Bormann, Pierce and Reiners, 1978) has documented the recovery of the system after a period of about six years during which vegetation was allowed to re-grow on the experimental plot. Some of the hydrological and chemical elements in the ecosystem had returned to pre-felling levels within three or four years after vegetation was allowed to re-grow (Figure 3.13).

Table 3.9
OUTPUT OF NUTRIENTS FROM HUBBARD BROOK FOREST ECOSYSTEM IN STREAMFLOW

A *Annual loss of material in dissolved and particulate forms*

Water year	Dissolved substances (kg/ha)	Total particulate (kg/ha)	% dissolved
1965–6	131·2	7·94	94
1966–7	147·3	41.91	78
average per year	139·3	24·91	85

B *Annual losses of dissolved material, classified by chemical elements (kg/ha/yr)*

Substance	Symbol	1965–6	1966–7	Average
Aluminium	Al^{+++}	2·4	2·8	2·6
Ammonium	NH_4^+	0·9	0·4	0·65
Calcium	Ca^{++}	9·0	10·7	9·85
Magnesium	Mg^{++}	2·7	2·9	2·8
Potassium	K^+	1·3	1·7	1·5
Sodium	Na^+	6·3	6·8	6·55
Bicarbonate	HCO_3^-	0·4	2·0	1·2
Chlorine	Cl^-	4·3	4·7	4·5
Nitrate	NO_3^-	6·5	5·9	6·2
Sulphate	SO_4^{--}	47·2	51·4	49·3
Silicon dioxide	SiO_2	31·3	36·8	34·05
Dissolved organic carbon	C	18·9	21·2	20·05
TOTAL		131·2	147·3	139·3

Source: Bormann, Likens and Eaton (1969)

Net losses of nutrients from ecosystems can arise if parts or all of the ecosystem are cropped or otherwise exploited as a resource. Rennie (1955) has attempted to evaluate nutrient removal from forest sites via the various components of timber thinning and clear felling, by collecting together published data from 42 studies. His analysis focused on calcium, potassium and phosphorus. The analysis shows that timber-exploited forests differ quite substantially from unexploited virgin forest in nutrient cycling, because of the continuous and unavoidable removal of nutrients from the site in the timber resources.

Although in absolute terms tree removal may be relatively small, the continued productivity of an ecosystem depends heavily on the ability of the underlying soil to replenish nutrient losses; and it is the ironic reality of forestry management that much afforestation has taken place on nutrient-deficient marginal upland soils. Rennie suggests that 'the nutrient uptakes of timber-producing forest of either coniferous or hardwood species are

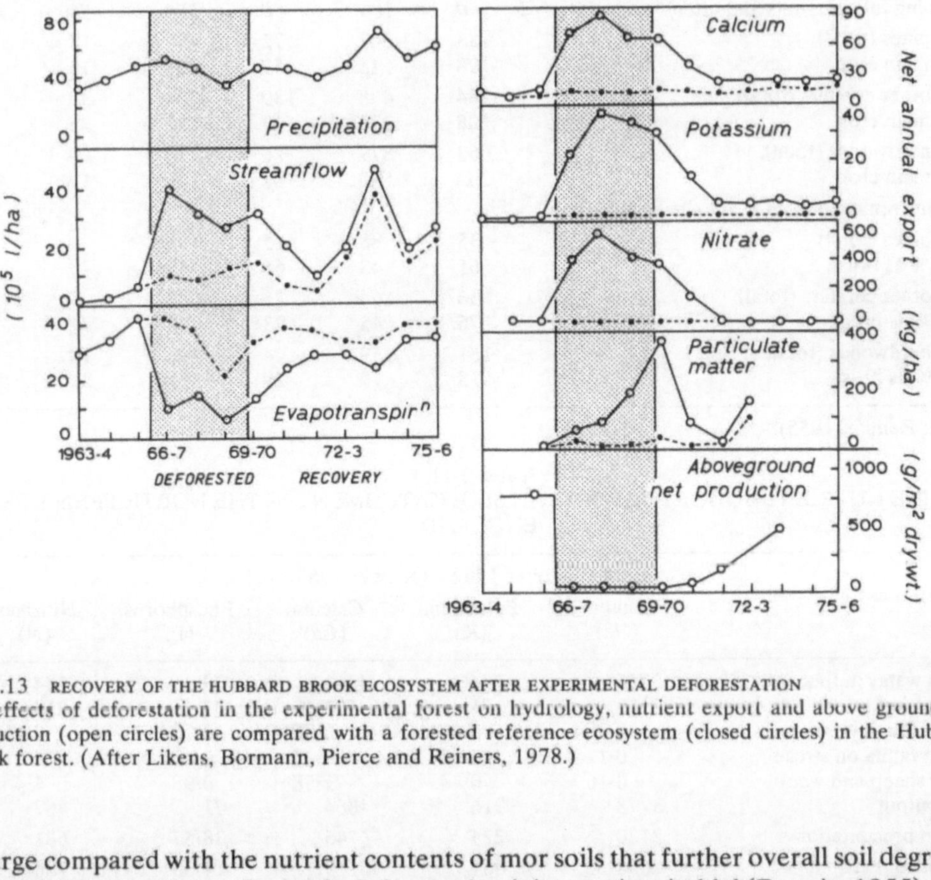

FIG. 3.13 RECOVERY OF THE HUBBARD BROOK ECOSYSTEM AFTER EXPERIMENTAL DEFORESTATION
The effects of deforestation in the experimental forest on hydrology, nutrient export and above ground net production (open circles) are compared with a forested reference ecosystem (closed circles) in the Hubbard Brook forest. (After Likens, Bormann, Pierce and Reiners, 1978.)

so large compared with the nutrient contents of mor soils that further overall soil degradation and, sooner or later, diminished site productivity, are inevitable' (Rennie, 1955). The effects of forestry practice on nutrient availability and export are summarized in Table 3.10.

An alternative ecological resource management situation is the grazing of moorland or other upland ecosystems by sheep. Crisp (1966) measured inputs and outputs of nutrients in a blanket bog system at Rough Sike in the English Pennines over a year to establish the nutrient budget and evaluate nutrient losses. Over the year net nutrient losses were relatively high. Losses per hectare ranged from 0·17 to 0·40 kg/yr for phosphorus to nearly 50 kg/yr for calcium, and the nitrogen data show an annual output of twice the input. Crisp therefore considered and measured the components of the nutrient cycles to assess the relative importance of various possible loss pathways. Certain factors such as the drift of

Table 3.10
FOREST EXPLOITATION AND THE AVAILABILITY AND EXPORT OF NUTRIENTS

(kg/ha)	Calcium (C)		Potassium (K)		Phosphorus (P)	
Available nutrients in calluna mor soil:						
(a) top 28 cm	27		17		19	
(b) top 50 cm	36		32		23	
Total nutrient uptake by forest growth			(years of growth)			
(excluding foliage, roots and litter):	*50*	*100*	*50*	*100*	*50*	*100*
(a) pines (total)	133	203	72	91	17	21
main crop	98	115	57	56	13	12
(b) other conifers (total)	264	438	139	234	30	41
main crop	208	274	113	152	25	28
(c) hardwoods (total)	463	879	126	225	29	50
main crop	302	520	82	129	19	28
Nutrients removed from forest site by thinning:						
(a) pines (total)	35	88	15	35	4	9
% as brush	65	43	66	49	75	55
(b) other conifers (total)	56	164	26	82	5	13
% as brush	76	45	73	44	77	52
(c) hardwoods (total)	161	359	44	96	10	22
% as brush	15	19	29	38	21	27

Source: Rennie (1955)

Table 3.11
OUTLINE NUTRIENT BUDGET FOR ROUGH SIKE CATCHMENT IN THE NORTH PENNINES OF ENGLAND

	(kg/ha) October 1962 – October 1963				
	Sodium (Na)	Potassium (K)	Calcium (Ca)	Phosphorus (P)	Nitrogen (N)
Stream water output	3755	744	4461	33	244
Peat erosion	23	171	401	37	1214
Drift of fauna in stream	0·004	0·011	0·003	0·010	0·118
Drift of fauna on stream	0·11	0·38	0·07	0·43	4·6
Sale of sheep and wool	0·16	0·44	1·58	0·98	4·4
Total output	3778	916	4864	71	1467
Input in precipitation	2120	225	745	38·57	681
Net loss for catchment	1658	661	4119	14·33	786
Net loss per ha	20·01	7·97	49·68	0·17–0·40	9·48

Source: Crisp (1966)

fauna either in or on the streamwater proved to be very small components of the total loss for most elements (Table 3.11), and although ecological resource exploitation in the form of selling sheep plus wool from the ecosystem was practised this had relatively little direct effect on mineral losses.

Since most pollutants in the environment have a chemical basis whether they are water pollutants, air pollutants or chemicals sprayed on to or dug into soils for agricultural purposes, it is inevitable that pollution can have a radical impact on nutrient cycling in ecosystems. Although this theme will be considered in detail in Chapter 7, it is convenient to point out that chemical pollution, particularly that associated with the introduction of toxic material into formerly natural ecosystems, can disrupt ecosystem stability by upsetting equilibrium states in nutrient cycles, by altering the range and types of ecological niche in the host ecosystem and by passing through food webs so that they become concentrated in the upper trophic levels where they can often have lasting and dramatic effects (Figure 3.14).

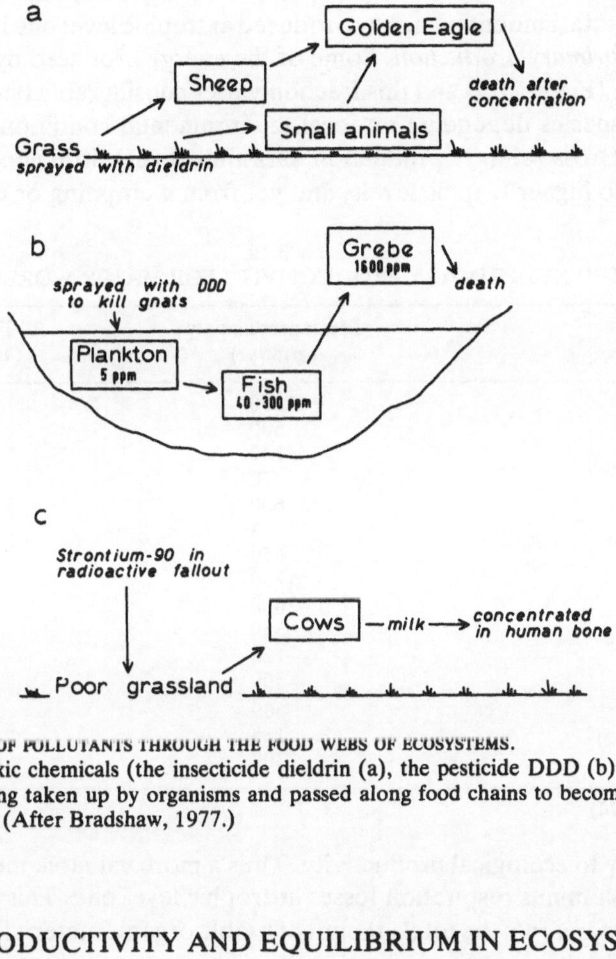

FIG. 3.14 MOVEMENT OF POLLUTANTS THROUGH THE FOOD WEBS OF ECOSYSTEMS.
Three examples of toxic chemicals (the insecticide dieldrin (a), the pesticide DDD (b) and the radionuclide Strontium-90 (c)) being taken up by organisms and passed along food chains to become concentrated in the highest trophic levels. (After Bradshaw, 1977.)

3.3 PRODUCTIVITY AND EQUILIBRIUM IN ECOSYSTEMS

3.3a Ecosystem productivity

One of the most important characteristics of an ecosystem from an ecological point of view as well as from a management or exploitation perspective is its *productivity*. This can reveal much about the 'state of health' of an ecosystem; different systems can be directly compared on the basis of productivity; and the significance of various environmental limiting factors can be assessed.

The process of 'building' organic matter in an ecosystem is closely dependent on the availability and movement of energy through the system, and the movement is essentially uni-directional with one single major source – the sun. There are thus two key factors in determining ecological productivity – the amount of solar radiation which is available to the primary producers at trophic level one for photosynthesis, and the efficiency with which the autotrophs convert the solar energy into usable forms by photosynthesis. All of the electrical and thermal energy used by plants is converted into chemical energy and so, at least in theory, a measurement of the amount of sugar produced in plants at trophic level one should provide an index of energy uptake by the plants, and thus a measure of productivity. The total amount of energy produced at trophic level one in an ecosystem is a measure of *gross primary production*. Some of the energy produced by photosynthesis is lost by respiration (Figure 3.5) and this fraction varies considerably between species, and within the same species depending on local environmental conditions such as weather (Billings, 1964). Gross primary production data thus reveal little about the amounts of energy available to higher trophic levels, and yet from a cropping or extraction point of

Table 3.12
ESTIMATED NET PRIMARY PRODUCTIVITY FOR MAJOR WORLD BIOMES

Vegetation unit	Mean productivity ($g/m^2/yr$)	Total for area (10^9 tonnes/yr)
Forests:	1290	64·5
Woodland:	600	4·2
Tundra:	140	1·1
Desert scrub:	70	1·3
Grassland:	600	15·0
Desert:	3	—
Cultivated land:	650	9·1
Fresh water:	1250	5·0
Reefs and estuaries:	2000	4·0
Continental shelf:	350	9·3
Open ocean:	125	41·5
Upwelling zones:	500	0·2
Total continental:	669	100·2
Total oceanic:	155	55·0
WORLD TOTAL:	303	155·2

Source: Simmons (1974)

view this is the key to ecological productivity. Thus a more valuable measure is the gross primary production minus respiration losses at trophic level one. This is the *net primary production* and it represents the total amount of usable organic material produced per unit of time at trophic level one. It is conventionally measured in $g/m^2/yr$ or $g/m^2/day$. Lewis (1974) has measured primary production in a trophic lake (Lake Lanao in the Phillipines) and he established that net primary production is 1·7 $g/m^2/day$, whereas gross primary production is 2·6 $g/m^2/day$. Net production in this case was 65 per cent of the gross production, and respiration losses amounted to 35 per cent at trophic level one.

Because productivity is controlled ultimately by the amount and spectral composition of solar radiation there are marked differences in productivity through space across the earth's surface. The net primary productivity of most major ecosystems has either been measured or estimated, and Table 3.12 summarizes a number of natural ecosystems. The highest productivities are found in the reef and estuary ecosystems, with natural forests and fresh water a close second. The lowest net productivities occur in the open ocean areas and perhaps predictably in desert environments such as desert scrub and open desert. Whilst Table 3.12 suggests that overall mean net productivity levels on land are more than four times those in ocean areas as a whole, when these productivities are weighted according to the total areas of the earth's surface represented in the two classes the ocean areas far exceed land in area, and so total net primary productivity of land is only twice that in the oceans. It is interesting to note from Table 3.12 that the productivity of cultivated land overall is 650 g/m^2/yr, which is not substantially higher than that of natural grassland (600 g/m^2/yr). This suggests that perhaps cultivation is a somewhat inefficient form (in an ecological sense) of land use when the amount of time, effort and resources devoted to it are taken into account. Table 3.12 only shows average productivity values for major vegetation types, however, and these disguise both marked regional variations within vegetation types and local variations due to environmental and management variations. The productivities of some cultivated ecosystems are summarized in Table 3.13. The data

Table 3.13
ESTIMATED NET PRIMARY PRODUCTIVITY FOR VARIOUS NATURAL, SEMI-NATURAL AND CULTIVATED ECOSYSTEMS

Ecosystem	Productivity (g/m^2/yr)
Natural and semi-natural:	
Giant ragweed (Oklahoma):	1442
Spartina saltmarsh (Georgia):	3285
Pine plantation, 20–35 yr old (England):	2190
Deciduous plantation, 20–5 yr old (England):	1095
Desert (Nevada):	40·15
Seaweeds (Nova Scotia):	723
Cultivated:	
Wheat (world average):	343
Rice (world average):	496
Potatoes (world average):	402
Sugar cane (world average):	1726
Mass algal culture, outdoors:	4526

Source: Odum (1959)

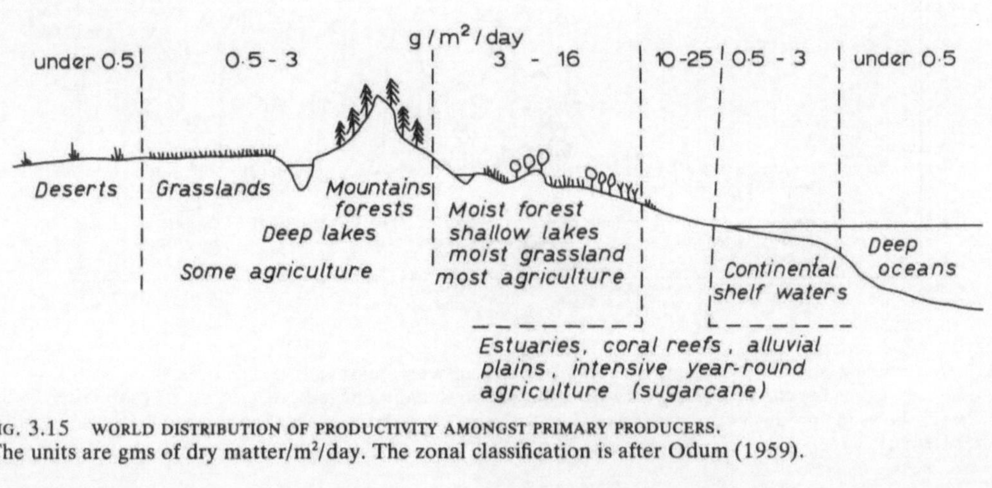

FIG. 3.15 WORLD DISTRIBUTION OF PRODUCTIVITY AMONGST PRIMARY PRODUCERS.
The units are gms of dry matter/m^2/day. The zonal classification is after Odum (1959).

for cultivated systems are striking. Rice and potatoes – the staple diet for many parts of the world – are clearly inefficient crops to grow, with productivities in the order of 400 to 500 $g/m^2/yr$ (less than natural grassland!); whereas sugar cane is ecologically efficient as a crop and mass algal cultures are extremely efficient in converting solar energy into food.

It is convenient to consider variations in primary productivity on two scales – the regional and the local. On a regional perspective Odum (1959) recognizes three main levels of productivity in the world (Figure 3.15). There are first of all regions of high ecological productivity such as shallow water areas, moist forests, alluvial plains and areas

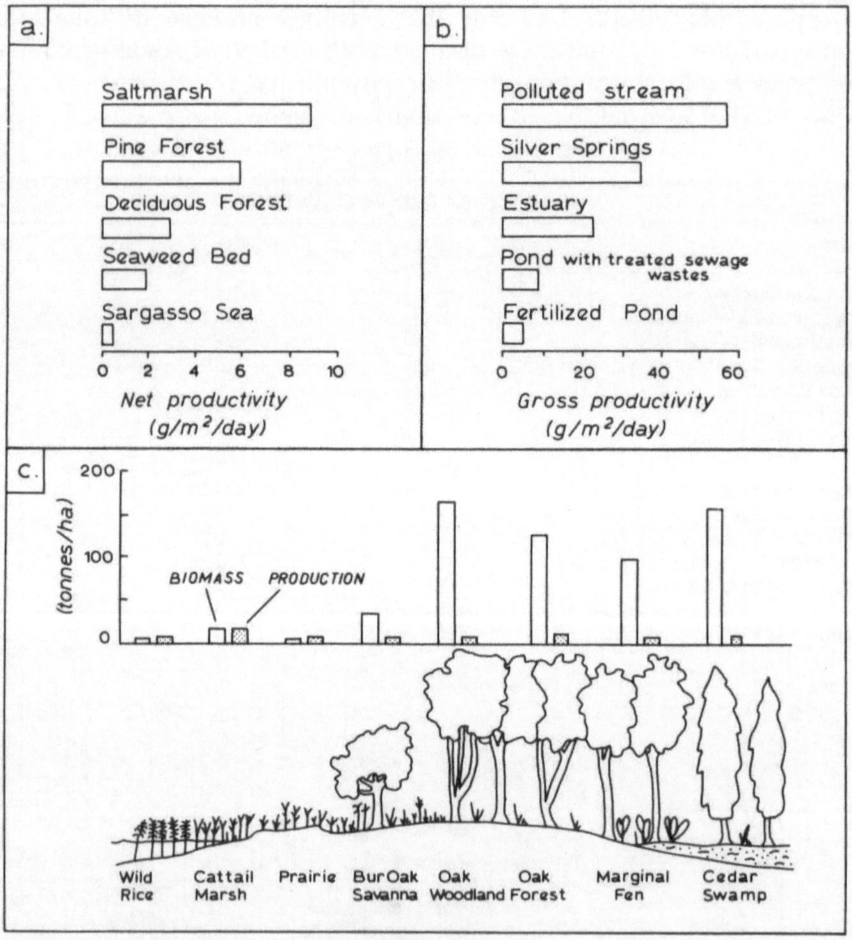

FIG. 3.16 SOME EXAMPLES OF VARIATIONS IN PRODUCTIVITY BETWEEN DIFFERENT ECOSYSTEMS.
Natural ecosystems (a) can be compared with some polluted and managed ecosystems (b) (both after Gates, 1972); (c) shows the productivity variations within different ecosystems of the Anoka Sand Plain region, USA. (After Reiners, 1972.)

of intensive cultivation. These are complemented by areas of intermediate productivity such as areas of grassland, shallow lakes and most farmland: and areas of low productivity such as arctic wastelands, deserts and deep ocean areas. Variations in productivity at the regional and global scale will be considered further in Chapter 5.

At the local scale there are a large number of published studies of ecosystem productivity and energy production from both natural and managed or polluted ecosystems (Figure

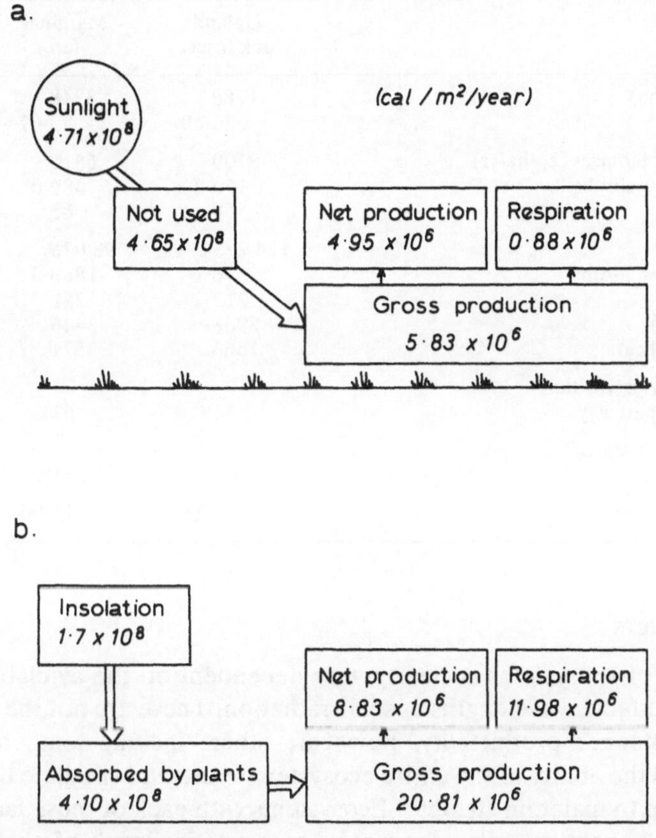

FIG. 3.17 ENERGY AND ECOLOGICAL PRODUCTIVITY.
(a) Energy relationships within the perennial herb grass vegetation of an old-field community in Michigan. (After Phillipson, 1966.)
(b) Energy relationships within Silver Springs ecosystem, Florida. (After Odum, 1957.)

3.16) which illustrate wide variations in productivity even within reasonably small areas (Figure 3.16c). One of the best documented ecosystems is an old-field community in Michigan studied by Odum (1959) and illustrated in Figure 3.17a. Over 98 per cent of the solar energy available to the system was not used, and only 1·24 per cent was available for plant growth and respiration. Gross primary production was thus a mere 1·24 per cent of the initial solar input. Respiration losses amounted to 15 per cent of the gross production and this left $4{\cdot}95 \times 10^6$ cal/m^2/yr as net primary production. The food energy potentially available to herbivores and carnivores was thus only 1·05 per cent of the original solar input. Odum (1957) has also studied the production levels in Silver Springs, Florida by the same approach; and net production in that ecosystem was even less, at 0·52 per cent of the original insolation (Figure 3.17b). Further information on productivity variations at the local scale is offered by Reiners (1972) who studied the structure and energetics of three forest ecosystems in Minnesota (Table 3.14).

Table 3.14
STRUCTURE AND ENERGETICS OF THREE FOREST ECOSYSTEMS IN MINNESOTA, USA

	Upland oak forest	Marginal fen	Cedar swamp
Total density (stems/ha):	1788	3348	2755
Basal area (m^2/ha):	26·49	25·07	42·22
Annual production of biomass (kg/ha/yr)	8700	6512	10 139
–herbaceous and low shrubs:	150·2	489·0	179·6
–tall shrubs:	58	65	–
Total biomass (kg/ha):	124 273	98 075	159 406
–herbaceous and low shrubs:	396·1	1880·9	542·2
–tall shrubs:	212	281	–
–dead wood (trunk):	8988	3446	3566
–dead wood (branches):	1663	571	317
Percentage of primary production entering decomposer pathways:	52%	62%	48%
Total detritus as percentage of above-ground biomass:	61%	331%	577%
Biomass/production ratio:	14·28	15·06	15·72

Source: Reiners (1972)

3.3b Limiting factors

The productivity of ecosystems is clearly very dependent on the availability in suitable amounts and at suitable wavelengths of solar radiation. These are not the only important factors which influence productivity, however; others include temperature, nutrient supply, water and the ability of the entire ecosystem to use and circulate the material and energy which help to maintain stability. Ecosystems with each of these factors present in suitable forms and quantities often show quite high relative levels of productivity. Quite often, however, one or more factors is in 'short supply' and this can inhibit efficient and productive ecological development. Such a factor is termed a *limiting factor*. For example in many desert areas there are plentiful supplies of nutrients, sunlight and temperature to support a reasonably healthy vegetation, but vegetation is generally sparse because water is a limiting factor.

Charles Darwin (1859) was one of the first to point out that although animals often have the capacity to increase their numbers within an area, in fact they do not do so. He identified four checks to continuous increases in populations – the amount of food available; the effect of predation by other animals; the effects of physical factors such as climate; and the effects of disease on populations. The 'Concept of Limiting Factors' can be traced back to the 1840s when a German agricultural chemist – von Leibig – found that the yield of a crop could be increased only by supplying the plants with more of the nutrient which was present in the smallest quantities (Billings, 1964). This suggested to von Leibig a 'Law of the Minimum' whereby productivity, growth and reproduction of organisms would be constrained if one or more limiting factors was present in less than a minimum threshold quantity, and yields would increase in direct proportion to the amount of that limiting factor added. Ecologists more recently have conceived a 'Law of the Maximum' because productivity and growth can be constrained as much by an excess of some limiting factors (such as heavy moisture conditions, over-abundance of some critical chemical elements) as by a shortage.

Blackman (1905) attempted to rationalize the concept of limiting factors and he proposed that there is a spectrum of significance of any one limiting factor, with upper and lower critical threshold conditions and an optimum. In theory, therefore, each limiting factor has three critical values (Figure 3.18) – a *minimum* condition below which the

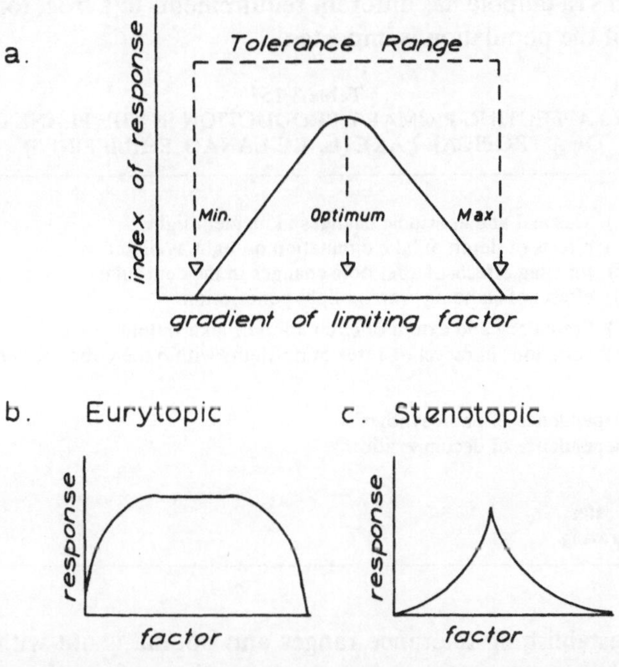

FIG. 3.18 TOLERANCE RANGES AND LIMITING FACTORS.
The generalized form of the tolerance response curve (a) can be contrasted to the eurytopic (b) and stenotopic (c) forms. See text for discussion.

phenomenon ceases altogether; the *optimum* condition where the phenomenon is exhibited at its most pronounced level; and the *maximum* condition above which the phenomenon ceases again. Blackman (1905) did point out, however, that it is extremely difficult to identify these critical states with any degree of certainty. Figure 3.18 summarizes the influence of a limiting factor on population response (measured by rate of respiration, rate of growth or rate of movement). It suggests that a population can only exist over a portion of the total range of variation of a given limiting factor (the *tolerance range* for that factor), and that the population is better adapted to certain parts of the range (the *optimum*) than to others. The range over which the adaptation is better is of basic importance because for any one limiting factor some organisms will show an optimum reaction to a wide range of variations of the factor (the *eurytopic curve* shown in Figure 3.18b), whereas other organisms may have a very narrowly defined optimum for the same factor (the *stenotopic curve* in Figure 3.18c). Inevitably the more narrowly defined the optimum the more critical is that particular limiting factor for that organism, because a small change in the limiting factor might trigger off a large, and perhaps irreversible, change in that organism's reaction (such as in reproductive or survival rates). On the other hand with the same species a broad optimum for a given environmental factor means that the organism can tolerate a wide range of variations without severe reactions. A given population might be eurytopic for some factors and stenotopic for others, and so it becomes important to establish the population reaction to environmental controls. A population's response to a given factor might also change through the life-cycle of the individual organisms (a tadpole has different requirements to a frog, for example) and so the age structure of the population is important.

Table 3.15
LIMITING FACTORS AFFECTING PRIMARY PRODUCTION IN THE PLANKTON COMMUNITY OF A TROPICAL LAKE (LAKE LANAO, PHILIPPINES)

A	Resource supply:		
	(i) Light	(a)	seasonal and aperiodic changes in incident light
		(b)	effects of depth of lake circulation on light availability
		(c)	time-lag effects of aperiodic changes in incident light
		(d)	effects of standing crop on light penetration
	(ii) Nutrients	(a)	Frequency and extent of circulation of lake waters
		(b)	variations in recycling rates of nutrients within the euphotic zone
B	Temperature:		
	(i) Temperature dependence of photosynthesis		
	(ii) Temperature dependence of decomposition		
C	Biomass removal:		
	(i) Sedimentation rates		
	(ii) Zooplankton grazing		

Source: Lewis (1974)

The process of establishing tolerance ranges and optima is not without problems or sources of potential inaccuracy. Quite often the determinations are done under laboratory conditions, using an 'artificial' environment where each limiting factor can be controlled and varied at will. Problems do arise, however, when attempts are made to transfer laboratory results to the real world situation because the latter is multi-dimensional and multi-variate and thus it does not lend itself to faithful replication under laboratory conditions. Blackman was clearly aware of this problem when he wrote that 'special research might at least show how far the recorded optima for assimilation and respiration are real metabolic truths, and how far they are illusions of experimentation' (Blackman, 1905). One reason for the problem is that although one limiting factor can be apparently optimal for a given organism that organism will not react at the optimum level if one or more other factors are still limiting (either beyond or below the tolerance range). Blackman (1905) offered the axiom that 'when a process is conditioned as to its rapidity by a number of separate factors, the rate of the process is limited by the pace of the *slowest* factor.' Organisms within natural ecosystems thus often exhibit very complex responses to environmental variables, and an example of the multi-dimensionality of the problem can be seen in the list of variables found to be important by Lewis (1974) in a study of primary production in the plankton community of a tropical lake (Table 3.15).

A second general problem with limiting factors is that many studies ignore biotic limiting factors – that is, the interactions between the organisms themselves. These types of factors are particularly important for animals, and they fall into two broad groups (Clapham, 1973) – the interaction between different populations of organisms; and a series of instinctive control mechanisms that are internal to the population itself (Table 3.16). A number of ecologists favour a *behavioural approach* to animal populations and they cite the now well-established principle that a population started from one pair of parents can be expected to grow to a predictable size, and to maintain that size if environmental conditions remain unchanged. Wynne-Edwards (1965) maintains that population growth is a density-dependent process, and that animal populations are controlled by a homeostatic process of automatic self-regulation, based on the balance between population density and available resources (the *carrying capacity* of that environment). The process of self-regulation in animal populations is related to a series of behavioural devices such as the territorial system adopted by birds and other organisms (whereby competition for territory leads to a mosaic of feeding and breeding plots within a given area); the social hierarchy (literally the 'pecking order') of many higher animals, especially those which prefer to live in groups; and the relegation of unsuccessful candidates in territorial battle to a 'non-breeding' surplus or reserve. Each of these mechanisms admits a limited number of individuals to share food and territory resources, and excludes all extras above the effective carrying capacity of the given habitat.

Table 3.16
FACTORS THAT INFLUENCE ECOSYSTEM PRODUCTIVITY AND ECOLOGICAL POPULATION BEHAVIOUR

Abiotic factors: determine the interactions between the organic and inorganic parts of the system.

(a) Physical factors: such as temperature, the quantity and spectral quality of light, depth of water, rainfall and climate.

(b) Chemical factors: such as pH and availability of nutrients.

Biotic factors: determine the interactions between organic parts of system.

(a) Interaction between different populations: such as mutualism, parasitism and predation.

(b) Instinctive control mechanisms that are internal to the population: such as social organization, territoriality and social hierarchies.

Source: Clapham (1973)

Because the ecosystem is composed of both biotic and abiotic elements and each interacts with the other, it is practically impossible to alter any single factor within an ecosystem without affecting the rest of the system. This holistic nature of the functional ecosystem was considered by Friedrich (1927), who suggested a 'Principle of the Holocoenotic Environment' to characterize the integrity of the system. Billings (1952) analysed the environmental factors controlling plant growth and distribution and he concluded that a holocoenotic principle is fundamental to any understanding of the environment/plant relationships. This high level of inter-relationship between all parts of the ecosystem means that if a particular limiting factor is removed or replaced there often follows a chain reaction of adjustments within the system. In some cases the adjustment can be so marked that it modifies the very nature of the ecosystem itself – and in very severe cases one ecosystem might even be replaced by another. The original limiting factor has thus triggered off ecosystem adjustment, and so Billings (1964) termed this the 'Principle of Trigger Factors'. Most trigger factors fall into one of two distinct types. First are the factors which were at one time limiting but which are no longer so, such as the application of fertilizers to improve nutrient levels in agricultural systems. The second type is the factor which has been introduced into an ecosystem but which was not there before and which has neither a well-developed cyclical mechanism within the natural ecosystem nor a real ecological niche. Examples of this second type include DDT and radioactive fallout material, both of which can trigger off radical ecological changes.

Paine (1966) has presented interesting evidence of the principle of trigger factors which suggests that the overall diversity of animal species within an ecosystem is related to the number of predators in the system and their efficiency in preventing single species of prey from monopolizing some important limiting factor. For example in the marine inter-tidal zone the normal limiting factor is space; and if predators capable of preventing monopolies of space are missing or are experimentally removed then the system becomes less diverse overall. A second example is the series of changes in ecology, hydrology, erosion and energy and nutrient flows within part of the Hubbard Brook experimental forest triggered off by deforestation (Figure 3.19).

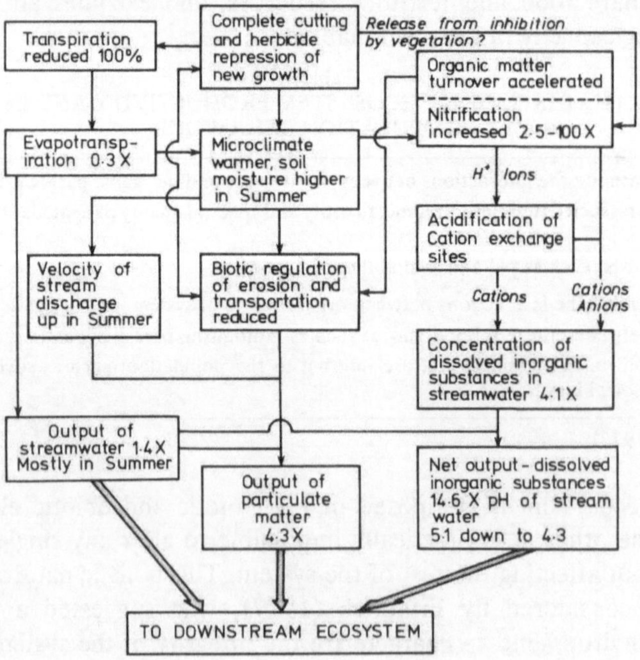

FIG. 3.19 CHANGES IN ECOLOGY, HYDROLOGY, EROSION AND NUTRIENT FLOWS IN HUBBARD BROOK ECOSYSTEM, TRIGGERED OFF BY DEFORESTATION.
(After Bormann, Likens, Siccama, Pierce and Eaton, 1974.)

A distinctly different approach to the problem of evaluating the relationships between environmental factors and ecosystems was demonstrated by Jenny (1961). He attempted to generalize the relationships between ecosystem properties and environmental factors by a *state factor equation* approach similar to that developed for the description of soil by Dokuchayev (1898) and himself (Jenny, 1941). The approach considers that ecosystem properties (l), soil properties (s), vegetation properties (v) and animal properties (a) within a reasonably homogeneous area (of unstated size) are related to, or are a function of, three 'state factors':

Lo, the initial state of the system when genesis begins,
Px, external flux potentials (that is, environmental gradients), and
t, the age of the system (time).

Thus in general terms the state factor equation can be written as:

$$l, s, v, a = f(Lo, Px, t)$$

Jenny (1961) further considered the specific attributes of the three state factors involved; and he sub-divided the initial system state (*Lo*) into parent material (p) and slope (r), and the external flux potentials (*Px*) into regional climate (*cl*) and a regional biotic factor (o) which covers all species which migrate or are carried into an ecosystem. Thus an extended form of the equation can be written:

$$l, s, v, a = f(cl, o, r, t \ldots \text{etc.})$$

There is some inter-dependence between the state factor variables because although r and p are not considered by Jenny to be time-dependent, *cl* and o may be regarded as a function of time. Although Jenny (1961) developed an approach which rationalizes the variety of environmental factors which might have an influence on ecosystem properties, relatively little interest has been shown in it because it is too generalized to allow the equations to be tested or appropriate descriptions of the 'state factors' to be quantified.

3.3c Equilibrium in ecosystems

Ecosystems are characterized by a complex web of relationships between individual organisms (created by flows of energy and nutrients) and by a mosaic of feed-back loops whereby a change in one part of the system may lead to repercussions in many other parts of the same system and in adjoining systems. It is widely accepted that ecosystems that have evolved over long periods of time can reach a *steady-state* condition characteristic of open systems in general, whereby the impact of local disturbances from the environment can be stabilized by ecosystem constraints and by the process of self-adjustment.

Brock has visualized a steady-state condition in mature ecosystems which he defines as 'a time-independent condition in which production and consumption of each element in the system are exactly balanced, the concentrations of all elements within the system remaining constant, even though there is continual change' (Brock, 1967). This clearly implies a two-phase time scale with, on the *short term*, constant ecosystem change on a seasonal basis (such as leaf fall and decomposition) and over a period of years – these would be natural ecosystem fluctuations within established thresholds; and on the *longer term* the steady-state condition of dynamic equilibrium. Confusion has arisen within ecology because of failure to isolate the two phases of this time scale, and also because of the difficulty of characterizing *stability* in a meaningful way.

Two main concepts of stability are used. One is stability viewed as a constancy of species numbers within an ecosystem, or of the numbers of individuals of a species within a population. This is often referred to as 'no-oscillation stability', and it implies continuity and predictability within systems (Dunbar, 1973). The other is stability viewed as the ability of a system to maintain, or return to, its original state after an external impact or change. This is often referred to as 'stability-resistance' (Regier and Cowell, 1972). Margalef relates the two types of stability to constancy of the environment, because he maintains (Margalef, 1968a) that the first type of stability achieves a steady-state under essentially constant environmental conditions, whereas the second type gives greater resistance to changes external to the ecosystem. Hill (1975b), however, has pointed out that ecologists rarely differentiate between the two types and he stresses that 'resilience' concerns the ability of a system to adjust to stress and this is the fundamental property of stability.

A basic problem is how to measure stability or resilience. Two suggestions are to measure the rate of the initial response of the ecosystem to the imposed stress, under either field or laboratory conditions (Hurd, Mellinger, Wolf and McNaughton, 1971); or to measure the extent of transformation of the ecosystem from its initial state. This latter approach is not without inherent problems, however, because although a system might regain its former levels of biomass, net productivity and diversity, the proportion of various species after the stress may differ from that in the ecosystem's initial state. The time period required for recovery of equilibrium might also be too long for observational measurement. In considering the overall development of an ecosystem, Olson concluded that 'a "*climax*" condition, in the sense of a steady-state or zero net community storage of matter or energy, may not be attained until long after composition and average biomass of many living species have become relatively constant, or begun to oscillate around some average value' (Olson, 1963). The time period for this to occur appears to be very variable, and to range from about a year in some tropical forests to over a thousand years on some sand dune soils.

Much attention has been devoted to the factors controlling stability in ecosystems, and the hypothesis that stability is mainly a function of the complexity of the trophic structure of an ecosystem has found widespread support. Elton (1958) concluded that diversity in food webs tends to promote ecosystem stability because it increases the system's resilience to outside invasions, and it reduces oscillations in populations within the ecosystem. Elton has also provided six main lines of evidence in favour of the diversity/stability association (Table 3.17). MacArthur (1957) has also pointed out that stability increases as the number of links in the food web increase, and that a large number of interacting feeding links produces a wide variety of adjustments to stress within ecosystems and it provides alternative channels for energy flow. An example of a stable trophic structure is the stream ecosystem studied by Patrick (1972) which had a wide variety of species at each trophic level. Patrick concluded that 'the variation in kinds of organisms and in their ecological and food preferences probably gives stability to the system and ensures that a given stage in the food web is not eliminated over time' (Patrick, 1972). The strategic benefits of the populations within the stream ecosystem include the many different species at trophic level one; the many energy pathways; the fact that each individual makes only a small contribution to overall stability in the system because most individuals are small in size and have rapid turn-over rates (because of short lives and high reproductive rates); and the fact that many of the species are capable of performing a given function within the ecosystem (that is, the species are 'generalists' rather than 'specialists', and their ecological niches overlap substantially).

Table 3.17
MAIN LINES OF EVIDENCE FOR THE THEORY THAT DIVERSITY LEADS TO STABILITY IN NATURAL ECOSYSTEMS

(a)	Mathematical argument that models of simple predator/prey relationships generally show large fluctuations
(b)	In laboratory experiments it is often difficult to maintain small populations of few species in simple habitats in balance; either the population sizes fluctuate or one or more species is liable to become extinct
(c)	Natural habitats on small islands seem to be more vulnerable to ecological invasion by foreign species than habitats on continents
(d)	Large-scale invasions or outbreaks most often happen on cultivated or planted land, where habitats and communities have been artificially simplified by man
(e)	Contrast the differences in stability between tropical communities (complex ecosystems – very stable) and arctic communities (simple ecosystems – very variable and unstable)
(f)	Research during the 1950s on pest control in orchards (half-way between natural woodland and arable open-field systems) showed that attempts to control orchard pests often revealed many problems of inherent instability

Source: Elton (1958)

There has been some debate over the relationship between diversity and stability. Odum has stressed that

> the question of diversity in nature is a particularly important one because man's theory of environmental management, at least up to the present time, is that reduced diversity leads to greater control of nature for the benefit of man. Nature's theory seems to hold that diversification results in greater biological control and stabilization of the environment (Odum, 1964).

One line of evidence that challenges the diversity/stability model is the experimental studies of Hurd and his colleagues (1971) who added inorganic fertilizer (NPK) to two successional old fields of similar size and of different ages, in New York; and they then monitored the ecological repercussions. An increased diversity of plants at trophic level one and of animals at trophic level two produced decreasing population stability at trophic level three, and so increased diversity in this case did not lead to increasing stability, but quite the converse. A second line of evidence is that some ecosystems may contain a single species high in the food web which can strongly influence the entire stability of the structure, and in these cases the stability of the overall ecosystem depends on the effects of stress on the key species rather than on overall species diversity. Paine (1969) has documented an ecosystem on a rocky inter-tidal zone where the removal of a dominant carnivore (the starfish *Pisaster achraleus*) led to significant changes in the ecosystem in terms of population density, species composition and general appearance of stability.

There are, nonetheless, many instances of ecosystem stability being closely related to ecological diversity. Tundra ecosystems, for example, are extremely simple in terms of animal communities and this often leads to large-scale population oscillations triggered off by small environmental stresses (Dunbar, 1973). The tropical rain forest ecosystem is also widely cited as a system with low resiliance to large-scale disturbance because the primary tree species are incapable of re-colonizing extensive areas of cleared land (Gomez-Pompa et al., 1972). Connell (1978), however, has pointed out that two opposing views on the nature of equilibrium in ecosystems like tropical rain forests or coral reefs exist. One is that stability usually prevails, and when a community is disturbed it quickly returns to its original state (natural selection adjusts species to the system). With this viewpoint ecological communities are seen as highly organized, co-evolved species assemblages in which ecological efficiency is maximized, life history strategies are optimized, populations are regulated and species composition is stabilized. The alternative view is that ecosystem equilibrium is rarely attained and disruption is so common that species assemblages seldom reach an ordered state. With this viewpoint, communities of competing species are not highly organized by co-evolution into systems in which optimal strategies produce very efficient associations whose species composition is stabilized. Connell (1978) maintains that the assemblages of tropical rain forests and tropical coral reefs conform to the non-equilibrium model where disturbances are frequent but the species present may be so alike in competitive ability that diversity is maintained by chance replacements (Figure 3.20).

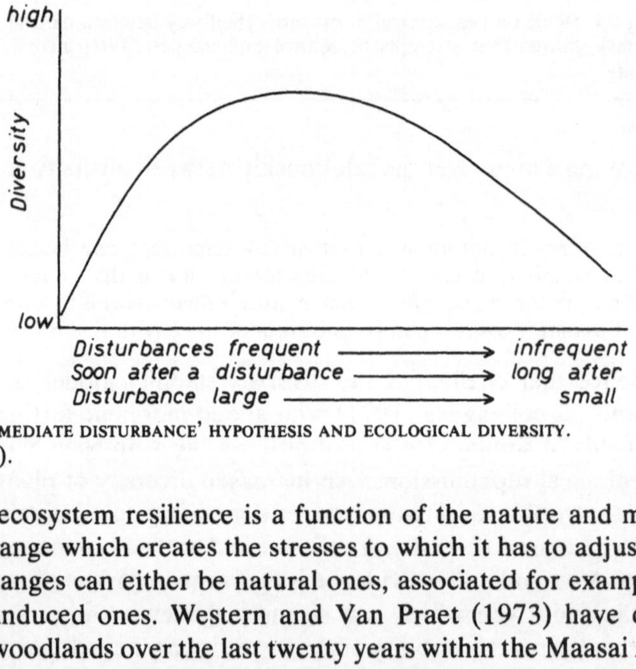

FIG. 3.20 THE 'INTERMEDIATE DISTURBANCE' HYPOTHESIS AND ECOLOGICAL DIVERSITY. (After Connell, 1978).

In many ways ecosystem resilience is a function of the nature and magnitude of the environmental change which creates the stresses to which it has to adjust. In essence the environmental changes can either be natural ones, associated for example with climatic change, or man-induced ones. Western and Van Praet (1973) have documented the extensive loss of woodlands over the last twenty years within the Maasai Ambelosi Game Reserve of East Africa in relation to environmental changes. A change in rainfall regime led to an upward rise of ground-water levels in the order of three to four metres, and this brought soluble salts into the rooting layers of the soils by capillary rise. This in turn led to wholesale death of trees within the ecosystem through physiological drought, and this triggered off successive changes in the entire structure and stability of the ecosystem.

Many ecosystem changes are associated with man-induced changes which often trigger off ecological changes to which many organisms cannot adapt. The large-scale removal of tropical rain forest, for example, has triggered off the destruction of soil structure and productivity in many areas. Large-scale pollution by oil, heavy metals, biocides and similar toxic elements which are either alien to natural ecosystems or normally present in much smaller quantities can also induce widespread and long-lasting ecosystem changes. Woodwell (1970) has rationalized changes in ecosystem equilibrium associated with pollution into four main areas of change – a simplified ecological structure; a reduced species diversity; a reduced standing crop of organic matter; and changes in ecosystem nutrient cycles.

Ecosystem stability is also of importance from the point of view of the potential of ecological resources for management and exploitation. Clapham (1976) has pointed out that distinct patterns of ecosystem resiliance and differential exploitability by man do exist, but they have not yet been worked out in detail. To this end he suggests a relationship between intensity of ecosystem response and intensity of management, which can be used to define a 'Coefficient of Exploitability'. As a first approximation of the nature and magnitude of spatial variations in this coefficient, Clapham (1976) calculated representative coefficient values from available empirical data, and clearly this is one applied area of ecosystem studies where research attention could be richly rewarded.

3.4 ECOSYSTEM ANALYSIS – PROBLEMS AND PROSPECTS

3.4a Problems in ecosystem analysis

Although many ecologists would agree with Colinvaux that 'the ecosystem concept is one of the most powerful ideas of ecology, a concept which allows us to examine the workings of the natural world in an objective and understanding way' (Colinvaux, 1976), there are a variety of basic problems which must be considered in using the concept of a dynamic, self-regulating ecosystem.

The problem of scale

In reality systems lie within systems as a nested hierarchy (hence Lindeman's (1942) stress that the ecosystem can be viewed 'within a space-time unit of any magnitude'), and yet for description and analysis a system must be abstracted from reality by fitting boundaries to it. In this way an inherently open system is 'closed' for analytical convenience. Whilst this process of closure gives flexibility for analysis it does at the same time lead to problems of definition. An ecosystem can be of any size from a goldfish bowl to the entirety of the earth's oceans. Clearly the nature of an ecosystem depends on the scale of resolution adopted in the analysis. A scale continuum exists between the single oak tree, the oak forest and the whole biosphere, for example. In practice the size of system adopted for measurement and analysis depends generally on either the particular interest of the observer, or convenience (both conceptual and practical). Closely allied to the problem of scale is the problem of actually defining boundaries for a given system.

Definition of boundaries

Nature is a continuum, and with few exceptions sharply defined natural boundaries are rare. But for ecological analysis, measurement, modelling and application, boundaries have to be fitted to a study ecosystem. In some cases, such as the Hubbard Brook ecosystem, the boundaries can be placed around convenient landscape units such as the drainage basin – but animals migrate between basins and the flow of energy and nutrients is rarely confined to topographic units of such convenience. Brock (1967) points out that ecosystem boundaries are often conceptual rather than physical ones, and that they are 'constructed' simply to create a barrier through which to consider the imports and exports of matter and energy. He also stresses that 'unless the . . . boundary is placed to contain within it a system in steady state, great difficulties in the subsequent analysis may evolve' (Brock, 1967). Evans (1956) has stressed the difficulties of determining the limits of a given system because a web of pathways of loss and replacement of matter and energy frequently connect one ecosystem with another. He maintains that 'this has led some ecologists to reject the ecosystem concept as unrealistic and of little use in description or analysis' (Evans, 1956). Identification of ecosystem boundaries is essential, however, because of the need to include all representative organisms within the study system, and all factors of the abiotic environment (nutrients, wind, light and so on). Hence boundaries are generally fitted to include all functioning processes (such as nutrient cycling) within the study system. Island ecosystems are quite popular for their convenient and abrupt boundaries, particularly remote islands with little contact with contiguous terrestrial ecosystems. Smith (1977) was able to describe the trophic structure and food webs for a small sub-Antarctic island (Marion Island) because it proved to be a good ecological laboratory (because of geographical and biological isolation, and freedom from interference by man) and because direct study was simple because of well-defined nutrient cycles and relatively uncomplicated trophic relationships.

Measurement of ecosystem forms and processes

It is often extremely difficult to measure accurately the ecosystem components which are important and it then becomes both necessary and convenient to devise a conceptual framework (such as Odum's *Energy Flow Diagram*) to help in the identification of key properties to measure. The process of tracing feeding relationships within an ecosystem, for example, is problematical in the extreme; and Phillipson (1966) outlines three main approaches to this particular problem. One is the direct observation of feeding habits, but this is a time-consuming approach and it is unlikely that all possible feeding links will actually be observed and recorded. An alternative is to collect samples of all species present in an area or habitat, and to analyse in the laboratory the gut content of each organism to establish 'who has eaten what'. Again the basic problem arises that not all feeding links will be established in this way, and the approach only really identifies the most recent food source for each organism, which might not be representative of a long-term preferred source. The third approach involves the labelling of lower trophic levels with some form of tracer (such as radioactive isotopes) whose transfer through the food web as it passes from one organism to another can be measured instrumentally. An example appears in Figure 3.21. This approach appears to solve many of the problems of incomplete recognition of food web links and of averaged feeding patterns, but the intentional introduction of toxic radioactive material into formerly natural food webs seems ethically unjustifiable to some ecologists. This latter approach has also been used in non-laboratory conditions where the movement of toxic or alien chemicals through food webs has offered a valuable opportunity for studying and tracing feeding relationships in ecosystems (Figure 3.14). Measurement problems also arise because of the migration of animal species between ecosystems either in the course of regular seasonal migration or of permanent migration induced by some longer term environmental change. Thomas has pointed out that 'ecosystem is a fashionable word, implying that all organisms interact and are in balance; it ignores migration – our insects migrate, our fishes migrate, our birds migrate and even our mammals migrate' (Thomas, 1975).

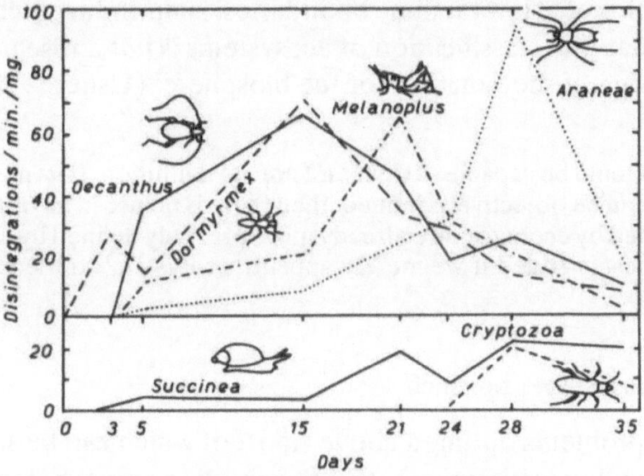

FIG. 3.21 RADIOACTIVE TRACER STUDY OF FOOD WEB STRUCTURE.
The results of a simple experiment in which a radioactive tracer (P^{32}) was used to isolate a food chain in an intact old-field community. The figure shows the build-up of P^{32} activity in the biomass of six major vertebrate populations following the labelling of a single species of dominant plant. (After Odum, 1962.)

Problems of ecological modelling

There are also problems inherent in attempting to model ecosystems, either mathematically or conceptually. For example Odum (1960) has pointed out that by 1960 there were available growing numbers of measures of the structure and function of ecosystems (such as biomass, production rates and trophic relationships) which allow quantitative comparisons between different ecosystems, and yet there was a basic need for further synthesis of the available data into generalized theories. He advocated the use of an electrical circuit analogy for nutrient cycles – a physical model constructed with batteries and wires wherein the flows between components could be measured with milliammeters. Various other types of ecosystem modelling strategies include mathematical modelling (for example the work of Halfon, 1976), conceptual modelling (such as Jenny's state factor equation approach – see page 99) and laboratory replication models of complex real world ecosystems (such as the microcosm approach advocated by Beyers, 1963).

Man-modified ecosystems

Attempts to identify basic differences between natural and man-modified ecosystems also raise problems because it is practically impossible to find any ecosystem of any size that has not been affected in some way or another, and to some extent by human impact, because of the very unity of the biosphere (see Chapter 2), and because of the long history and widespread distribution of human activities. It is perhaps convenient to distinguish less modified natural systems from highly modified anthropogenic systems, but this particular problem applies to any studies or approaches in ecology and it does not relate simply to the concept of ecosystems.

The ecosystem in theory and practice

Perhaps more basic than most of these other operational problems is the debate over whether the ecosystem concept is more valuable in theory than in practice, because of the inherent problems of scale, interaction, boundaries, man-induced change and the like. Usher has stressed that 'the classification of ecosystems is *not* a research tool, and cannot advance our knowledge of the functions of the biosphere' (Usher, 1973). Brock is of the opinion that

> the word *ecosystem* should be capable of clear and precise definition. If its meaning is not restricted to something which can be objectively defined, then there is no use in having the word, since most natural systems studied by ecologists are already more precisely defined by existing words (such as lake), and nothing new is added if we merely append 'ecosystem' to the existing words (Brock, 1967).

3.4b Values of the ecosystem approach

Despite the various problems outlined above (most of which can be solved with carefully structured analysis and measurement), the ecosystem approach has a number of very significant values, and it has a potentially large contribution to make to the study of the environment from both an ecological and a management point of view. The main benefits of the approach can be seen in terms of convenience of scale, flexibility of framework, fundamentality of the unit and assets of the systems approach *per se*.

Convenience of scale

The ecosystem is a convenient scale on which to consider plants and animals and their interactions with each other and with the environment. Odum has observed how 'ecologists rally around the ecosystem as their basic unit just as molecular biologists now rally around the cell' (Odum, 1964), and the *International Biological Programme* has made a concerted and co-ordinated effort to increase understanding of how entire ecosystems function and to model behaviour of ecosystems. For example the United States Ecosystem Analysis Program has concentrated on six major ecosystems – grasslands; deciduous and coniferous forests; deserts; arctic tundra; and tropical forests (Hammond, 1972).

Flexibility of framework

Ecosystems also provide a convenient framework in which to view ecology and interaction between various parts of the environment (both biotic and abiotic). Odum (1968) maintains that each of six major themes (Table 3.18) related to both the structure and function of the organic world is integrated within the ecosystem approach. Colinvaux has suggested that 'the ecosystem concept is one of the most powerful ideas of ecology, a concept which allows us to examine the workings of the natural world in an objective and understanding way' (Colinvaux, 1976); whilst Caldwell concludes that 'no more promising means for relating the biological and social science contributions to an understanding of human life in its environmental context has yet been found' (Caldwell, 1966). Because it offers such a unifying framework within which to consider both biotic and abiotic aspects of environmental management, Ripley and Buechner (1967) refer to the ecosystem as a point of synthesis of the 'human-society-plus-environment complex'.

Table 3.18
MAIN PROPERTIES OF THE STRUCTURE AND FUNCTION OF ECOSYSTEMS

ECOSYSTEM STRUCTURE:	
(a)	composition of the biological community (such as species numbers, biomass and life history)
(b)	quantity and distribution of abiotic materials (such as nutrients and water)
(c)	range or gradient of conditions of existence (such as temperature or light)
ECOSYSTEM FUNCTION:	
(a)	range of energy flow through the system (ECO-ENERGETICS).
(b)	rate of nutrient cycling (ECO-CYCLING).
(c)	regulation by physical environment and by organisms (ECO-REGULATION)

Source: Odum (1968)

Fundamental ecological unit

The ecosystem also appears to be a more meaningful unit on which to base strategies of environmental management than the community, whose virtues have been pointed out by Moss and Morgan (1967). Evans (1956) argues that the ecosystem is the basic unit of ecology for four main reasons. The ecosystem involves the movement and accumulation of matter and energy through the medium of living things and their activities; and the system is characterized by a large number of regulatory mechanisms which limit the numbers of organisms present, influence their physiology and behaviour and control the qualities and rates of movement of matter and energy. The concept of ecosystems applies at all levels from individual plants to the entire biosphere; and ecosystems are, by definition, open systems rather than closed ones.

Systems analysis and modelling

A further value of ecosystems is that techniques of systems analysis can be readily applied (Chorley and Kennedy, 1971). In particular mathematical modelling of system behaviour and response can be valuable in characterizing system changes in response to either internal or external stresses, and in predicting the likely ecosystem changes associated with human manipulation or exploitation. Halfon (1976) has demonstrated that modelling techniques can give valuable results, but he also warns that much study now needs to be devoted to objective methods of modelling, and in particular to solving operational problems of systems identification and optimal solutions for systems aggregation procedures. Modelling can be valuable in predicting the impacts of management strategies. For example Botkin, Janck and Wallis (1972) have outlined a dynamic computer model of forest growth in which future changes in the state of a forest can be predicted on the basis of the present state of the system plus random future changes. The model has been calibrated for the Hubbard Brook ecosystem, and it simulates the end process of competition between species and between individuals, of secondary succession (see Chapter 4) within the forest, and of changes in vegetation related to altitude. The model has predicted a peak standing crop of the forest at about 200 years after clear-cutting, with relatively stable (but nonetheless fluctuating) species composition afterwards. This prediction is clearly of considerable value for site management and resource planning.

Applied ecosystem analysis

The ecosystem concept is of considerable value for evaluating the effects of human activities on the environment. Odum (1974) has pointed out that by about 1960 the theory of ecosystems was quite well developed, but not in any way applied. The applied ecology of the 1960s consisted of managing components of ecosystems as more or less independent units. Thus forest management, wildlife management, water and soil resource management were widely practiced, but ecosystem management was not. Odum adds that 'practice has now caught up with theory. Controlled management of the human population together with the resources and life support system on which it depends as a single integrated unit now becomes the greatest, and certainly the most difficult, challenge ever faced by human society' (Odum, 1974). Marquis (1968) is even more forceful in advocating the need for careful manipulation of ecosystems. He is of the opinion that

> man will be interested in how to design segments of ecosystems, assemble segments together into ecosystems, and regulate those ecosystems to better satisfy human goals. He will develop models of ecosystems to enable him to evaluate the stability of ecosystem behaviour, its sensibility to various disturbances, and the reliability of its many parts (Marquis, 1968).

Many ecosystems are now to some extent controlled by man, and Clapham (1976) has identified three main characteristics of such man-dominated systems. One is that the distribution and function of human ecosystems are controlled by economic, social and political factors rather than by factors of the natural environment. Furthermore, the stability of the ecosystem rests on a narrower resource base in natural systems than in human ecosystems because resources can be, and are, transferred into the latter. Finally there are different rates of change in ecosystem structure and operation between natural and human ecosystems. In the former, changes in equilibrium are generally associated with genetic alterations or with long-term climatic or other environmental change; whereas in the latter ecosystem change is often associated with social and cultural change.

The most striking human ecosystems are the completely *urban ecosystems*. Hughes (1974) has pointed out that urban systems differ from natural ones in two very basic ways. First there are massive energy flows within the urban system (energy flow through the commercial centre of a major city may be a thousand times that in surrounding rural areas). Secondly the materials cycle is broken in urban systems, because materials need to be constantly replaced from outside the city (thus the urban system is closely dependent on other systems for resources) and waste disposal in urban systems generally involves the discharge of concentrated materials into other non-urban ecosystems (ironically often those on which the city depends for sustenance). The major components of the urban ecosystem are shown in Figure 3.22.

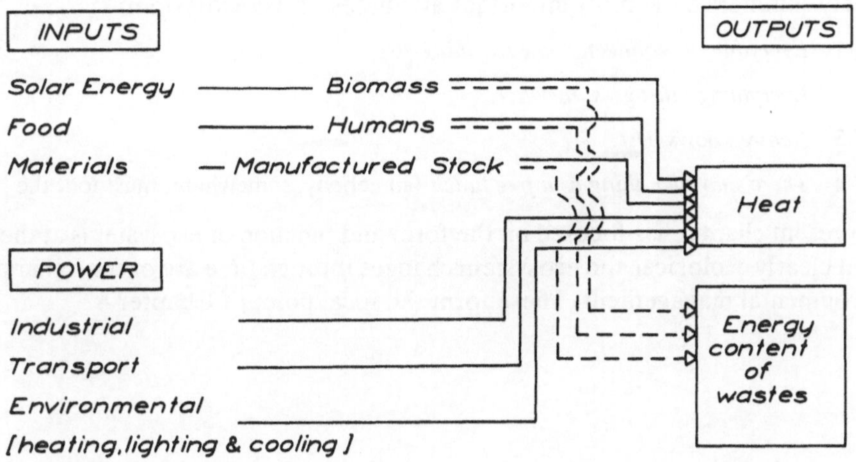

FIG. 3.22 THE URBAN ECOSYSTEM.
Major flow pathways (solid lines) are distinguished from minor ones (broken lines). (After Hughes, 1974.)

Ecosystems and geography

A final value of the ecosystem concept and approach is that it has much to offer to geography. Stoddart (1965) has identified four main properties which make ecosystems of value to geographical studies:

(a) *Monistic nature* – the ecosystem approach views environment, plants, animals and man, and their interactions, together simultaneously.

(b) *Structured form* – ecosystems have clearly defined structures which make investigation of flows within the system and components of the system conceptually simpler.

(c) *Functional entities* – the system involves a framework of interaction between the main components, and between them and the flows of matter and energy, which link the whole system together and make it work.

(d) *General systems* – the ecosystem is an open system, tending towards a steady-state condition. This is a valuable conceptual benefit.

Tomlinson has also concluded that 'the ecosystem approach provides a framework within which any geographer can work; there is no need for the break between physical and human geography – it merely becomes a question of emphasis on a particular part of the system' (Tomlinson, 1972).

In conclusion, the ecosystem approach stresses the unity of all parts of the environment, including both biotic and abiotic elements; it allows the links and pathways in flows of energy and nutrients to be traced and compared; it facilitates measurement and comparison of the components of different types of ecological communities, and identification of equilibrium and non-equilibrium states; and it highlights basic ecological principles which should be carefully evaluated when guidelines of ecological management and exploitation are compiled. Commoner (1972) has formulated four 'Basic Laws of Ecology' which effectively summarize the most important attributes of the ecosystem:

LAW 1 *Everything is connected to everything else.*

LAW 2 *Everything must go somewhere.*

LAW 3 *Nature knows best.*

LAW 4 *There's no such thing as a free lunch* (somebody, somewhere, must foot the bill!).

This present chapter has focused on the form and function of ecosystems at the present time, but clearly ecological and ecosystem changes through time are of central importance to environmental management. These form the focal point of Chapter 4.

4
Ecological Changes Through Time

> Many ecologists have . . . come to realize that the past is a key to understanding the present, and they are beginning to look at current research and management problems in the light of historical experience (Sheail, 1971).

The structure, function and stability of ecosystems vary considerably in both time and space. Since these are the two fundamental dimensions of ecological change it is important to consider each individually. Ecological changes through time occur, and are of fundamental significance, on a wide range of time scales from the level of seconds and minutes (for physiological and behavioural patterns of plants and animals) to the many millions of years of evolutionary change documented in the fossil record. A number of books have been devoted specifically to the history of British vegetation, including those by Godwin (1956), Hawksworth (1974b), Lousley (1953), Pennington (1969) and Perring (1970). The concern in this chapter is not to review specifically the ecological history of any particular area, nor to evaluate the wide range of techniques currently available for establishing ecological histories in various environments (reviewed, for example, by Kellman (1975) and Randall (1978)), but rather to consider the significance of inherent and man-induced change over a range of time scales to the management of the environment in general and ecosystems in particular. Charles Darwin (1859) was aware that in nature there is a frequent ousting of one form by another either over long periods of time (involving the extinction of species) or over relatively short periods of time (involving purely local eliminations). More recently Margalef (1968b) has stressed that the processes of *ecological evolution* and *ecological succession* are basically the same. It is useful, nonetheless, to consider these two time scales independently.

4.1 EVOLUTION AND ECOSYSTEMS

4.1a The theory of evolution

The publication of Darwin's *The Origin of Species by Means of Natural Selection; or the Preservation of Favoured Races in the Struggle of Life* in 1859 influenced the thinking and perspectives of both scientists and laymen alike, both then and since, perhaps more so than almost any other book. In the book Darwin marshalled wide-ranging evidence on the principles and mechanisms of evolution by natural selection. This radically affected the whole of science in dispelling the myths of catastrophism and of a world whose existence could be measured in terms of, at the longest, thousands of years; and it also placed a very clear perspective on the fundamental role of competition between and within species, over both space and time, in the development of plants and animals. The theory of evolution has commanded much attention since the mid nineteenth century, and much has been written on the subject. Here is not the place to consider the whole realm of evolution, but attention can be focused on several aspects of the topic because evolution stresses the inherently dynamic nature of ecosystems and of the organic components of the biosphere.

Three particular aspects of evolution are important from the environmental management point of view. First, it provides a *broad perspective* and a valuable time scale of ecological development through time, as well as a broad background of natural ecological change against which it is possible to evaluate the magnitude and nature of recent ecological changes. Some conservationists falsely argue that ecosystem changes *must* be minimized; but this ecological-preservation approach is both unnatural and unwise in the light of natural ecological changes through time. Ager has argued that

> conservation, though laudable the perspective, is . . . in the long run . . . as futile as Canute trying to stop the tide. We may preserve rare species in zoos, just as we preserve fossils in museums, but extinction for the species is as inevitable as death for the individual. It is the natural fate of every living organism, and requires no special cause (Ager, 1976).

Clearly any optimal conservation strategy must accept change as inevitable. Static forms of conservation might readily produce good results in the short term, but clearly in the long term they are doomed to failure.

Secondly and closely related to this, evolution offers a convenient framework for considering *extinction of species*, and the allied natural fluctuations of genetic variety and species diversity through time. The inevitability of extinction has been stressed by Martin, who concluded that

> extinction is *not* an abnormal fate in the life of a species. When all the niches in a biotic community are filled, extinction must occur as rapidly as the evolution of new species. The fossil record of the last ten million years bears witness to this fact, for it is replete with extinct animals that were sacrificed to make room for new and presumably evolving species (Martin, 1967).

Whilst from an ecological point of view extinction is inevitable and acceptable, from an environmental management point of view the crucial questions are 'What order of magnitude of extinctions is "natural" and ecologically acceptable?'; 'How can human use, exploitation and management of ecosystems induce accelerated or entirely anthropogenic extinctions?', and 'What can and should be done to alter the rates and directions of extinctions or threatened extinctions at the present time (and is it wise to consider doing so)?'

Thirdly, *evolution has not stopped* – it is an on-going dynamic ecological process, and the adaptation of individuals, species and ecosystems to changing environmental conditions whether natural (such as long-term climatic changes) or man-induced (such as through loss of habitat) is both expectable and ecologically beneficial. The key factors from an environmental management point of view include the time lag between environmental change and ecosystem or species change; the magnitude of ecological change in relation to environmental change; and the feedback mechanisms within ecosystems and within individual organisms which bring about the ecological changes.

Traditional views (in Western society) on the origin of the many plants and animals of the biosphere rely heavily and understandably on religious doctrine which states that life owes its origin to an act of divine creation. Thus the Book of Genesis narrates *The Creation* in which dry land, grass and trees, whales and every living creature were created within a week of divine construction. This was a notion of an 'instantaneous biosphere', which did not change through time. By the eighteenth century, however, recognition of change in animal and plant species was demanded by findings of fossil extinctions within the stratigraphic record, and a theory of evolution was in its infancy when Lamarck (1744–1829) challenged the belief in the fixity of species. By the mid nineteenth century, evolution had captured the imagination of both Charles Darwin (1809–92) and Alfred Russell Wallace (1823–1913), simultaneously but independently. Their thoughts were presented to the scientific world in two papers read together at the same meeting of the Linnean Society in London on 1 July 1858.

The five most important threads in Darwin's argument on evolution have been isolated by Robinson (1972):

(1) Individuals of a species vary, and this variability is small but significant.

(2) Particular characteristics make some individuals more efficient and better able to stand up to conditions in their struggle for life.

(3) Any advantageous characteristics that an organism possesses would give it a better chance of survival and reproduction.

(4) If some advantageous characteristic were transmitted to and through the organism's offspring, this would give succeeding generations superiority in the struggle for existence.

(5) Any disadvantageous traits would handicap individuals in their struggle – they would be more likely to be eliminated by competition from the better-equipped members of their race.

Thus Darwin (1859) clearly saw a struggle between organic forms, which could lead to a winner (*evolution*) and a loser (*extinction*), especially when the struggle took place between closely related species. His basic argument is that gradual modification and diversification of organisms stems from a process of natural selection and adaptation to environmental factors, both being slow but on-going over long periods of time. Once some variation in a species begins, and this proves to be a benefit in the struggle for survival, this line of variation becomes progressively more pronounced with each succeeding generation.

DeVries, a Dutch botanist, challenged Darwin's notion of progressive evolution, and he stressed that it was possible for offspring occasionally to differ so markedly from the parents that a 'new species' is developed within a single generation (Robinson, 1972). This more or less spontaneous evolutionary change was described as a *mutation*, and it was understood to give rise to inheritable variations in a species which can be spread by inter-breeding.

The key to understanding the mechanism of evolutionary changes lay in work on heredity carried out by Gregor Mendel, a monk in Moravia (now Eastern Germany) (Riley, 1967). Mendel experimented with pea plants and noted how certain strongly contrasting characteristics (such as tallness and shortness, and colour) are transmitted to succeeding generations as the result of cross-fertilization. The basic building-block for the process, and Mendel's basic unit of inheritance, is the *gene*. Mendel deduced that each of the cells involved in organic reproduction carries only one of a pair of genes; selection of the gene takes place at random; the two genes (one from each parent) meet and join at fertilization; one of the pair of genes is *dominant* over the other and it determines the outward characteristics and appearance of the offspring as it grows up. Later research has demonstrated that each of the strings of genes are *chromosomes*; and that chromosomes can be damaged in a number of ways (such as through excessive exposure to X rays and to particles emitted by atomic disturbances), causing gene mutations (Riley, 1967).

The process of mutation is central to modern views on Darwinian evolution, of which Dobzhansky (1950) has summarized the five basic elements.

(1) The mutation process furnishes the raw materials of evolution.

(2) The process of sexual reproduction produces countless gene patterns.

(3) The possessors of some gene patterns have greater fitness than the possessors of other patterns in available environments.

(4) Natural selection increases the frequency of superior, and fails to perpetuate the adaptively inferior, gene patterns.

(5) Groups of gene combinations of proven adaptive worth become segregated into closed genetic systems, called *species*.

He stresses that natural selection does not *change* an organism – it offers an opportunity for the organism to react to a change in the environment by adaptive transformation. Whether or not this reaction occurs depends on the availability of genetic materials supplied by mutation. Gause (1934) has pointed out that two or more species with similar life-styles cannot co-exist indefinitely in the same habitat, because one will inevitably prove more efficient than the others and it will crowd out or even eliminate its competitors. Harper (1967), however, has argued that natural diversity of species implies that the 'struggle for existence' is not regularly forced to decide between 'stronger and weaker brethren'; often the struggle in an area is evaded or does not occur because of variations in feeding habits and habitat preferences in animals, and because mixtures of plant species often form stable associations which possess self-stabilizing properties.

Whilst it is not appropriate here to consider in detail the actual mechanisms of evolution and adaptation, from a management perspective it is important to consider the *scale* at which evolution operates. Traditional evolution theory explains traits of species in terms of the advantages to *individuals*, and community-scale evolution, if it exists at all, is regarded as coincidental. Wynne-Edwards (1962), on the other hand, has argued strongly the case for *group selection* based on selection between groups within a specific population rather than on selection between individuals; and Margalef has gone further, in emphasizing that 'evolution cannot be understood except in the frame of ecosystems' (Margalef, 1968a). The *ecosystem* as a unit of natural selection has been evaluated by Dunbar (1972), who has defended the thesis that ecosystems as a whole (sub-divided into sub-systems as is convenient and necessary) can act as units of natural selection, and that ecosystems compete for survival in the same sense that individuals do. In the case of ecosystem-scale evolution the ultimate objective of the process is stability of the system, which promotes survival of the component species as well as the ecosystem overall. Dunbar argues that those ecosystem changes which lead to more resistant states (that is, which increase system stability) are assimilated into the system, and although the ecosystem derives its properties from its component species, the evolution of those species is controlled by the operation of the ecosystem itself. Darnell has also argued that 'the functional ecosystem is the fundamental selectional unit of evolution, and that evolution proceeds by mutual adjustment of ecological entities into harmonious systems with some degree of permanence' (Darnell, 1970). Wilson (1976) has presented a similar case for considering the *community* to be the basic unit of evolution, and he presents a model whereby a species cultivates its community so as to maximize its own fitness. If evolution occurs at the ecosystem or community levels, clearly ecological management strategies should recognize this and they should aim to harmonize induced changes within and between ecosystems so as not to deflect the course of evolutionary changes in adverse or perhaps irreversible manners.

4.1b Evolutionary history of plant and animal species

One important outcome of the early work on evolution was the quest to trace the ancestral history of individual species of plants and animals. Tracing the course of evolution proved to be important because it fostered the development of biological classification systems, especially that suggested and widely used by Linneaus. The course of evolution is best documented in the fossil record, and so palaeontology has made valuable contributions to ecology.

Bakker (1971), for example, has considered the accumulated evidence on Brontosaurs – the 'greatest land animals of any age', which reached lengths of up to a hundred feet, and weighed probably over 20 tons whilst alive. Brontosaurs had long been considered (and generally restored in museums) as swamp or lake-dwelling herbivores. Bakker has, however, suggested a terrestrial habitat because of the physiology and structure of the creatures – tall (not built for water propulsion), with no molar teeth but a powerful gastric mill (stones embedded in the stomach lining) for grinding tough food; a body configuration more elephant-like than hippo-like; and a long neck for browsing high in shrubs and trees.

Much attention has been given by palaeontologists to establishing the sequence of evolutionary changes in plants and animals throughout the fossil record. Interpretations generally fall into two distinct schools, termed by Gould (1971):

(a) *Phyletic gradualism:* this school sees all branches of the animal and plant world as slowly evolving side by side, over long periods of time, even though some groups evolve more rapidly than others.

(b) *Punctuated equilibria:* this school views the organic world as being normally in a state of equilibrium, but this is periodically interrupted or punctuated by short, sudden happenings or catastrophes.

Ager has suggested that the fossil record suggests evolutionary change like the life-style of a soldier, that is 'consisting of long periods of boredom and short periods of terror' (Ager, 1976). The fossil record is thus not characterized by gradual evolution so much as by episodic activity, with the sudden evolutionary advance of one group of species at the expense of others. Consequently the fossil record shows evidence of a large number of ecological 'take-overs', such as the mass extinction of dinosaurs. Mass extinctions seem to be characteristic of the fossil record, and Ager (1976) and others have suggested a correlation between mass extinctions with major regressions of the sea, characterized by a cyclic pattern of extinctions with a periodicity of approximately sixty million years (Table 4.1).

Classical theory on evolution distinguishes two broad types of evolutionary change. First is the view that evolutionary changes are *autogenetic*, or directed from within the organisms themselves, and although this view was previously widely held it has relatively few supporters today. The alternative view is that the changes are *allogenetic* or *environmental*,

Table 4.1
THE CYCLIC PATTERN OF MASS EXTINCTIONS OF SPECIES

10^6 years before present	Event
600	start of the Cambrian faunal explosion
540	mid-Cambrian extinctions (trilobites, etc.)
480	?Ordivician extinctions and glaciation
420	?mid-Silurian regression; reef extinctions
360	Frasnian extinctions (such as coral reefs)
300	?beginning of Permo-Carboniferous glaciation
240	end of Permian extinctions (brachiopods, etc.)
180	end of Triassis extinctions (e.g. ammonites)
120	early Cretaceous floral change
20	end of Cretaceous extinctions (reptiles, etc.)
0	Pleistocene glaciations and post-Pleistocene extinctions

Source: Ager (1976)

and thus directed by external factors. This conforms to the theories of Lammarck on genetics and Darwin on evolutionary adaptation and selection, and the view is widely supported. A valuable contrast in evolutionary strategies between different types of environment has been identified by Dobzhansky (1950). In environments with harsh physical conditions (such as the Arctic) the struggle for existence is believed to be effectively *passive* in that physical factors of the environment destroy great masses of living organisms. In physically 'mild' environments (such as tropical areas), however, natural selection is seen to be a *creative* process, which may lead to the emergence of new modes of life and of more advanced types of organization.

4.1c Extinction of species

The theme of environmental stability and evolutionary strategies has recently been considered by Futuyma (1973), who pointed out that the evolution of specialized relationships between species will be most pronounced in constant environments (hence the high degree of inter-dependence between species in the tropical rain forest ecosystem). Because of this, in areas characterized by environmental fluctuations, unusual environmental changes may cause the extinction of inflexible species only; whereas comparable changes in a generally constant environment may cause extinctions not only of the relatively inflexible species but also of those species highly dependent on them. Futuyma envisages that 'with continual slow environmental change, groups of species will go extinct in concert rather than independently as expected of broad-niched species that have evolved in fluctuating environments' (Futuyma, 1973).

In essence extinctions arise due to the replacement of one species in a given ecological niche (see pages 73–5) by another. Generally a species will be replaced by a more competitive one (for that environmental niche), or by one whose tolerance response to the suite of prevailing environmental conditions is closer to the optimum. Ecologists use the term 'extinction' in two quite different senses:

(a) *species extinctions*: species extinction is said to have occurred when all of the members of a particular species throughout the world have died or been exterminated. This particular branch of evolutionary and genetic strain is henceforth lost for ever

(b) *local extinctions*: these are said to have occurred when all individuals of a given species within a specified area (generally of stated size) have been removed either by death or by enforced or induced permanent migration elsewhere.

Both scales of extinction are of importance from an environmental management point of view, and also from the ethical point of view; but clearly completed and threatened species extinctions are the more pressing and lamentable because of the irreplaceable loss of genetic variety and evolutionary potential.

Extinctions have been documented over a wide variety of timescales throughout the fossil record. Many species extinctions have occurred because of *environmental changes* which alter the prevailing environment beyond the tolerance ranges of affected species. Throughout the Permo-Trias, for example, dinosaurs resembled advanced mammals or birds, and they had distinct advantages over mammal-like reptiles of the day in their mobility. However, micro-fossil evidence pin-points a sudden drop in worldwide temperature at the Cretaceous/Tertiary boundary, and the sudden but prolonged cold stress led to extinction of dinosaurs which were too large to escape by hibernating in burrows or other micro-habitats available to smaller warm-blooded animals which survived extinction. The dinosaurs were thus unable to survive prolonged drops in body temperature (Bakker, 1972).

A second major cause of species extinctions may be the *outbreak of disease* or *pest infections*, under changing environmental conditions. An example is the spread of Dutch Elm Disease throughout Britain in the early 1970s, which has been aggravated by the rapid die-back of trees infected by the fungus (*Cerato cystis ulmi*) and the presence in recent years of a new aggressive strain of the fungus vector (Burdekin and Gibbs, 1972).

A third cause of extinctions has been *direct hunting* and the *persecution of species*. Towards the close of the Pleistocene, for example, over a hundred species of large American mammals (including mammoths and many species of horses) became extinct, but the fossil record shows no marked loss of small invertebrates, plants, aquatic organisms or marine life. Martin (1967) has evaluated several hypotheses for this 'selective' mass extinction, including:

(a) *ecological substitution* by other species of large carnivores competing for the same food resources (but the mammal extinction far exceeded possible replacement by new species that could easily have been accommodated by the prevailing habitat),

(b) *climatic change* and *accelerated competition* between large mammals before they could re-adjust to changes in vegetation and climate (but there is no fossil evidence of competitive stress, and the extinctions apparently occurred when environmental conditions were improving for many species),

(c) *human over-kill*; many of the extinctions coincide in time with the first archaeological records of fire (charcoal deposits) and with the distinctive stone tools of Early Stone Age hunters.

Martin (1967) concluded that the use of fire in fire drives (to drive whole herds of big game over cliffs) for hunting led to massive extinctions because, by the very crudeness of the technique, whole herds had to be decimated to kill the relatively few animals needed for food. The fire drives would presumably have been frequent events because the food would not remain fresh or edible for long periods of time under primitive culinary conditions. In a similar field, but with less academic conviction, Hogarth (1976) has examined the likely causes of extinction of the dragon, 'common' throughout Europe in late mediaeval times but extinct by the late eighteenth century. The three most likely causes are *commercial over-exploitation* (dragons were used for transport and as fireworks, and their products were thought to have valuable medicinal purposes); *slaying* (by saints, kings, knights and gods – generally in single but mortal combat) and *loss of credibility* (the most likely explanation because of the failure to survive of any actual specimens and because of a recent dearth of sightings). The total extinction of dragon species, however, has been debated by some ecologists; and the perpetual and recently accelerated quest for identification of the large but infrequently sighted inhabitant(s) of Loch Ness in Scotland (Scott and Rines, 1975) promises to rekindle the flickering flames of this debate.

A contemporary example of extinction associated with direct exploitation of species (but this time a threatened rather than actual extinction) involves the coastal redwood (*Sequoia sempervirens*) along the west coast of North America. Already the redwood forests are much smaller and more decimated than formerly, and two management strategies currently being adopted with remnant areas of forest have increased the extinction probabilities of the redwood and aroused the passionate concern of conservationists. Many of the older mixed age class stands of redwood are to be extensively felled for logging (except in some preserved park areas) and the general environment of the forests is now being radically affected by changes in fire and water regimes within the ecosystems and by grazing, cultivation, road building and a host of allied land-use changes. Namkoong and Roberds (1974) point out that the old growth redwoods in particular are threatened and so survival of the species will depend on young, single-aged stands of the redwoods. They recommend management strategies to control mortality and increase successful reproduction among the redwoods to increase the survival probabilities of the species. A further example of local extinction occasioned by exploitation concerns the edelweiss plant, formerly but no longer common on dry, rocky Alpine slopes above the timber line, which has been depleted by the direct impact of over-collecting (Moyal, 1976).

A fourth major cause of extinctions has been *man-induced environmental change*, through habitat removal and by altering the equilibrium states within ecosystems (see Chapter 6). These changes appear to have accelerated of late in both diversity and magnitude. Because many of these environmental changes are either occurring at the present time or occurred in the recent past, there is a wide range of documented examples from many parts of the world, and many of these relate to local rather than global, and threatened rather than actual extinctions. Duffey (1968), for example, has related the extinction of the Large Copper Butterfly (*Lycaena dispar Haw, batanus opth.*) at Woodwalton Fen National Nature Reserve in Huntingdonshire to the draining of the Huntingdon Fens over the last two hundred years; and Wells (1968) concluded that the changing distribution of plants such as the Pasque Flower (*Pulsatilla vulgaris*) has reflected the incidence of ploughing in the nineteenth century on chalk and limestone areas of lowland England. Dempster, King and Lakhani (1976) have traced the changing fortunes of the Swallowtail Butterfly (*Papilio machaon britannicus*) in Britain. Although the butterfly was formerly common throughout the East Anglian Fens, the area of suitable habitats was drastically reduced by drainage of the fenland meres (completed by 1850). The species became extinct at Wicken Fen by the early 1950s, and it now survives only in the marshes around the Norfolk Broads (though in declining numbers). The authors relate the local extinction of the butterfly to reductions in the status and performance of the main food source of the butterfly – Milk Parsley (*Pencadanum palustre*) – occasioned by loss and fragmentation of its habitat due to fenland drainage and the natural over-growth of carr vegetation. A third example is the relatively recent die-back of hedgerow trees (especially oaks) in north Humberside documented by Balmer, Rothwell and Boatman (1975), who concluded that the main causal factor has been the large-scale and widespread destruction of hedgerows and hedgerow trees within the last ten years which has substantially increased the exposure of the remaining trees and now threatens them with die-back.

Whilst extinction is clearly an inherently natural ecological process, there are many signs that man-induced extinctions, or threatened extinctions, on the global scale are increasing in number through time. Silverberg's (1973) data show that in the period 1801 to 1850 only two kinds of mammals became extinct in the world; between 1851 and 1900 this had risen to some thirty-one mammals and a large number of bird species; and by the period 1901 to 1944 there had been forty recorded extinctions. The trend is continuing, because in the early 1970s the World Wildlife Fund published a list of over a thousand species of plants and animals currently on the verge of extinction; and the Smithsonian Institute submitted to the United States Secretary of the Interior a list of 2099 native North American species, sub-species and variants now in danger of extinction and in need of protection. In Europe the available data are equally alarming – Ribault (1978) stresses that of 2069 endemic plants and flowers only 694 are *not* now threatened with extinction.

Extinction of species is clearly an area where environmental management has a vital role to play, in differentiating between 'natural' extinctions and 'accelerated' or 'anthropogenic' extinctions. One basic problem is that both evolution and extinction are on-going processes, and so identification of man-induced changes from 'background' natural changes is difficult. The problem is made more difficult in the light of recent findings on contemporary evolution of species.

4.1d Contemporary evolution

Dyer (1968) has stressed the need to study recent evolution for both academic and applied reasons. The former include the benefits of examining the evolutionary changes as they occur (this thus allows insight into the processes themselves, and into their immediate controls), and of analysing the formation of new species; and the latter include the need to apply this knowledge to problems of conservation, environmental disturbances, agricultural improvement (such as the production of improved strains and new varieties of crop plants) and the spread of drug-resistant pathogens. Several cases of recent evolution of species can be considered.

One concerns the *domestication of animals*. Dobzhansky (1955) has shown that by evolutionary changes a general tendency arises in the course of the domestication of animals, whereby the longer a particular species has been domesticated the more numerous are its domestic varieties (Table 4.2). Thus, for example, only five domesticated varieties of budgerigar have evolved since 1840, whereas over 200 varieties of dogs have evolved since perhaps 10 000 BC.

Table 4.2
AGE OF DOMESTICATION OF PRINCIPAL DOMESTIC ANIMALS AND VARIETIES

Species		Domestication	Number of varieties
Pigeon	(Columba livia)	prehistoric	140
Donkey	(Equus asinus)	prehistoric	15
Guinea Pig	(Cavia porcellus)	prehistoric	25
Dog	(Canis familiaris)	10 000–8000 BC	200
Cattle	(Bos taurus)	6000–2000 BC	60
Pig	(Sus scrofa)	5000–2000 BC	35
Chicken	(Gallus domesticus)	3000 BC	125
Horse	(Equus caballus)	3000–2000 BC	60
Cat	(Felis manisulata)	2000 BC	25
Duck	(Anas platyrhynchos)	1000 BC	30
Rabbit	(Oryctolagus cuniculus)	1000 BC	20
Goose	(Anser anser)	1000 BC?	20
Peacock	(Pauo cristatus)	early historic	4
Guinea Hen	(Numida numida)	prehistoric, and AD 1500	1
Canary	(Serinus canaria)	AD 1500	20
Budgerigar	(Melopsittacus undulatus)	AD 1840	5
Maral	(Crevus maral)	AD 1850	1

Source: Dobzhansky (1955)

A second case concerns 'Industrial Melanism', that is the appearance and spread of dark forms of a number of species (such as moths) as an evolutionary response to industrialization and its environmental pollution effects. Dyer (1968) has summarized the evidence on the spread of melanism, as a diffusion process, of the Peppered Moth (*Biston betularia*) in England since the melanic form was first discovered in Manchester in 1848. Speckled varieties of the moth, previously dominant, were progressively replaced by dark coloured ones (by a Darwinian form of 'natural selection', triggered off in the presence of a major dominant dark coloured gene). Soon after 1848 the appearance of the dark melanic forms was recorded in counties near to Lancashire – such as Cheshire (1860), Yorkshire (1861), Staffordshire (1878) and Westmorland (1870). Eastern counties (Norfolk, Suffolk and Cambridgeshire) all recorded their first melanic forms between 1892 and 1895, and by 1897 there are reports of sightings in southern England. Dyer points out that the diffusion of the melanic form follows a normal population growth curve – in that the initial spread would be extremely slow if one single mutational event was involved; this would be followed by a period of rapid spread; the rate of increase would then slow down considerably as the frequency of the melanic form approached 100 per cent. The Peppered Moth is not the only casualty of industrial melanism, however, because within the past hundred years in industrial areas of Britain and North West Europe some 48 species of moth have become melanic and a further 250 have gone markedly darker in colour.

4.2 SUCCESSIONAL DEVELOPMENT OF ECOSYSTEMS

One striking feature of vegetation development in most environments is that groups of plants (not only those genetically related to each other) often grow together. Groupings of plants which are adapted to local habitat conditions are referred to as *Plant Associations* or *Plant Communities*. Vegetation thus often has an apparent orderliness and continuity. George Perkins Marsh clearly had this in mind when he wrote:

> Nature, left undisturbed, so fashions her territory as to give it almost unchanging permanence of form, outline and proportion, except when shattered by geological convulsions; and in these comparatively rare cases of derangement, she sets off at once to repair superficial damage, and to restore, as nearly as practicable, the former aspect of her dominion (Marsh, 1864).

The orderliness in vegetation relates to two inter-related ecological factors – the continuity of vegetation response to spatial variations in environmental factors (see Chapter 5), and the continuity of vegetational development through time.

Studies of vegetational development through time have been made in many areas, and these generally show that development is a slow process involving a large number of small sequential changes, and that ultimately a relatively stable and self-perpetuating plant community is often established. This entire sequence of directional, sequential change was termed *Vegetation Succession* by Clements (1916), who saw vegetation development as an orderly and predictable sequence which developed along definite pathways, towards predictable end situations. Clements rationalized the successional development of vegetation into a five-phase model:

PHASE 1 *Nudation*: the initial creation of a bare area.

PHASE 2 *Migration*: the arrival of plant seeds, etc.

PHASE 3 *Ecesis*: the establishment of the plant seeds, etc.

PHASE 4 *Reaction*: competition between the established plants, and their effects on the local habitat.

PHASE 5 *Stabilization*: where populations of species reach a final equilibrium condition, in balance with local and regional habitat conditions.

Throughout this successional sequence vegetation change is continuous and sequential. Succession is thus composed of a series of transitional vegetation communities 'en route' to an equilibrium community. Clements termed the transitional stages *Seres*, and the final equilibrium state the *Climax Vegetation*. Different vegetation communities develop on different types of habitat, and so the seral stages can be identified by appropriate prefixes – thus hydroseres occur on wet sites such as swamps and lake edges; xeric seres occur on bare rock with little freely available moisture; and psammic seres occur on moving sand such as on sand dunes. Whatever the form of the seral sequence, the whole sequence of seral communities is termed a *Prisere*.

A considerable amount of attention has been devoted to the concept of succession since the pioneer formulation of concepts by Clements (1916). Whittaker (1953) has stressed the evolutionary significance of succession, and Horn has offered a definition of succession as 'a pattern of changes in specific composition of a community after a radical disturbance or after the appearance of a new path in the physical environment for colonization by plants and animals' (Horn, 1974). Regier and Cowell (1972) identify three main prerequisites for succession. One is the creation of a 'new' habitat; and the second is a situation where all biotic variables acting within the habitat remain within the ranges to which many organisms have already become adapted (that is, within their 'tolerance range'); and the third is that many such organisms have access to the environment.

Many examples of priseres and of successional sequences have been traced, and following Clementian terminology these describe two basic forms of succession:

> *Primary Succession*: the developmental sequence on newly exposed bare areas which have not supported vegetation cover previously – such as newly emerged oceanic islands, ablation zones in front of glaciers, dune systems in the process of development, recently deposited river alluvium, and man-made habitats such as pit heaps.

> *Secondary Succession*: the developmental sequence on areas which have in the past supported vegetation which has either in part, or in whole, been destroyed – such as by human interference in the form of trampling, fire, intentional removal of vegetation, and road construction activities; or by natural processes in the form of lightning fires, the influence of drought or floods and severe storms.

Horn defined secondary succession as 'the process of re-establishment of a reasonable facsimile of the original community after a temporary disturbance' (Horn, 1974). Clearly, from an environmental management point of view, the notion of secondary successions is most valuable, since it offers scope for natural regeneration of vegetation after human impact, and thus for assisting in the natural repair of degraded environments.

4.2a Primary succession

Primary successions have been studied in a wide variety of habitats, and some examples will serve to highlight the major types of vegetational and environmental change which characterize normal priseral development. One classic example of primary successions is the study by Cooper (1923) of vegetational changes between 1916 and 1921 in recently de-glaciated areas in the Glacier Bay area of Alaska. Three major stages in the succession were identified (Table 4.3), and the study also offered an insight into rates of change at different periods throughout the succession. It started with a very slow initial rate of colonization, followed by marked acceleration in the rate of change occasioned by increased effective reaction between thicket species and the habitat (dense spruce forest had become established within 125 years of the melting of the ice). There was subsequently a reduction in the rate of change when spruce became dominant, so that the entry of hemlock into the association was a very slow process.

Successional studies have also been based on new environments such as recently emerged oceanic islands and areas affected by volcanic activity. An interesting example of the former is the ongoing study into the evolution of organic life on Surtsey reported by Fridriksson (1968, 1975). The island appeared on 15 November 1963 as a result of a submarine volcanic explosion, and the initially circular cone of ash and cinder has subsequently been modified by volcanic events on and around the main island. Biological observations of life forms on the island have been carried out at regular intervals, and the results show a detailed picture of the early stages of successional development on a previously sterile island. Within six months of the major eruption, various strains of bacteria and a few moulds (carried by air and sea) had arrived, along with one fly, several birds (including seagulls searching for food washed up by the waves) and several living plant fragments which had drifted into the shore. Amongst the first plants to arrive by drifting were diatoms and bacteria (by August 1964); and sixteen algal species had colonized the coast by 1967. The first bottom invertebrates on submarine slopes around the island had arrived by November 1964; and by 1967 some twelve to fourteen species of marine animals had invaded the inter-tidal zone. By 1967 also twenty-six species of insects had arrived on the island, mostly carried by air in favourable winds; and the first vascular plants were seedlings of sea rocket (*Cakile edentula*) found growing on the sandy beach by June 1965.

Table 4.3
SUMMARY OF SUCCESSIONAL STAGES AT GLACIER BAY, ALASKA

Major stage	Characteristic species	Sequence of change
PIONEER STAGE:	herbs and mat-forming shrubs	(i) initial colonization by Rhacomitrium mosses (ii) gradual increase in perennial herbs, then (iii) establishment of Dryas (with mat-forming tendencies)
THICKET STAGE:	willow and alder	(i) dwarf, creeping willows enter vegetation community (ii) willow and dryas shade out mosses and herbs, these then disappear (iii) shrubs such as willow (*Salix*) and alder (*Alnus*) become dominant, these enrich soil with nitrogen
FOREST STAGE:	coniferous forest	(i) appearance of sitka spruce (*Picea*) followed later by (ii) mixture of spruce and hemlock (*Tsuga*)

Source: Cooper (1923)

The development of vegetation cover on the island of Krakatoa after the 1883 eruption provides a classic case of primary succession, and it offers the prospect of tracing a longer sequence of changes than the Surtsey example. The result of the eruption in 1883 was to bury the whole island of Krakatoa under up to thirty metres of ash and pumice, and hence destroy and remove all vestiges of the former luxurious tropical forest vegetation. Natural succession subsequently led to recolonization so that within thirty years there had developed a distinctive shore vegetation and an inland community, and within fourteen years the centre of the island was covered by tall grass several metres high. Subsequently the succession, as traced by Richards (1952), appeared to be heading towards re-establishment of the former climax vegetation structure (Table 4.4). The rapid speed of the succession on the island has been attributed to the nature of the underlying material, which is pervious and thus penetrable by roots, and it is rich in plant nutrients. Succession on volcanic habitats has also been traced in Papua by Taylor (1957), who found that – unlike the Krakatoa case – many volcanic sites were pioneered by tree species which rapidly formed a close-canopy woodland community, although the succession towards climax was still expected to take several centuries to complete.

Recently created habitats provide further abundant scope for elucidating successional sequences. For example, developing sand dune systems at Blackeney Point and Southport were studied in detail by Salisbury (1922, 1925) over a 300-year time scale, based on examination of old topographic maps and present-day vegetation; and a series of samples taken along this scale allowed analysis of the rate of soil development and plant succession.

Table 4.4
PRIMARY SUCCESSION ON KRAKATOA, AFTER THE 1883 VOLCANIC ERUPTION

Year	Total number of plant species	Coast	Lower slopes	Upper slopes
		Coastal woodland climax	Lowland rain forest climax	Submontaine forest climax
1934	271		Mixed woodland largely taken over from Savanna	Woodland with smaller trees, fewer species taking over
1932				
1928	214			
1920				
1919			Scattered trees in grassland, single or in groups with shade species beneath. Thicket development in ravines	
1908	115	Wider belt of woodland with more species, shrubs, coconut palms	Dense grasses up to 3m high, woodland in ravines	
1897	64	Coastal woodland develops	Dense grasses	Dense grasses with shrubs interspersed
1886	26	9 species of flowering plants	Ferns and scattered flowering plants, blue-green algae beneath on ash	
1884			No life	
1883	0	Volcanic explosion; all life killed; hills deeply gullied by rain		

Source: Richards (1952)

Soil changes through time included the progressive leaching of calcium carbonate, and a fall in pH of the sand; and vegetation changes followed the Clementsian sequence of migration, ecesis, reaction and eventual stabilization. Man-made habitats such as disused pit heaps also offer scope for the natural development of successions – and these are valuable because a well-defined time scale for the changes can often be established. Brierley (1956) studied vegetation development on pit heaps of various ages, and he considered the significance of factors such as the wind-blown arrival of plant seeds, the availability of light for photosynthesis, and the extent of root competition on pits of various ages. Hall (1957) has characterized the basic differences in flora between pit heaps and the adjacent areas, and related these to the opportunity given to pioneer communities to occupy relatively bare ground and to the peculiarity of the pit heap environment (especially in terms of soil acidity) in relation to neighbouring 'natural' soils.

4.2b Secondary succession

Secondary successions have been studied under a variety of situations from purely natural change along a spectrum of human interference to purely man-made change.

Natural environmental changes which can trigger off secondary successions of formerly natural ecosystems include the impact of *climatic extremes* such as storm and tidal action, and the impact of natural lightning. Johnson (1976) has described natural regeneration of Corsican Pine plantations (planted since the 1870s) on dune areas along the North Norfolk coast, which were devastated by a storm surge in 1953 which inundated all low-lying areas and killed the trees. After the area was re-drained, succession involved invasion by scrub species which increased after reduction in the rabbit populations in the area (after the advent of myxomatosis in 1954/5), and which is now being replaced slowly and irregularly by natural regeneration of trees and shrubs. The impact of *fire* has triggered off secondary succession in a great many areas, and whether the fire is natural or man-induced the ecological changes can be dramatic and long-lasting. Swan (1970) evaluated the response of four plant communities in New York State to accidental burning – these were northern hardwoods, oak woods, goldenrod/poverty grass field communities, and little bluestem fields. The most gradual successional changes were observed in the oak woods, which proved to be the most stable after fire, with a lower percentage of dead trees (16 per cent) than the northern hardwoods (35 per cent), a higher percentage of saplings sprouting (87 per cent versus 43 per cent), a higher average number of sprouts per sapling (4.4 versus 2.5), and a higher percentage of ground layer species remaining unchanged or increasing in frequency (87 per cent versus 67 per cent). Although burning has adverse effects on some successions, in others it is a basic environmental factor to which the ecosystem is adjusted and successional ecosystem changes follow if the fire is suppressed or eliminated. For example, Day (1972) maintains that although the introduction of effective fire-protection programmes in Southern Alberta's Rocky Mountain Forest has encouraged a progression towards later successional stages than the pine-dominated assocation favoured by a combination of natural 'dry' electrical storm fires and fires started by man, an active programme of controlled firing would be essential in the long term to ensure favourable habitats for fauna in the area and the preservation of the wilderness value of the Mountain Forest. Controlled burning is also a basic ingredient of heather management programmes, because it restores heather to a more juvenile phase of growth so that there is more nutritious herbage available to grazing animals, and so that annual productivity of green shoots is at a maximum (Grant, 1968).

Heather burning is generally carried out for improved grazing, but the impact of *grazing per se* on secondary succession and vegetation re-generation can also be very marked. The effects of grazing can perhaps best be evaluated by considering the vegetation changes which follow from a cessation of grazing and the removal of grazing pressure. Such is the case of vegetation changes on chalk down grasslands since the pressures of rabbit grazing were reduced with the advent of myxomatosis in South East England in 1954. Based on a study of repeated observation of vegetation along fixed transects in seven National Nature Reserves, Thomas (1960) documented a series of vegetation changes since 1954 (Table 4.5). A second example of grazing is the study by Peterken and Tubbs (1965) of phases of regeneration of unenclosed woodlands in the New Forest since 1850, in relation to fluctuations in grazing and browsing pressure (Figure 4.1). Three main periods of active regeneration in the deciduous woodland were noted. First was a period (A) between 1648 and 1763; followed by period B between 1858 and 1915, and a more recent period (C) since 1938. The three main herbivores in the New Forest are deer, cattle and ponies, and although the numbers of these have changed dramatically through time the authors point

Table 4.5
SUMMARY OF MAIN CHANGES IN CHALK DOWN GRASSLAND VEGETATION SINCE THE ADVENT OF MYXOMATOSIS

(a)	Increased presence of bush-like Juniper (formerly heavily damaged by rabbits); and likelihood of rapid colonization of much downland by scrub, shrubs and trees
(b)	Reduced numbers of those plant species encouraged by rabbits. In general this represents a benefit to agriculture because many species (such as *Atropa belladona*) are poisonous to stock
(c)	Changed proportions of grass species – especially the rapid loss of *Festuca rubra*; this leads to development of unstable downland turf
(d)	Better supplies of young leaves of grass and clovers for game birds; increased and diversified plant growth (such as the appearance of some orchids)
(e)	Increase in woody plant species such as brambles and gorse bushes

Source: Thomas (1960)

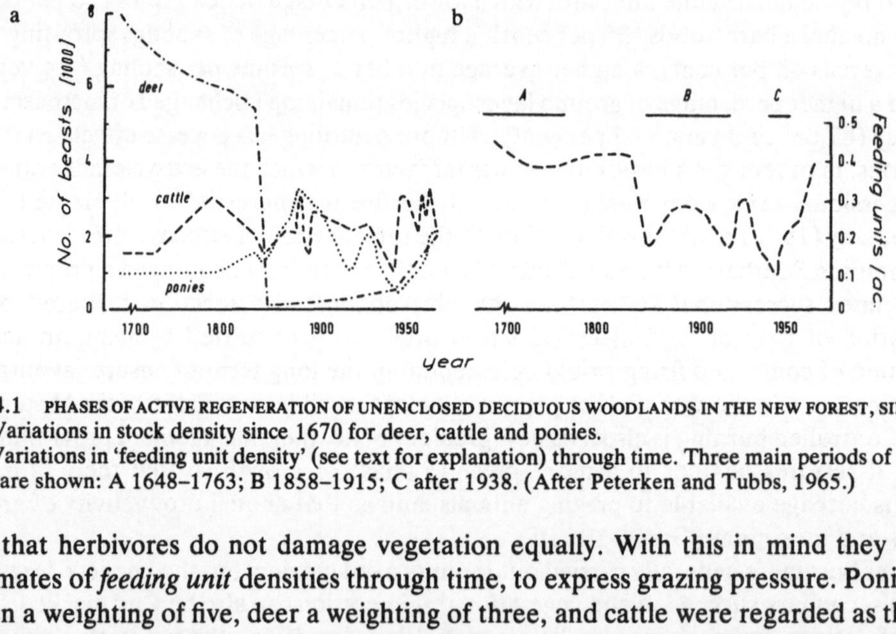

FIG. 4.1 PHASES OF ACTIVE REGENERATION OF UNENCLOSED DECIDUOUS WOODLANDS IN THE NEW FOREST, SINCE 1850.
(a) Variations in stock density since 1670 for deer, cattle and ponies.
(b) Variations in 'feeding unit density' (see text for explanation) through time. Three main periods of regeneration are shown: A 1648–1763; B 1858–1915; C after 1938. (After Peterken and Tubbs, 1965.)

out that herbivores do not damage vegetation equally. With this in mind they derived estimates of *feeding unit* densities through time, to express grazing pressure. Ponies were given a weighting of five, deer a weighting of three, and cattle were regarded as the basic unit of one. There does seem to be some correlation between reduced grazing pressure and woodland regeneration in the New Forest (Figure 4.1b), especially for period B which follows the almost complete removal of deer after 1851, following the 'New Forest Deer Removal Act'. Grazing pressures have since reverted to high levels, and Peterken and Tubbs conclude that 'the implication for the future management of the New Forest is that, if browsing pressure continues to rise at the present rate, successful regeneration in the unenclosed woods will become impossible in the next few years' (Peterken and Tubbs, 1965).

Direct human impact – rather than indirect via grazing – can also have marked effects on vegetation regeneration and secondary successions. The North West Territories of the Canadian Arctic have been exploited in recent years in the search for oil and gas reserves, and Hernandez (1973) has studied the impact of disturbances of both soil and vegetation cover on natural recolonization of disturbed areas. Many of the effects which are spatially most widespread derive from the clearance of lengthy transect lines for seismic surveys of sub-terranean deposits, and often the thawed surface soil layers are bulldozed for access and for emplacement of the geophones. After six years of natural recolonization of such a line the plant cover was only 43 per cent, and this comprised different species to the 'natural' areas. For example, grasses were common, dwarf shrubs were rare, lichens nearly absent, moss cover was greatly reduced compared to the surrounding areas. The sequence of invasion and succession in such harsh environments is clearly a slow one – and this has marked implications for the overall long-term stability of such ecosystems (see pages 99–103), and for the potential of large-scale exploitation in these sensitive systems.

Historical land-use changes associated with cycles of colonization and depopulation also produce large-scale opportunities for secondary succession. The case of St Anthony's Wilderness, near Philadelphia, has been presented by Forrester (1974). The region was

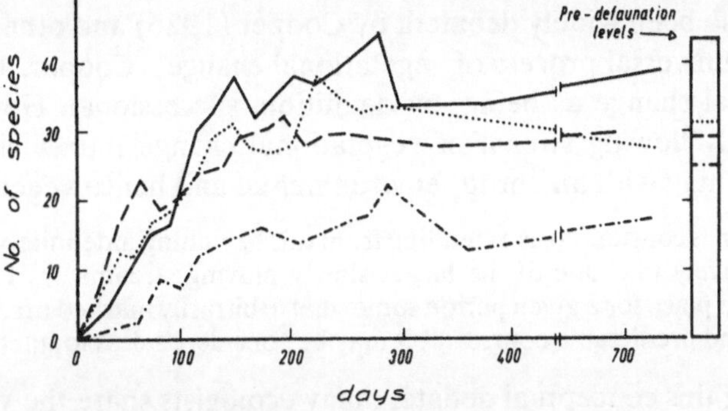

FIG. 4.2 EXPERIMENTAL ZOOGEOGRAPHY OF ISLANDS.
The time-colonization curves for four small mangrove islands at Florida Keys, showing patterns of re-colonization by arthropod communities after initial extermination by methyl bromide fumigation. (After Simberloff and Wilson, 1970.)

first opened to settlers in the 1750s, and the discovery of coal in the area in the mid nineteenth century led to dramatic population expansion, which was accentuated by the advent of the railroad. Coal mining soon declined in the area, however, and it was replaced by lumbering. By 1900 all of the local hillslopes had been stripped of their tree cover, and soon the resources were depleted so the population abandoned the area. Since abandonment the process of secondary succession has recommenced, and the area is now a second generation wilderness, with the return of trees and natural vegetation.

One interesting case of secondary succession, which concerns animals rather than plants, is the case of the 'experimental zoogeography of islands' described by Simberloff and Wilson (1969, 1970). In the experiment the entire arthropod population of six small mangrove islands in Florida Keys were exterminated by methyl bromide fumigation (Wilson and Simberloff, 1969), and the process of re-colonization was subsequently monitored by frequent censuses on the islands. After one year a number of species had roughly re-attained their original levels on five of the islands, and 'time colonization' curves could be established for each island (Figure 4.2). These curves show two peaks (Simberloff and Wilson, 1969):

(a) a large number of species is present initially, but this is prior to the establishment of populations belonging to the dominant constituent species;

(b) this was followed by a decrease in numbers, as densities of the constituent populations approached the levels they had been at prior to the de-faunation;

(c) there then occurred a lower 'peak' level, at which species interactions contributed significantly to species extinction rates.

The first peak was termed the 'non-interactive species equilibrium' and the second was termed the 'interactive species equilibrium'. After two years the species numbers had not changed greatly, but there were significant changes in the species composition on the islands because of high rates of population turn-overs (Simberloff and Wilson, 1970).

4.2c General characteristics of succession

Although used extensively since the pioneer work by Clements (1916), the concept of a 'succession' has also been widely debated, by Cooper (1926) and others. Cooper viewed succession as 'the universal process of vegetational change' (Cooper, 1926), and thus he saw *all* vegetational change as being, by definition, successional. He preferred to subdivide a continually flowing stream of vegetational change into what he termed 'unit successions', beginning with an abrupt environmental and hence vegetational change.

> The unit succession is considered as extending from this beginning indefinitely into the future until it merges insensitively into one of the larger slowly moving streams . . . It is the vegetational sequence in a given place for a given period somewhat arbitrarily marked off . . . it commonly runs a rather definite and predictable course . . . it may be considered developmental (Cooper, 1926).

Notwithstanding this conceptual debate, many ecologists share the view of Odum that 'ecological succession is one of the most important processes which result from the community modifying the environment' (Odum, 1962). Whittaker (1953) has reviewed the concepts of succession and climax, and he isolated four major ecosystem changes which accompany successional development. These are a progressive increase in community complexity and diversity; a progressive increase in the structure (massiveness) and productivity of ecosystems; an increasing maturity of soils; and an increasing relative stability and regularity of populations within the ecosystem. These provide a convenient framework for considering general characteristics of succession.

Ecological diversity

Species diversity generally increases during succession because of the increasing number of habitat niches available in the later stages of seral development. An interesting illustration of such changes in diversity is provided by Odum (1962), based on a laboratory experiment in which changes in a series of culture flasks containing plankton communities were monitored through various successional stages (Figure 4.3). The experiment started with a 'climax', relatively old plankton community, characterized by a high diversity of species and a low ratio of production to biomass (thus gross production tends to equal community respiration). To this was added a fresh culture medium of old tissues, which triggered off the successional changes. During early succession the community had low species diversity because of the dominance of a few plankton species, and production exceeded respiration. Through time, however, there occurred a gradual and progressive change towards the climax (steady-state) situation of high diversity and low production/biomass ratio. Regier and Cowell (1972) have suggested a generalized model of changes in

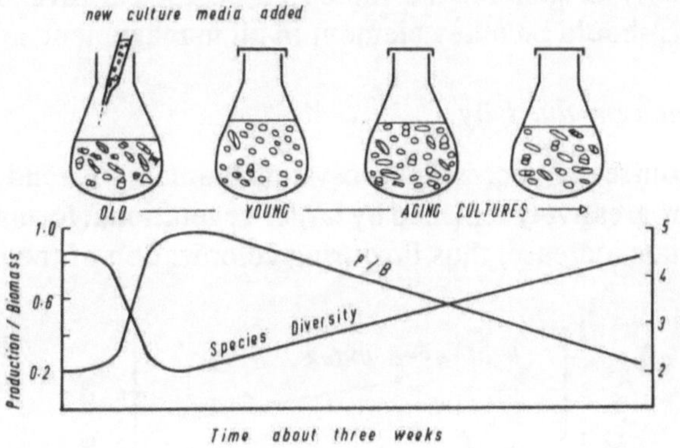

FIG. 4.3 EXPERIMENTAL ILLUSTRATION OF CHANGES IN ECOLOGICAL DIVERSITY THROUGH TIME, DURING THE COURSE OF A SUCCESSION.
Based on laboratory cultures. (After Odum, 1962.)

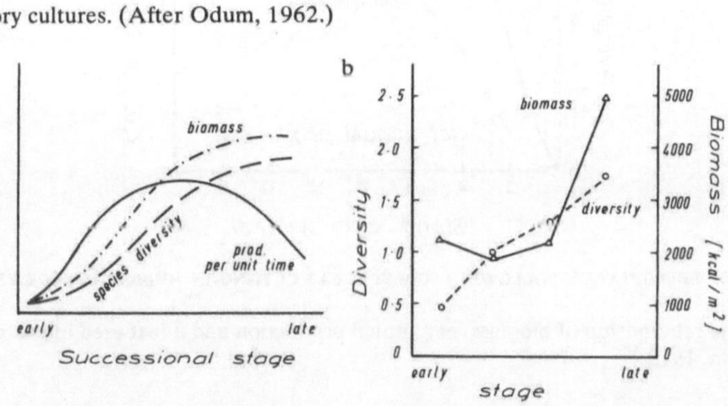

FIG. 4.4 GENERAL PATTERNS OF CHANGE IN BIOMASS AND SPECIES DIVERSITY THROUGH SUCCESSION.
(a) The general model of biomass, production and species diversity as a function of successional state in, and external stress on the ecosystem.
(b) An illustration of the model, based on an ungrazed salt-marsh at Milford Haven, South Wales. (After Regier and Cowell, 1972.)

biomass and species diversity as a function of successional state of the ecosystem (Figure 4.4), which demonstrates increasing diversity through succession. Early stages tend to be dominated by a few species of highly opportunist organisms which are not effective competitors with most longer-lived species; and these are replaced through the succession. The application of the model is confirmed by observations from an ungrazed salt-marsh in Milford Haven in South Wales (Figure 4.4b).

The general notion of progressively increasing diversity throughout the entire course of a succession has been challenged by Loucks (1970), however, on the basis of observations of forest communities of different ages in Wisconsin. He demonstrated a peak in forest community diversity about 100 to 200 hundred years after initiation of secondary succession, and a reduced diversity later in succession. This appears to be related to periodic disturbances (such as fires) with average intervals of between 50 and 200 years, which recycle the system and maintain a periodic wave of peak diversity. On a management theme, Loucks concluded that 'any modification of the system that precludes periodic, random perturbations and recycling would be detrimental to the system in the long run' (Loucks, 1970). Clearly in general the increased ecological diversity through natural ecological successions should be a key element in all management strategies.

Ecosystem structure and productivity

Clearly through the course of a succession ecosystem biomass will tend to increase, as small pioneer species are progressively replaced by larger vegetational forms, and as the number and diversity of habitats increase, thus favouring colonization of the ecosystem by more,

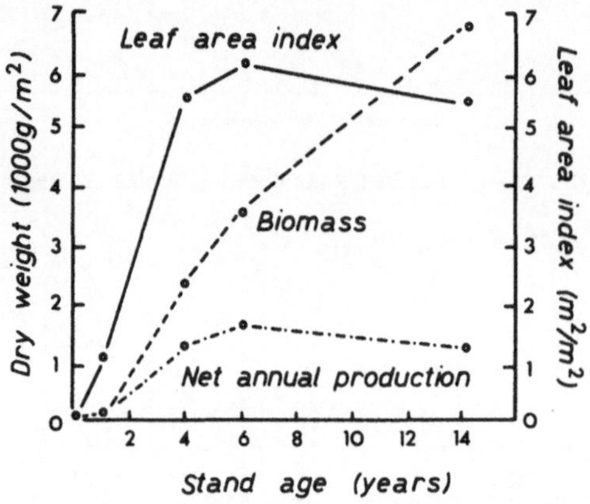

FIG. 4.5 PATTERN OF REVEGETATION FOLLOWING FOREST CLEAR CUTTING IN HUBBARD BROOK EXPERIEMENTAL ECOSYSTEM.
The figure shows the relationship of biomass, net annual production and a leaf area index to stand age. (After Marks and Bormann, 1972.)

and generally larger, faunal species. Productivity will in general increase also, at least during early succession (Figure 4.4a). What is important, however, is the relationship between the two. The Odum model (Figure 4.3) shows a progressive reduction in the ratio between production and biomass through succession; and Regier and Cowell (1972) stress that under 'climax' conditions a high percentage of the biomass at any period in time is locked up in relatively large, long-lived tissues and bodies of a relatively few kinds of organisms – thus the production/biomass ratio is low. Specific examples illustrate these tendencies. Thus biomass increases overall through succession in the Milford Haven salt-marsh (Figure 4.4b); and observations on the rate of revegetation following forest clear-cutting in the Hubbard Brook forest (Marks and Bormann, 1972) shows increasing biomass through time over the first 14 years of secondary succession, increasing production for the first six years with little overall change subsequently (Figure 4.5), and a production/biomass ratio which diminishes throughout the period.

Although these studies show increasing productivity throughout the succession, the time scale in each case is quite restricted; and Lindeman (1942) has pointed out that when complete priseres are considered, smooth growth of productivity is rarely found in natural successions because successional seres consist of a number of stages and these are reflected in changing productivity levels. The hypothetical productivity changes through the development of a hydrosere from an initial deep lake through to eventual terrestrial climax vegetation are shown in Figure 4.6 and the details are summarized in Table 4.6

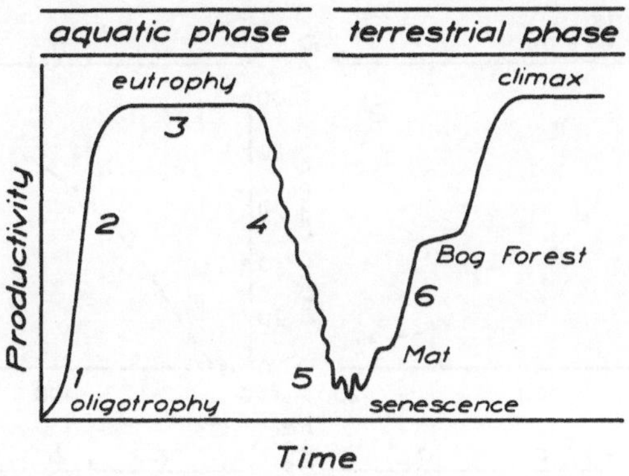

FIG. 4.6 SEQUENCE OF VARIATIONS IN ECOLOGICAL PRODUCTIVITY THROUGH TIME, OVER A LONG-TERM DEVELOPMENT BY SUCCESSION OF A HYDROSERE.
See Table 4.6 for explanation. (After Lindeman, 1942.)

Table 4.6
SEQUENCE OF HYPOTHETICAL PRODUCTIVITY CHANGES DURING DEVELOPMENT OF A HYDROSERE

Stage	Productivity changes
1	Relatively short-lived pioneer stage – deep lake. Productivity is relatively low due to the low nutrient status of the water (*Oligotropic lake*). Stage equilibrium
2	Influx of nutrients from drainage basin in streamflow – leads to increased productivity and the beginning of sedimentation in the lake
3	By now the nutrient limiting factor is fully recovered; excess nutrients stimulate the rapid growth of organic matters, and this exceeds the amount which can be oxidized by respiration, prediction and bacterial decomposition. Reduced oxygen supply to lower lake levels affects the bottom fauna – leads to a *Eutrophic lake*. Stage equilibrium
4	Oxygen supply becomes increasingly a limiting factor of productivity – lake sedimentation reduces mean lake basin depth
5	Lake is eventually completely filled with sediments (*Senescene*). (a) There is an increase in the shallow shore-line area populated with pondweeds. (b) There is an increased marginal invasion of terrestrial stages
6	Productivity begins to increase, punctuated by stages of stage equilibrium; the common form would be: (a) Mat vegetation (partly floating?) of sedges, grasses, heaths and some mosses. (b) Increased productivity with succession to Bog Forest Stage – this might be permanent or succeeded by a regional climax vegetation

Source: Lindeman (1942)

Changing soil characteristics

A key element in the Clementsian theory of succession is that progressively, through time, the changing vegetation communities can modify their local habitat and environment more and more. In terrestrial ecosystems it follows, therefore, that soil characteristics should change markedly throughout the development of a prisere. Crocker and Major (1955) studied the development of certain soil properties on recently exposed surfaces of different ages in the Glacier Bay area of Alaska, through a succession stretching from initial pioneer stage, through an alder-dominated stage, and across a transition to a spruce dominated stage. Changes were observed in soil properties of bulk density (Figure 4.7a); soil reaction; organic carbon content (Figure 4.7d); calcium carbonate; total nitrogen (Figure 4.7c);

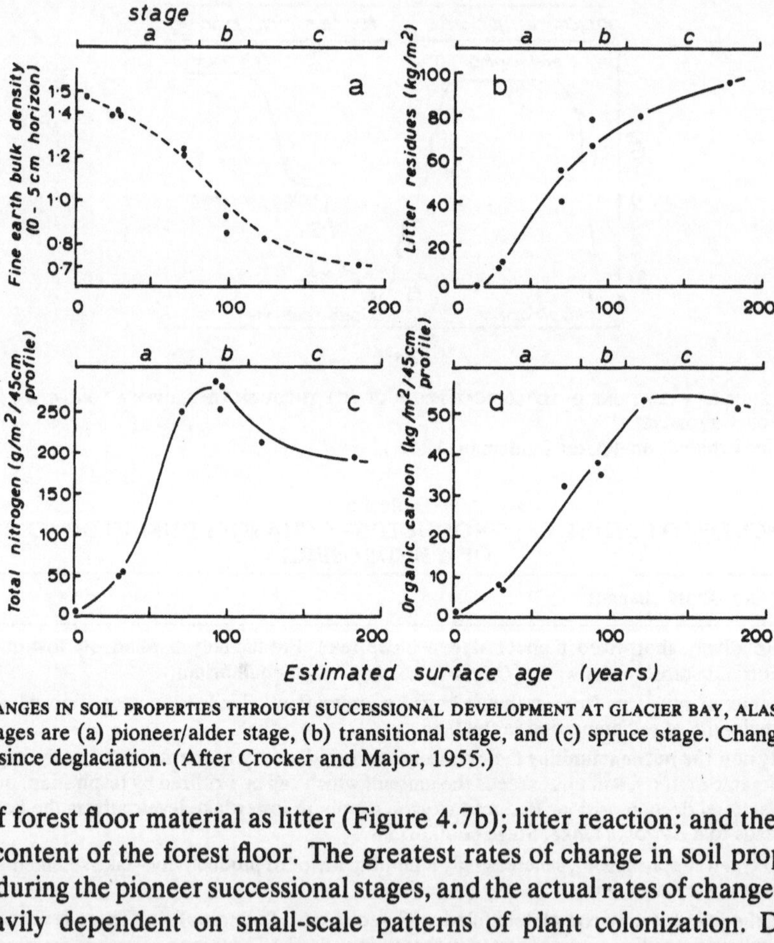

FIG. 4.7 CHANGES IN SOIL PROPERTIES THROUGH SUCCESSIONAL DEVELOPMENT AT GLACIER BAY, ALASKA.
The three stages are (a) pioneer/alder stage, (b) transitional stage, and (c) spruce stage. Changes are plotted against time since deglaciation. (After Crocker and Major, 1955.)

amount of forest floor material as litter (Figure 4.7b); litter reaction; and the carbon and nitrogen content of the forest floor. The greatest rates of change in soil properties were recorded during the pioneer successional stages, and the actual rates of change were shown to be heavily dependent on small-scale patterns of plant colonization. During early succession, therefore, spatial patterns of soil properties were very marked, whereas the subsequent development of alder thicket led to a greater spatial uniformity of soil properties.

Although changes in soil characteristics through time in the Glacier Bay study were quite small after about 100 to 150 years (Figure 4.7), the time taken to reach 'stability' depends strongly on the nature of the initial soil material. Barren dune sand thus takes longer to reach equilibrium than does glacial till; and from Olson's (1958) analysis of soil changes on southern Lake Michigan sand dunes it appears that up to 1000 years of successional development is required before soil properties such as carbon and nitrogen content, soil moisture status and cation exchange stabilize.

Ecosystem stability

Whittaker's (1953) conclusion concerning increasing ecosystem stability through succession has been echoed by a number of ecologists (such as Odum, 1969), and it is based on Clements' (1916) views on vegetation as an organism. Clements maintained that 'the developmental study of vegetation necessarily rests upon the assumption that the unit or climax formation is an organic entity. As an organism the formation arises, grows, matures and dies' (Clements, 1916). Just as the association between ecological diversity and stability (see pages 99–103) is widely debated, so the link between the two as they both change through time (during succession) has attracted conflicting views. Horn (1974) ascribes this in part to the common failure to distinguish between what he terms 'stability' and 'dynamic stability'. The former refers simply to the overall absence of ecological change within a system, and Horn maintains that this generally shows a small increase through succession. The latter refers to the speed at which the community rebounds after some temporary disturbance, and he maintains that this generally decreases during succession.

Time scales

Attention has also been focused on the time scales involved in succession, and on rates of change within priseres. These provide valuable diagnostic indicators of potential ecosystem stability, susceptibility to induced changes, and the time scales required (in management strategies) for natural self-generating 'repair' of damaged ecosystems. Cooper (1926) outlined a general tendency for rates of ecosystem change to be slow at the beginning of successions (because initial habitat conditions make it difficult for plants to establish themselves in numbers and in bulk sufficient to have marked effects on the environment); to increase through succession (because larger numbers and increased diversity of plant species lead to greater change, because of greater reaction between organic and inorganic components of ecosystems); and then to decrease as the climax stage is approached (this is a self-limiting tendency associated with a reduction in species diversity at climax). This sequence of changes in rates of ecosystem change helps to account for the 'S-shaped' curves of changing biomass and species diversity through time in Regier and Cowell's model (Figure 4.4a), and the 'S-shaped' curve of biomass changes through time in Figure 4.5. The pattern of soil changes observed by Crocker and Major (Figure 4.7) also conforms to this generalized sequence of changes in rates.

The time required for attainment of the climax state was considered by Odum (1962), who concluded that it is related to community structure. Thus successions in open-water ecosystems tend to be relatively brief (measured perhaps in terms of weeks) because the community can only modify the physical environment to a small extent. With the forest ecosystem, on the other hand, succession is much longer because a large biomass accumulates through time, and the community continues to change in species composition and to regulate its physical environment. Ecologically adverse sites such as sand dunes and recent volcanic lava flows have on the whole very lengthy successions, which often require at least a thousand years before the climax vegetation structure can be attained.

4.2d Climax vegetation

The end point of the process of succession, according to Clements (1916), is the *climax community*. When this is established, small-scale environmental changes cease, the existing species are capable of reproducing in situ, no further aggressive colonists invade the area, and in theory the most diverse and complex communities possible in that location exist there (Kellman, 1975). The vegetation under such conditions is often referred to as the *Climatic Climax Vegetation*, because climate is the most important single environmental factor which controls the structure and composition of the climax community. Tansley pointed out that the climax community 'will be dominated usually by the largest and particularly by the tallest plants which can arrive on the area, and can flourish under the

Table 4.7
CLIMAX AND ASSOCIATED STAGES OF VEGETATION SUCCESSION AS DEFINED BY CLEMENTS (1936)

CLIMAX:	the final stage of vegetational development in a given climatic unit	
PROCLIMAX:	'all communities that simulate the climax to some extent in terms of stability or permanence, but lack the proper sanctions of the existing climate' Proclimax states include	
	SUBCLIMAX:	'the stage preceding the climax in all complete seres, primary and secondary',
	DISCLIMAX:	'results from the modification or replacement of the true climax, either as a whole or in part, or from a change in the direction of succession' (such as through disturbance by man or by grazing animals),
	PRECLIMAX:	similar to the subclimax state, but here the likelihood of succeeding to the climax state is low because compensation by local factors will rarely (if ever) be overcome within the existing climate,
	POSTCLIMAX:	a community which, for some reason, has progressed beyond the climax stage (that is, it is post-climax to lower life forms)

particular conditions which it presents' (Tansley, 1935). For this reason most climax communities are dominated by tree species (such as the Tropical Rain Forest), and low grading shrubs and herbaceous plants can only be dominant where they are not shaded out by tree canopy above (such as in Arctic Canada, on desert fringes, and similar areas).

In Clements' (1916) original formulation the climax state exists when vegetation development is in equilibrium with, and at an optimum for, a given climatic environment; and various deviations from the strictly climax state can arise as *proclimax states* (Table 4.7). In Clements' view the climax is the major unit of vegetation (synonymous with the *biome* concept of more recent workers – see Chapter 5), it is considered as the more or less permanent final stage of succession, and its visible unity is related to the life-form of the dominant species. The climax concept has environmental management implications, because Whittaker defines climax as 'a partially stabilized community steady-state adapted to maximize sustained utilization of environmental resources in biological productivity' (Whittaker, 1953).

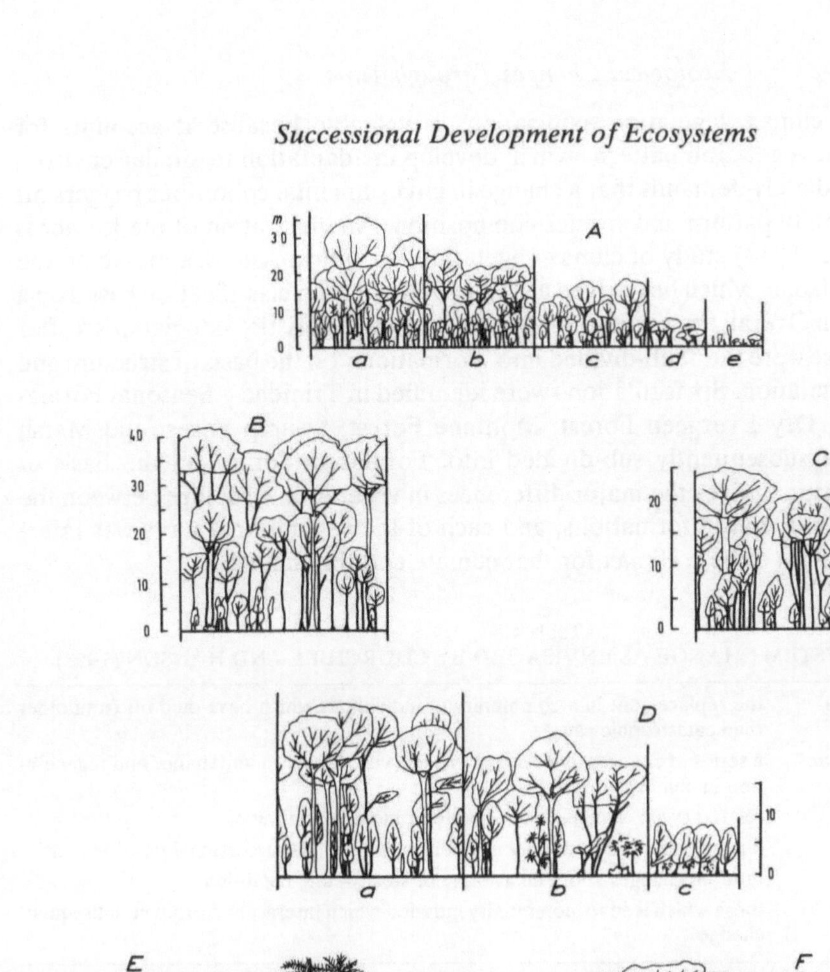

FIG. 4.8 VEGETATION STRUCTURE OF CLIMAX VEGETATION ASSOCIATIONS IN TRINIDAD.

Key to the vegetation units:

A	Seasonal Formations:	(a)	evergreen seasonal forest
		(b)	semi-evergreen seasonal forest
		(c)	deciduous seasonal forest
		(d)	thorn woodland
		(e)	cactus scrub
B	Rain Forest		
C	Xerophytic Rain Forest		
D	Montane Forest	(a)	lower montane rain forest
		(b)	montane rain forest
		(c)	elfin woodland
E	Swamp	(a)	swamp forest
		(b)	palm swamp
F	Marsh	(a)	marsh forest
		(b)	marsh woodland
		(c)	palm marsh

(After Beard, 1944.)

The concept of climax vegetation associations is valuable because it accounts for similarities between vegetation patterns which develop in adaptation to similar environments, and it immediately demands that a change in environmental conditions triggers off a change in community pattern and species composition. An illustration of the former is provided in Beard's (1944) study of climax vegetation in tropical America, based on the vegetation of Trinidad, in which he derived a system of vegetation classification based on a three phase division. Overall the vegetation was divided into 'Floristic Associations' (after Clements), and these were then sub-divided into 'Formations' on the basis of structure and life forms of the vegetation. Six formations were identified in Trinidad – Seasonal Formations, Rain Forest, Dry Evergeen Forest, Montane Forest, Swamp Forest and Marsh Forest. These were subsequently sub-divided into 'Formation Series' on the basis of habitat. Figure 4.8 summarizes the major differences in vegetation structure between the various formations and within formations, and each of the vegetational units was interpreted by Beard (1944) to be a climax for that climate and that habitat.

Table 4.8
TYPES OF ECOSYSTEM CHANGE AS ENVISAGED BY CHURCHILL AND HANSON (1958)

Non-Cyclic Replacement:	the replacement in a community of individuals which have died off from other than catastrophic causes
Phasic Cycle Replacement:	a series of changes in a cyclical system (vital for self-maintenance and regeneration in communities)
Inter-community Cycles:	related cyclic changes between two or more communities
Directional Change:	a reasonably ordered sequence of change (such as successional development)
Fluctuation:	random changes about an average or steady-state condition
Spatial Changes:	those which lead to 'community mosaics' which interact and produce subsequent change

An illustration of environmental change triggering off vegetation change is provided by Vuilleumier (1971), who provided a synthesis of available evidence that many components of the flora and fauna of South America have been greatly affected by climatic shifts, particularly during and since the Quaternary. He concluded that 'in the broad region of South America that lies within the tropics, a series of humid-arid cycles drastically and repeatedly altered vegetation patterns during the Quaternary. Both mountain and lowland rain forest were fragmented during dry periods, and were able to re-expand during humid phases' (Vuilleumier, 1971). Contraction of rain forest into isolated and fragmented remnants led to evolutionary changes which contributed significantly to the development of the wide species diversity witnessed in the rain forest environment today.

The validity of the concept of climatic climax vegetation has been debated on a number of grounds, however. Tansley (1935) questioned the exclusive significance of climate in the Clementsian framework, and he pointed out that a great many other physical factors can affect 'climax' vegetation – including soil, slope, aspect, exposure, human and animal influences. Tansley thus favoured the notion of distinguishing self-induced (*Autogenic*) from externally-induced (*Allogenic*) successions. In considering African vegetation, Keay (1974) has concluded that 'the stable climax may be a rather theoretical concept', however useful, because it reflects a *dynamic* (rather than *static*) stability, in as much as animal variations in the seasons and the natural ageing and death of the component plants and animals lead to changes, and more drastic natural influences (such as tornadoes) produce more marked changes. Churchill and Hanson (1958) have evaluated some major areas of disagreement with respect to the applicablity of climax concepts to any area, and in particular to arctic and alpine areas. They argue that changes in ecosystems are inherent and inevitable; and they rationalize possible ecological changes into six discrete types, each complementary and non-exclusive (Table 4.8). Churchill and Hanson conclude that each of these types of change are within the framework of climax, however. Cooper (1926) has concluded that although the climax concept is useful it is rather subjective, and often too rigidly defined. In his view 'it is not possible to mark it [climax] off absolutely from the period of more active succession', and 'it is conceivable that the period which we name the climax might continue so long that a direct, gradual transition might occur, governed externally by climate, and in extreme cases by evolutionary production of new species' (Cooper, 1926). Inherent instability of climatic and other environmental factors over sufficiently long periods of time to allow complete succession to climax has also been suggested by Burgess (1960), who pointed out that the evidence accumulating from Quaternary Studies in many areas of the world indicates that at times there must have been very rapid replacements of one characteristic vegetation by another.

Many of the problems of applying the climax concept stem from the difficulties of identifying the climax stage of succession. Clements (1936) offered several diagnostic criteria including:

(a) the life form of the vegetation (dominated by the largest or tallest plants which could flourish in a given area under given climatic conditions?),
(b) the presence of certain dominant species (indicator species) through all, or nearly all, associations,
(c) faunal indicators (but these were thought to be of less value than floral indicators because of their greater mobility between areas),
(d) dependable records of previous vegetational structure and composition (though these are rarely available for sufficiently long periods of time).

Problems also arise because of the many possible forms of deviation from the pure climax state (Figure 4.9) because of either natural or man-induced factors. The successional sequence can often be suppressed before the ultimate climax state is reached (by grazing, burning, various land-use management practices) and so the 'prevailing climax' at a site may differ somewhat from the potential 'climatic climax' for that site. So common and widespread are 'suppressed successions' that some authorities view the climatic climax as an ideal end product of succession which is rarely if ever found in nature. Terminologies for non-climax states have been debated. For example Godwin (1929) preferred to use the term 'deflected succession' rather than Clements' (1916) 'sub-climax'; and pre-occupation with this form of debate has probably retarded rather than assisted the advancement of successional studies.

Eyre (1968) has rationalized the various forms of non-climax association by the use of a valuable figure (Figure 4.9) which illustrates how a succession (*prisere*) can either progress directly to the *climatic climax community*, or be deflected by some arresting factor into a *subclimax community*. The subclimax can 'succeed' to the climax via a series of *subseres* if the arresting factor is removed. On the other hand the climatic climax vegetation may be either drastically affected by felling, cultivation and the like, to produce a bare soil or relic community (which can then progress once again to the climax via secondary succession), or otherwise deflected without being completely removed (such as through grazing) and so form a *plagioclimax community*. Relaxation of the limiting factors can lead to succession via subseres back to the climatic climax community again.

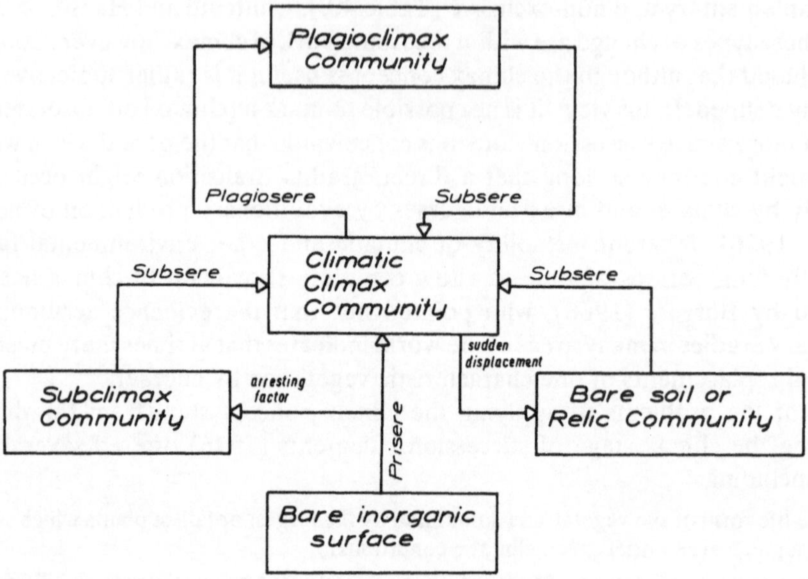

FIG. 4.9 VARIATIONS ON THE CLIMATIC CLIMAX STATE OF SUCCESSIONAL DEVELOPMENT. (After Eyre, 1968.)

4.3 RELEVANCE OF EVOLUTION AND SUCCESSION TO ENVIRONMENTAL MANAGEMENT

There are clearly a wide range of contexts in which the two basic time scales of ecological and ecosystem change (evolution and succession) are of relevance to environmental management. Perhaps of fundamental importance is the fact that changes through time are both inevitable and inherent in ecosystems, and these help to preserve ecosystem stability and productivity. Thus any efforts to manage or control ecosystems, for either conservation or exploitation, should be planned with this in mind. Conservation schemes which do not provide either scope within the ecosystem, or stimulation from the environment (and adjacent ecosystems) beyond the immediate boundary of the ecosystem, are thus ecologically unwise.

The significance of evolution to environmental management is clear from the examples of threatened and actual species extinctions quoted earlier in this chapter, and from the accumulating evidence that evolution and adaptation within individual species is going on at the present time (hence the importance of industrial melanism in moths, for example). At the evolutionary time scale *any* extinctions, whether or not the species has any commercial, aesthetic or scientific 'value' is regrettable because 'extinction is forever' and once a strain of genetic material is lost from the total gene-pool in the biosphere it can never be re-created. There is thus growing concern that stocks of the genetic material of threatened and endangered species be held 'in reserve' in seed banks, botanical gardens, zoological gardens and the like to prevent their permanent loss.

The significance of ecological changes on the successional time scale is equally real, and there are a number of situations in which the principles of succession are of central importance in environmental management and in conservation in general. For example Odum (1962) has stressed the conflict which arises between the needs of man for early stages of successional sequences as food sources (because of the relatively large net primary productivity available for harvesting in the early stages of a prisere), and the needs for climax stages as stable ecosystems which are capable of withstanding environmental and management changes. He suggests that in environmental management there is clearly a need for both productive and stable environments, and he stresses the need for a sound mixture of early and mature successional stages even within relatively small areas. Some conservationists advocate that mature successional stages should be preserved for quite the opposite reason to that offered by Odum (1962). Thus Horn has commented that

> conservationists and advocates of wilderness preserves have often cited the conventional generalization that diversity conveys stability, arguing that diverse natural communities should be conserved for their stabilizing influence on the simple artificial communities invented by man. One could equally argue that if complex systems were inherently stable they should need *no* protection. The opposite view, that diverse communities and climax communities are inherently fragile, is a much more powerful reason for their requiring protection from human disturbance, perhaps even from well-intentioned management (Horn, 1974).

Succession is also important because it is possible to view certain external controls on ecosystem development as deflecting factors which create sub-climax vegetational states. Regier and Cowell (1972) have considered two special cases of the reversal of normal successional processes. One concerns *exploitation* of ecological resources (such as through the harvesting of fish resources) where generally the largest and/or the most highly prized species are harvested first and this can readily affect successional status especially if the exploitation is recurrent and large-scale. The second case concerns *pollution* which can affect succession because in general the species which manage to survive pollution, or those which actually benefit from it (see Chapter 7), tend to be small, short-lived and opportunist by life-style. The species characteristic of stability and climax tend to disappear first.

There are situations, however, where management is aimed specifically at maintaining sub-climax communities. Green (1972) points out that this is the case in Britain with chalk grassland, dune grassland and heathland. He suggests that an important feature of these types of community is low fertility, especially in terms of nitrogen availability, and that this is directly related to their floristic diversity. Successional stages produce a build-up in fertility, loss of floristic diversity and often a reduction in the wildlife and/or amenity value of the habitat. As a result traditional forms of management on such communities (such as through grazing, burning and mowing) probably arrest seral development, partially by preventing a build-up of fertility within the ecosystem.

In summary, therefore, there are various situations in which ecological changes through time are important from a management point of view. Although many land-use changes (especially those which involve large-scale removal or destruction of natural ecosystems) are ecologically detrimental (see Chapter 6), it is quite possible that if new habitats are created (such as on motorway verges, railway embankments and in disused quarries) the secondary successions which will establish themselves on such sites could produce ecosystems which are more diverse and ecologically more interesting than the original ones. Similarly there are a wide range of management strategies which aim principally to deflect or suppress successions – such as the burning of heather moorland, the grazing of grassland, the thinning and coppicing of woodland and the clearance of undergrowth – and yet with the intervention of such practices the ecosystems remain ecologically productive, a wide range of habitats and micro-scale environmental gradients is preserved and the overall ecological value of the habitat is maintained.

Ecological changes through time are inherent and vital in all ecosystems, whether natural or man-made. Changes which are induced both from within the ecosystems and from without (in response to external environmental and biotic stimuli) tend to increase the stability and often the diversity of ecosystems – and these must surely be the focal points of any environmental management strategy which places ecological considerations utmost. However, many of the ecological and ecosystem changes through time are closely related to, if not conditioned by, ecological changes through space. The space–time continuum is of basic concern in ecosystems. Thus it is essential to consider ecological changes through space (Chapter 5).

5
Ecological Changes in Space

> The facts of plant (and animal) distribution . . . are themselves facts of geography in that they help to differentiate the earth's surface, besides being inescapable factors in the human environment (Edwards, 1964).

The geographer has traditionally been interested in spatial variations of environmental factors, and so consideration of ecological changes through space is well within the realms of concern of the geographer. Inevitably, present-day patterns of ecological activity reflect evolutionary and successional processes and the time scale over which ecosystems have developed, and so the subject matter of this chapter is a sequel to Chapter 4. Equally inevitably, distributions are conditioned by interactions between organisms, and between organisms and their environment, and so the structure and function of the ecosystem is important in a spatial sense (Chapter 3). In many ways, therefore, spatial distributions of organic matter, and of its characteristics (such as productivity and diversity), integrate the suite of ecological processes and relationships which are of fundamental importance to environmental management. The extent and nature of man-induced ecological change is also apparent in spatial distributions. Because of the unity and integrated nature of the Biosphere overall, and of individual ecosystems themselves, it becomes important to consider spatial aspects of natural and man-modified ecosystems as a pre-requisite to environmental management.

This chapter will focus on the distributional patterns and scale of spatial variations in ecological properties at a variety of scales from the global to the local, and on an evaluation of the need for and approaches to resource surveys and analyses, with special reference to the surveying and mapping of vegetation. This will provide a framework for considering (in Chapter 6) the notion of ecological resources.

5.1 SPATIAL DISTRIBUTIONS OF ECOLOGICAL PROPERTIES

From an environmental management point of view there has been growing interest in spatial variations and patterns of ecological change at all scales from the global to the very local. At the world scale the interest has been expressed through the *International Biological Programme* (Newbould, 1964), aimed at increasing our understanding of the overall amounts and spatial distributions of the organic resources of the biosphere about which little global-scale data existed previously (Baer, 1967). The results of the IBP studies are now available, and they provide an invaluable data base for attempts to evaluate the pattern of 'organic wealth' at both global and national levels (Eyre, 1978). At the other end of the scale spectrum, attention has been focused on micro-scale patterns of spatial variations in plant and animal distributions and in ecosystem processes, in efforts to resolve the optimum size and shape of reserves designed to conserve natural ecosystems (Hooper, 1971).

This interest in a spectrum of scales has been complemented by interest in a range of aspects of spatial distributions, of which the following provide some illustrative examples.

Species occurrence

The adoption of standardized and co-ordinated mapping schemes at both national and international scales has yielded information on the distributional patterns of individual species of plants and animals. This is of value in identifying the spatial range of the species, in isolating species with restricted ranges (perhaps those with increasing probability of imminent extinction?), and in providing base data for evaluating the environmental controls on distributions. In Britain, a standard format for mapping plant and animal distributions has been adopted widely, based on a standard 10 km grid square over the whole of the British Isles. The distribution of a large number of species of wild flowering plants has been mapped on this basis in the *Atlas of the British Flora* (Perring and Walters, 1962), and a time perspective on the distribution of some species has been given by dividing the available records on reported sightings into two time periods – before 1930 and subsequently. This has been of central value in identifying those species whose geographical range has contracted between the two periods, and the scheme has been of basic significance in highlighting those species which are most in need of protection. The task of compiling an accurate and up-to-date atlas of flora for a unit the size of Britain clearly depends largely on 'chance' findings and careful recognition by voluntary amateur observers, and so many locations and a number of species have no recorded observations. The task is ongoing, however, and it is co-ordinated by the Biological Records Centre at Monks Wood Experimental Station run by the Nature Conservancy Council. A supplement to the atlas was published in 1968 (Perring, 1968), and reports on distributions of newly-recorded plant species are published regularly in journals such as the *Journal of Ecology* and *Watsonia*. A number of national ecological groups and societies run National Distribution Schemes in conjunction with the Biological Records Centre – including the Botanical Society of the British Isles (flowering plants), the British Trust for Ornithology (birds) and the Mammal Society (mammals). Mapping of the distribution of fresh-water fish in the British Isles has also begun (Maitland, 1969), and the results, when available, should provide a sound basis for future planning and management of fish populations in the country. Some preliminary results are shown in Figure 5.1. Overall, therefore, the National Distribution schemes provide valuable information on the nature and location of threatened species which is of vital importance in any attempts to provide a comprehensive conservation strategy.

The interest in species distributions has not been confined to plants, nor has it focused solely on terrestrial habitats. Cramp, Bourne and Saunders (1974), for example, have collated data on the distribution of sea birds around Britain and Ireland and on the location of breeding colonies. This shows a concentration in sea bird distributions throughout the year in locations where warmer Atlantic waters pass into the English Channel, the Irish Sea and around the North of Scotland into the North Sea.

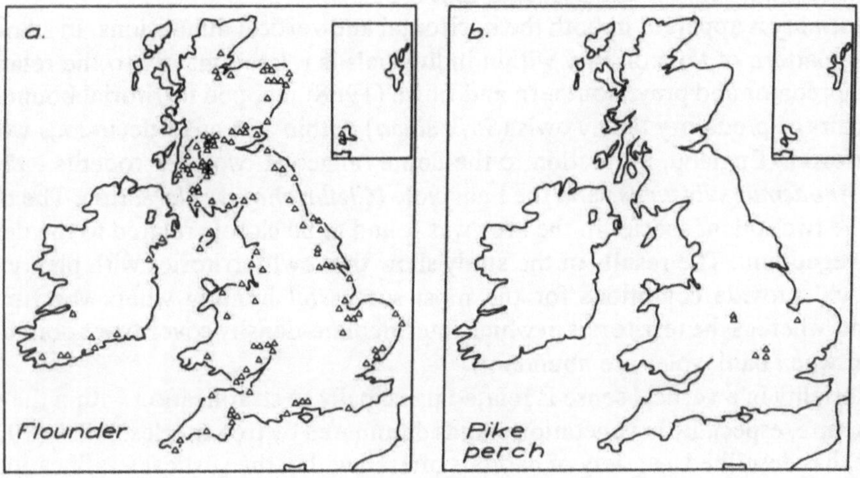

FIG. 5.1 PRELIMINARY RESULTS OF NATIONAL MAPPING OF FISH POPULATIONS IN GREAT BRITAIN.
The distributions of the flounder (*Platichthys flesus*) (a) and the pike perch (*Stizostedion incioperca*) (b) are shown, based on the 10 km grid squares of the National Distribution mapping scheme. (After Maitland, 1969.)

Species characteristics

Most studies of spatial variations have been devoted to mapping and explaining the range of rare species or of communities, based on the criteria of whether it is present or not. There have been relatively few attempts to consider the spatial distribution of characteristics of species within the zone of occurrence. One such study by Watson and Hanham (1977) focused on the variations in flowering colour of the butterfly weed (*Asclepias tuberosa*) throughout Oklahoma. The weed shows considerable variation in the colour of the inside of the flower across its range within the United States. In Oklahoma it is strikingly orange in the east and dominantly yellow in the west. The study also considered environmental controls for the observed pattern of variations, and it showed a significant association between variation in flower colour and atmospheric moisture, with temperature variations of secondary importance. Such studies are of value in increasing understanding of the environmental control of ecological appearance and of response to external stimuli, which could then be used in formulating appropriate environmental management strategies.

Territoriality in animals

Studies of local-scale movement patterns in animals and birds are valuable in isolating ecosystem boundaries, in understanding inter- and intra-species behavioural and feeding relationships, and in understanding the ecology of the individual species involved. From a management point of view they are essential in forming a basis for delimiting areas to be set aside as nature reserves, and in isolating ecological and spatial inter-relationships within ecosystems. Territorial patterning in animal and bird communities is related to factors such as the diversity of habitats available, the feeding and physiological requirements and limitations of the individual species, the tolerance range of given species to micro-scale environmental changes, and the competitive inter-relationships between species for food resources and personal space.

Territoriality is apparent in both the horizontal and vertical dimensions. In a horizontal sense the pattern of territoriality within individuals is related mainly to the relationship between predator and prey. Southern and Lowe (1968) mapped territorial boundaries of eleven pairs of predatory tawny owls (*Strix aluco*) within 104 ac of deciduous woodland near Oxford in England, in relation to the home ranges of two prey rodents – the wood mouse (*Apodemus sylvaticus*) and the bank vole (*Clethrionomys glareolus*). The distribution of the two rodent species in the area was found to be closely related to the density of ground vegetation. The results of the study show that owl territories with plenty of bare ground will provide conditions for the most successful hunting when wood mice are abundant; whereas the territories in which intermediate-density cover types dominate will be better when bank voles are abundant.

Territoriality in a vertical sense is related principally to stratification within the vegetation structure, especially in vegetation stands dominated by tree species. Allee (1926), for example, has described a variety of habitats offered within the vertical stratification of the tropical rain forest of Panama (Table 5.1), and he considered the animal communities in

Table 5.1
VERTICAL STRATIFICATION OF HABITATS WITHIN THE TROPICAL RAIN FOREST ECOSYSTEM

	Stratum	height above ground
(i)	air above forest	
(ii)	tree tops above main forest roof	above 125 feet
(iii)	upper forest canopy	75–100 feet
(iv)	lower tree tops	40–60 feet
(v)	small trees	20–30 feet
(vi)	higher shrubs	about 10 feet
(vii)	forest floor	
(viii)	subterranean	

Source: Allee (1926)

relation to environmental factors. Most animals live on the forest floor where the environment is relatively homogeneous, and there is a marked reduction in the number of species and individuals present in areas with low light conditions, with even fewer species in the upper forest levels (because of the fluctuating environmental conditions and the hardships of arboreal life). Figure 5.2 summarizes some of the major aspects of vertical stratification within a temperate woodland and the tropical rain forest.

Ecological diversity

Some of the relationships between ecological diversity and ecosystem stability were considered in Chapter 3, and since there are marked spatial variations in the occurrence of individual species, and different species respond in different ways to a given series of spatial changes in environmental conditions it is logical to anticipate marked distributional patterns of ecological diversity. Pielou (1975) has reviewed the nature and controls of ecological diversity at both local and global scales, and these will be considered more fully later in this chapter (see pages 150–2). The patterns of variation of diversity are of significance to environmental management because diversity affects stability and the resiliance of ecosystems to man-induced change (see Chapter 3), and because diversity reflects the inherent conservation value of a particular habitat, ecosystem or environment (see Chapter 8).

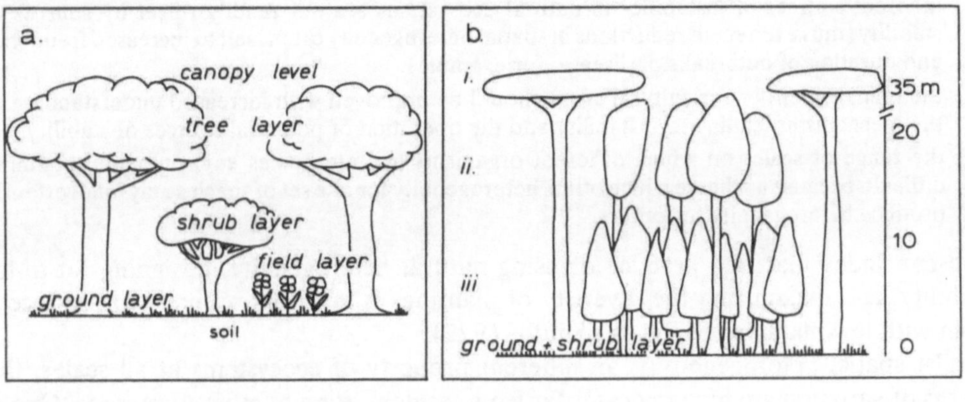

FIG. 5.2 VERTICAL STRATIFICATION WITHIN FOREST ECOSYSTEMS.
The figure shows general characteristics of stratification within temperate woodlands (a) and within the tropical rain forest (b).

Ecological productivity

The scale and nature of variations in primary productivity (both net and gross) between different types of ecosystem were illustrated briefly in Chapter 3, and the results of the IBP have facilitated analysis of spatial variations in productivity across the earth's surface (Eyre, 1978). These are very marked, even within similar environments. For example Ryther (1969) has pointed out the common assumption that the ocean is a single ecosystem in which food chains involving the same number of links and ecological efficiencies apply throughout, implying relative spatial homogeneity. In fact, as Ryther's analysis confirms, the spatial variations in primary productivity between different ocean areas, coupled with wide variations in ocean food – web dynamics, act to produce spatial differences in fish production 'which are far more pronounced and dramatic than the observed variability of the individual causative factors' (Ryther, 1969). Because of these different facets of spatial variations in ecological factors, the notion of identifying or isolating meaningful ecological regions or ecosystems is open to question. The problem of locating an ecosystem boundary somewhere along a continuous (and rarely abrupt) series of ecological changes between adjacent ecological systems (see Chapter 3) is compounded by this inherent spatial heterogeneity of ecological factors; and so perhaps the best approach to isolating ecosystems is to seek to optimize between maximizing the 'between-area' variations and minimizing the 'within-area' variations. Few criteria for consistent and ecologically (rather than statistically) based optimization have been considered, however.

Although it produces these types of operational problems, spatial heterogeneity of ecological factors is believed to encourage ecosystem stability (through diversity). The low success rate of many attempts to 'culture' ecosystems (that is, to isolate and maintain an assemblage of species) is believed by Smith (1972) to be in part a function of failure to recognize this inherent heterogeneity, which is at times reflected in failure to enclose 'safe' conditions for some of the cultured species, leading in turn to increasing overlap in activity amongst remaining species and reduced system stability overall. Smith (1972) also highlights three implications for environmental management behind spatial heterogeneity:

(a) inherent sources of instability in natural ecosystems are not readily offset by sources of stability (thus even small reductions in spatial heterogeneity often lead to increased frequency and duration of outbreaks of disease, for example)
(b) the management of agricultural crops should be improved with increased understanding of the mechanisms leading to instability and the operation of potential sources of stability.
(c) the range of scales on which different organisms operate makes environmental planning difficult, because a scheme which offers heterogeneity for one set of species may inadvertently produce heterogeneity for others.

Smith concludes that 'the practice of using natural heterogeneity, designing lot-to-lot variability, and maximizing the diversity of plantings is much more likely to produce a system with low maintenance costs' (Smith, 1972).

Whilst spatial heterogeneity is an inherent property of ecosystems at all scales, the patterns of variation are in many cases far from random, for a number of reasons. One is that many environmental characteristics (such as climate) vary in relatively regular ways, and groups of species tend to have similar tolerance ranges for given environmental factors. Many distributions are closely related to evolutionary development and successional stage, and these tend to vary in relatively coherent fashions, except where abrupt environmental changes have occurred. In addition many distributions are influenced by the activities of man, and these tend to have distinct spatial patterns. A further reason is that boundaries between ecosystems are often located by optimizing within- and between-area ecological variations, and so spatial patterns are defined on a regular basis. There are thus quite often patterns of ecological variation which can be identified and mapped at a variety of scales from the global to the local. Whilst these patterns can be recognized in a range of ecological factors (such as the species present, the vegetation structure, the organic diversity and ecological productivity), in final analysis these all reflect the distributional ranges of individual species. Such ranges are affected by a variety of factors, including the age (in an evolutionary sense) of the species, its genetic variability, its physiological tolerance, its mode of reproduction and dispersal, and the arrangement of geographical boundaries relative to the movements of the species. In turn, the distributional ranges influence the degree of mixing of the gene pool of the species, the probability of geographical isolation, and the possibility of catastrophic extinction.

Information on the pattern of ecological variations is required for both academic and applied reasons. The former stem from the fact, as stressed by Edwards (1964), that 'the facts of plant (and animal) distribution are themselves facts of geography'. The latter include land-use planning, resource evaluation and the formulation of conservation strategies. It is convenient to consider the pattern of ecological variations on two distinct spatial scales – the global and the regional.

5.2 ECOLOGICAL VARIATIONS – THE GLOBAL SCALE

The roots of modern ecology have been traced (see Chapter 1) back to a period before World War One when ecologists were concerned principally with mapping and evaluating the distribution of vegetation in Britain and elsewhere; and since this time a variety of approaches to the identification of world-wide ecological patterns have been adopted.

The Climax Formation approach

This is based on the study, description and mapping of the major climax plant communities of the world, following Clements' (1916) views on vegetation development (see Chapter 4). The approach was championed by Weaver and Clements (1938), and the basic unit for mapping was the 'Regional Climatic Climax Vegetation' – that is, the *assumed* climatic climax vegetation for a given environment, as based on the prevailing regional climate. The main assumption is that within any climatic region the vegetation develops from various positions along successional pathways that converge towards a single climax. Consequently under this approach all climax, secondary successions, plagioclimax and allied communities are mapped together as the regional climax. The approach thus provides a map of climatically-possible vegetation, but whether or not this is actually present on the ground depends on local environmental and land-use history. Thus a true picture of actual vegetation is not offered, and so alternative mapping schemes were sought which were based on description of the existing plant life. Two main approaches evolved to meet this need, but Dasmann stresses that they 'are not mutually exclusive, but nevertheless produce differing results' (Dasmann, 1972).

The Floristic approach

This derives from the work of Braun-Blanquet (1932) who mapped the distribution of species and groups of associated species (communities) in Europe. The maps produced in this way show variations in plant taxonomy rather than in the appearance of the vegetation, and they show the actual vegetation present in the area. This regional scale of analysis was projected to the world scale by the work of Good (1947) and Gleason and Cronquist (1964) who identified *floristic regions* or *floristic provinces* across the world. Good identified six basic regions (Table 5.2), and the map of these (Figure 5.3a) is similar to the map of faunal regions suggested somewhat earlier by Wallace (1876) (Figure 5.3b and Table 5.3).

Ecological approach

This alternative classifies vegetation by its general appearance and structural relations rather than specifically on the basis of species composition or successional status; and the early pioneer mapping at the world scale was done by Warming (1909). The emphasis here is on the *vegetation formation*.

Biotic approach

There have been a number of attempts to classify and map world distributions of animals, the earliest being Sclater's (1858) study of bird distributions. The scheme most widely used

Table 5.2
THE FLORAL REGIONS IDENTIFIED BY GOOD (1947)

BOREAL REGION:	the largest of the regions; includes North America, Europe and Northern Asia. Each is bounded by a distinct barrier (such as a mountain range, ocean or desert)
PALAEOTROPICAL REGION:	includes three major sub-divisions – Africa; Indo-Malaysia; and Polynesia. Few plants are common to any of these areas
NEO-TROPICAL REGION:	most of Central and South America
SOUTH AFRICAN REGION:	the Cape region and Kalahari desert
AUSTRALIAN REGION:	Australia and Tasmania. Plant varieties are dominated by *Eucalyptus* (over 600 species)
ANTARCTIC REGION:	New Zealand, the southern tip of South America, and the ocean islands of the Southern Ocean

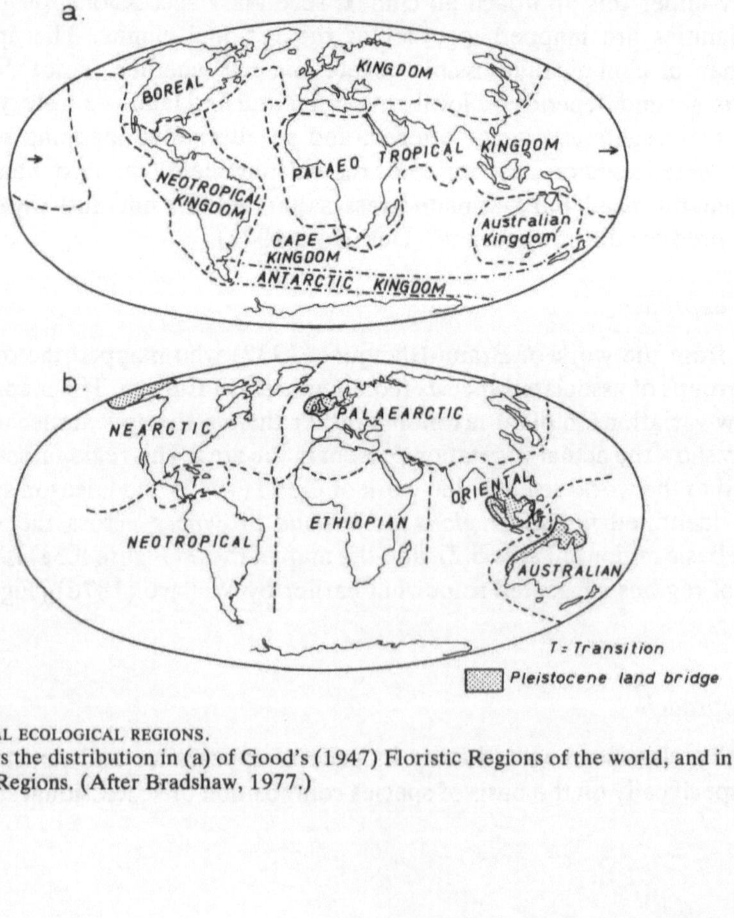

FIG. 5.3 GLOBAL ECOLOGICAL REGIONS.
The figure shows the distribution in (a) of Good's (1947) Floristic Regions of the world, and in (b) of Wallace's (1876) Faunal Regions. (After Bradshaw, 1977.)

in the past, and still generally accepted today, was developed by Wallace (1876) who distinguished a series of six major faunal (or zoogeographical) regions on the basis of differences in animal taxonomy (Table 5.3). Each of the major regions, shown in Figure 5.3b, can be divided into sub-regions on the basis of climate. Many ecologists and biogeographers now regard the Antarctic as an additional faunal region in the sense used by Wallace. There are clearly a number of similarities between Wallace's faunal regions and Good's floristic regions. Biotic distributions within smaller areas have also been

Table 5.3
THE FAUNAL REGIONS IDENTIFIED BY WALLACE (1876)

PALAEARCTIC REGION:	includes Europe and Northern Asia. Home for 136 families of vertebrates (3 unique), 100 genera of mammals (35 unique) and 174 genera of birds (57 unique)
NEARCTIC REGION:	North America and Greenland. Many similarities with the Palaearctic (the two were linked by land bridges during the Tertiary and the Pleistocene). Home for 122 families of vertebrates (12 unique), 74 genera of mammals (24 unique) and 169 genera of birds (52 unique)
ORIENTAL REGION:	India and the Far East; tropical area dominated by peninsulas and islands. Home for 164 families of vertebrates (12 unique), 118 genera of mammals (55 unique) and 340 genera of birds (165 unique)
ETHIOPIAN REGION:	tropical Africa south of the Sahara, and Saudi Arabia. Home for 174 families of vertebrates (22 unique), 140 genera of mammals (90 unique) and 294 genera of birds (179 unique)
AUSTRALIAN REGION:	possibly connected by land to the Oriental region. Home for 141 families of vertebrates (30 unique), 72 genera of mammals (44 unique) and 298 genera of birds (189 unique)
NEOTROPICAL REGION:	South America. Home for 166 families of vertebrates (44 unique), 130 genera of mammals (103 unique) and 683 genera of birds (576 unique)

Source: Bradshaw (1977)

mapped. For example Dice (1952) classified North America into *Biotic Provinces* on the basis of distinctive differences between floral and faunal distributions. More recently both Darlington (1957) and Muller (1974) have considered the whole realm of identifying faunal distributions.

Biome approach

Whilst the mapping of both plant and animal distributions has proceeded along the lines outlined in these four basic approaches, attention has also focused on the need to classify the earth's surface, and its diversity of organic forms, by a more general scheme which includes both plant and animal life together. Thus Clements and Shelford (1939) developed a biome approach which was subsequently applied in mapping the ecology of North America (Shelford, 1963). *Biomes* are characterized by a prevailing regional climax vegetation and its associated animal life. Thus although it integrates biotic and faunal components of the ecosystems, the approach as advocated by Clements and Shelford still suffers from the problems of the Climax Formation approach, in that the vegetation actually growing in an area may differ substantially from the potential climatic climax vegetation of that area.

Nonetheless, as Robinson stresses, 'the biome approach provides the most convenient unit for the study of ecological relationships' (Robinson, 1972), and it thus provides a basis for identifying large-scale regional ecosystems. This possibility has been realized within the *International Biological Programme*, which used as a basis for global data collection six major biomes – desert, grassland, tropical forest, deciduous forest, coniferous forest and Arctic/alpine tundra (Hammond, 1972). Dasmann (1972) has also stressed the value of a biome approach for providing spatial data necessary in the formulation of large-scale conservation strategies, because it is readily applicable on the global scale and it takes into account both plants and animals. Because the Clements and Shelford (1939) approach emphasizes ecological similarities rather than taxonomic differences between biomes, however, Dasmann (1972) suggested an alternative scheme based on combining the Clements and Shelford approach to biome analysis and Wallace's approach to identifying faunal regions. The four main steps in his approach are as follows:

STEP 1: identify *faunal regions* in the style of Wallace,
STEP 2: sub-divide each of these into *biomes*,
STEP 3: these biomes can then be sub-divided into *biotic provinces*,
STEP 4: in detailed studies these biotic provinces can be further sub-divided by vegetational, floristic or faunistic differences.

Productivity approach

Spatial patterns in productivity at the global scale might be anticipated because the overall productivity of ecosystems is controlled ultimately by the quantities and quality of solar radiation available to and utilized by the primary producers (see Chapter 3), and these have marked spatial patterns around the earth's surface. The aim of much recent ecological research has been to provide measurements of productivity levels in different types of ecosystem, and between ecosystems in different environments (Newbould, 1964). This has been a central theme in both the *Man and the Biosphere* and the *International Biological Programme* studies. Attempts can also be made to collate available data to produce maps of world variations in ecological productivity (such as those produced by Eyre, 1978), although much spatial interpolation is necessary to produce such apparently detailed maps. Nonetheless the productivity approach does provide a valuable means of considering global ecological resources, for evaluating the ecological efficiencies of various land-use and land-management schemes, and for evaluating possible orders of priority to adopt when devising species and ecosystem conservation plans. The approach also offers a yardstick by which to measure the ecological losses associated with land-use changes from natural vegetation (such as Tropical Rain Forest) to agricultural or urban land.

Diversity approach

Wallace's study of animal distributions led him to conclude that 'animal life is, on the whole, far more abundant and varied within the tropics than in any other part of the globe' (Wallace, 1876); and a number of workers have since focused attention on the pattern of variation of ecological diversity at the global scale (for example Pianka, 1966). Fischer (1960) illustrated marked 'diversity gradients' in which diversity of species are closely related to latitude for a range of organisms such as ants, American snakes, nesting birds, snake species in Argentina and marine species (Figure 5.4), and he stressed that 'the correlation of floral and faunal diversity with latitude is one of the most imposing biogeographic features on earth' (Fischer, 1960). In most cases the diversity of species increases towards the tropics, although in some organisms (such as some groups of algae), species diversity decreases towards the tropics. Both Dobzhansky (1950) and Fischer (1960) account for the generalizations shown in Figure 5.4 as functions of evolution, in that biota in the warm humid tropics are likely to evolve and diversify more rapidly than those on higher latitudes (because of more constant environmental conditions and relative freedom from climatic 'disasters').

If the relationship between ecological diversity and stability (see Chapter 3) is upheld, then clearly this latitudinal pattern of variation of diversity suggests a parallel distribution of ecological stability, with minimum stability towards the poles (hence the unstable and fragile Arctic ecosystems ?, see pages 99–103) and greater stability towards the equator

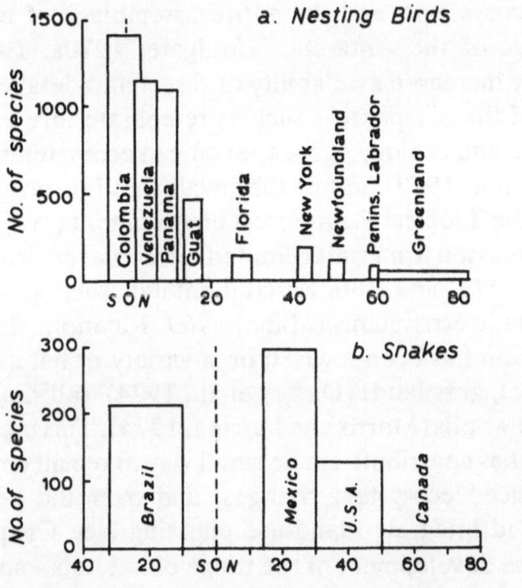

FIG. 5.4 DIVERSITY GRADIENTS OF VARIOUS SPECIES, IN RELATION TO LATITUDE.
Showing the distribution of diversity amongst nesting birds (a), American snakes (b) and ants (c). (After Fischer, 1960.)

(the Tropical Rain Forest?). Clearly this has global-scale environmental management implications for the exploitability of ecosystems and for priority areas for ecosystem conservation.

5.3 ECOLOGICAL VARIATIONS – THE REGIONAL SCALE

The various approaches to evaluating ecological variations at the global scale have been made possible largely by the increasing number of studies of spatial variations at more local scales. Information on ecosystem structure and composition, ecological productivity, species composition and diversity, and the environmental controls of these, is accumulating from many areas, both as a result of co-ordinated international research (such as the IBP) and of independent studies. Thus the last decade has witnessed considerable advances in understanding of the spatial and ecological relationships of terrestrial environments such as the arctic tundra (Bliss et al., 1973) and the Fennoscandian tundra (Wielgolaski, 1975b) ecosystems, and the entire assemblage of terrestrial and aquatic ecosystems characteristic of the Antarctic (Holdgate, 1970a, 1970b). This has been supplemented by greatly increased availability of direct and detailed observations of the more temperate parts of the biosphere – such as temperate forest ecosystems (Reichle, 1973) and the vegetation and ecology of Mediterranean ecosystems (Castri and Mooney, 1973; Wright and Wanstall, 1977) and of the invaluable but progressively diminishing ecological resources of the Tropical Rain Forest environment (Whitmore, 1975). Observations have also been amassed from water-limited terrestrial environments such as desert ecosystems (Noy-Meir, 1973) and from wetland habitats such as peatlands (Moore and Bellamy, 1974) and aquatic ecosystems (Mann, 1969; Raymont, 1966).

Within Britain, attention has been focused on a variety of habitats – including heathlands (Gimingham, 1972), grasslands (Duffey et al., 1974), salt-marshes and sand-dunes (Ranwell, 1972) and oak woods (Morris and Perring, 1974). This increased understanding of these specific habitats has contributed in no small way to recent concern over the nature and extent of man-induced ecosystem changes, and over the need for wise wildlife conservation schemes and integrated land-use planning (see Chapter 8). Studies have repeatedly shown that the development of the range of habitats – such as the New Forest (Tubbs, 1968) – is conditioned by both the natural environment and the history of land use, management and exploitation in the area.

Studies of regional-scale ecological variations have thus shed light on two areas of central concern to environmental management – man's influence on ecosystems and ecological factors controlling ecological variations.

5.3a Man-induced ecological changes

In Britain as in many countries there appear to be relatively few areas where a fully 'natural' vegetation cover exists. These are becoming increasingly fragmented, and many observers fear that contraction in both size and numbers will continue unless checked by some conservation action (such as legislation). There are a wide variety of ways in which man can influence the vegetation of an area.

Purely destructive actions

Many aspects of land use, such as forest clearance, hedgebank removal and many agricultural practices, lead to direct destruction of vegetation. The development of savanna and grassland vegetation in southern Papua, New Guinea, has been interpreted by Eden (1974) as being a product of the combined effects of shifting cultivation and forest burning practices; and evidence from a wide range of areas demonstrates that fire has played a significant role in extending grassland vegetation into formerly forested areas.

Establishment of new types of vegetation

such as tree planting. Tubbs (1974) has traced the long history of forest clearance and associated changes in the agricultural economy in Britain from the Neolithic period onwards; and Harrison (1974) has highlighted the contracting size and numbers of heathlands in Southern England because of land-use pressure caused by increasing demand for permanent pasture and arable land, building land, forestry, mineral extraction and military training.

Introduction of species between areas

Species of plants and animals have been introduced into areas beyond their native zones for many centuries, and such introductions can at times have marked effects on local vegetation composition and structure. For example Jarvis (1973) has evaluated the intentional introduction of exotic plants into England during the period between 1550 and 1700, especially from North America, as a result of the increased variety of plants available to horticulturalists, a climate of fashion culture and taste which encouraged the cultivation of plant rarities and curiosities, and a contemporary tendency to enlarge the ornamental components in gardens, parks and forest groves.

Farming practices

Natural vegetation and ecosystems can be radically affected by various farming practices such as the restriction of growth to plants of economic importance and the effects of grazing. Miller and Watson (1974) have documented a series of ecological shifts related to grazing effects, whereby those plants which are most palatable to stock are grazed most frequently and these may be replaced by species able to withstand such grazing pressures. The general realm of farming effects on wildlife and natural ecosystems will be considered more fully in Chapter 6.

Habitat changes

Many ecological changes can be triggered off by habitat changes, such as through land drainage. Green (1973) has evaluated a series of changes in agricultural practice (related especially to increased field drainage and increasing use of artificial fertilizers) in England and Wales which have coincided with significant and relevant climatic fluctuations to produce changing habitat conditions, particularly in aquatic habitats, in recent years. Habitat changes will also be considered in greater detail in Chapter 6.

These sketchy examples show the scope for human impacts on formerly natural vegetation, and indeed Riley and Young have suggested that 'to assess the extent to which a plant community has been modified by man is one of the most challenging problems in studying vegetation' (Riley and Young, 1966). From the environmental management point of view an inventory of present vegetation distribution is essential, however, and although study of the factors controlling present-day distributions is not the prime criterion of such inventories, it is important to understand some of the basic causes of prevailing distributions.

5.3b Environmental controls of ecological variations

Because the ultimate control on ecosystem productivity is the amount and spectral composition of incoming solar radiation (see Chapter 3), and because early attempts at large-scale ecological mapping were dominated by the Clementsian notion of climatic climax vegetation, it is largely inevitable that much attention has been devoted to the importance of climatic factors in influencing ecological variations in space. Approaches to the large-scale mapping of ecological and climatic systems followed similar lines initially, and indeed Eyre (1964a) has pointed out that climatic classification originated largely in

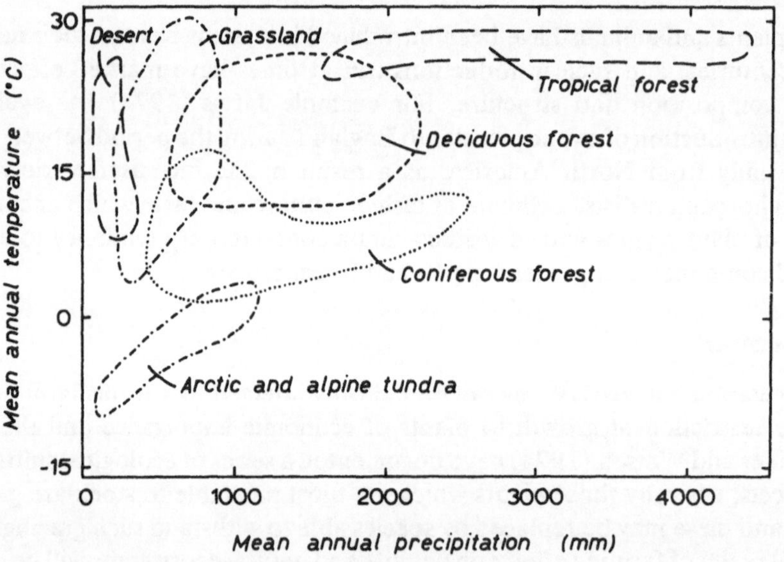

FIG. 5.5 CLIMATIC RELATIONSHIPS OF SIX MAJOR GLOBAL BIOMES.
(After Hammond, 1972.)

attempts to rationalize the vegetation pattern over the earth's surface. This led, inevitably, to gross over-simplifications, in that – as Eyre (1964a) pointed out – the 'climatic regions' derived by Köppen and Thornthwaite were often regarded as being *absolutes* rather than abstractions. This is not to say that there are not very marked associations between the distributions of major ecological variations and the pattern of climatic variation at the global scale. Figure 5.5 shows clearly a climatic basis for recognizing six major biomes as used in the ecosystem analysis programme of the IBP (Hammond, 1972). The early studies, however, might stand accused of using circular arguments in attempts to evaluate the pattern of association between climate and vegetation – because each was used as a basis for mapping the other! Subsequent attempts have sought to identify and account for observed associations between climate and vegetation in greater detail, and available studies fall into two broad types – those concerned with species distributions and those concerned with ecological productivity.

Species distributions

Hutchinson (1947) pointed out that previous work had shown restricted distributions of species to be largely a function of temperature, and he sought to rationalize this association by considering two aspects of temperature control over the distributions of individuals. One arises where the dispersal of those individuals is checked by temperature conditions too extreme for the survival of the individuals. For species spreading towards the poles from the equator the critical extreme temperature is the winter (cold) one; and the excessive heat of summer temperature is critical for species spreading towards the equator. The other arises where the distribution of individuals is affected by the critical temperatures necessary for reproduction and completion of the life cycles of the individuals.

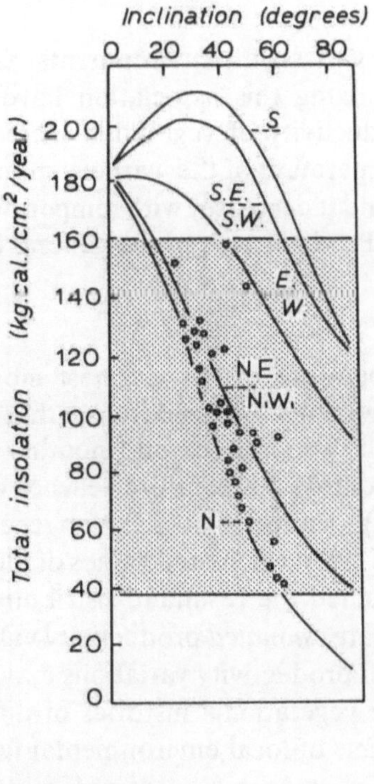

FIG. 5.6 THE BOYKO INSOLATION/EXPOSURE FACTOR.
The figure shows the total IE amplitude of *Laurus nobilis* in its border zone in Palestine (an area with between 600 and 800 mm of annual rainfall). (After Boyko, 1947.)

Species can spread towards the poles as far as the summer temperatures are suitable, and towards the equator as far as the requisite low temperatures can be maintained in winter. This enabled Hutchinson (1947) to recognize four distinct types of spatial zonation of species, each having a unique combination of summer and winter boundaries for poleward and equatorward limits, and each with a spatial configuration around the world which closely follows the distribution of summer and winter temperatures.

An alternative approach was adopted by Boyko (1947), who considered the value of plants as *climatic indicators*. For the geographical distribution of most species he noted that the limits towards the poles are generally conditioned by temperature whereas the limits towards the equator are conditioned by the prevailing water balance. Vegetation distributions also have a vertical dimension, and vertical zonation in mountain areas is controlled at the upper limits by temperature as a limiting factor, and at the lower limits by water balance. This framework enabled Boyko to identify two basic environmental factors controlling both macro- and micro-scale vegetational distributions – *insolation* (I), which reflects both temperature and water balance, and *exposure* (E) which is related in part to the orientation of the land surface at the study site. He then defined an *insolation/exposure factor* (IE) which 'can be regarded as decisive for the composition of most vegetation types in semi-arid and arid regions of low latitudes' (Boyko, 1947). Figure 5.6 shows the distribution of *Laurus nobilis* within the region of Palestine with annual rainfall of less than 800 mm, on slopes with annual insolation of below 160 kcal/cm^2.

Ecological productivity

The productivity of ecosystems varies with environmental factors (see Chapter 3), and several approaches to characterizing this association have been suggested. Paterson (1956) concluded that the productivity of vegetation increases with the length of the growing season, the average temperature of the warmest month, the annual precipitation and amount of solar radiation; and it decreases with temperature range. This enabled him to formulate an 'Index of Plant Productivity', of the general form:

$$I = \frac{Tm \quad P \quad G \quad S}{120 \quad (Tr)}$$

where Tm is the average temperature of the warmest month (°C), Tr is the annual temperature range between the warmest and coldest months (°C), P is the annual precipitation (cm), G is the length of the growing season (months) – defined as the number of months in which the average monthly temperature reaches or exceeds the threshold of plant growth (assumed to be 3 °C) – S is the solar radiation received (as a proportion of that received at the poles). Paterson (1956) calculated values of the index for different parts of the earth's surface, and he classified the resultant distribution into six 'Zones of Plant Productivity' (Table 5.4). These are *estimated* productivity values and Haggett (1972) has stressed that the actual pattern of productivity variations may differ markedly from these zonal tendencies because of the vegetational histories of different areas, the effects of human interference and the effects of local environmental factors not considered in the equation.

Nonetheless Paterson's (1956) study paved the way for other attempts to find environmental variables which correlate well enough with observed ecological productivity variations to be of value in predicting the latter for unmeasured sites. For example Rosenzweig (1968) concluded that *actual evapotranspiration* (AE) provides a measure of the simultaneous availability of water and solar energy in an environment at a given point in time, and these are reflected in the amount of plant activity (that is, productivity). By collecting data on net annual above-ground productivity (NAAP, measured in g/m^2) from a variety of environments and estimating annual actual evapotranspiration (AE, measured

Table 5.4
ZONES OF PLANT PRODUCTIVITY PREDICTED FROM CLIMATIC DATA BY PATERSON (1956)

Zone	Index of plant productivity	Main geographical range
A	over 50·00	equatorial belt (such as the Amazon Basin of South America, and the Congo Basin of Africa)
B	10·01–50·00	restricted to the tropics (mostly South America)
C	3·01–10·00	some of the most densely populated areas of the world (such as eastern United States, western Europe, much of India and South China)
D	1·01–3·00	mainly confined to cool temperate climates, with maximum rainfall in the summer period (such as central and eastern Europe)
E	0·26–1·00	adjacent to Zone F, relatively narrow zone around deserts but large areas of North America and Soviet Asia
F	below 0·25	cold, dry areas (such as ice caps, northern taiga fringes and mid-latitude deserts)

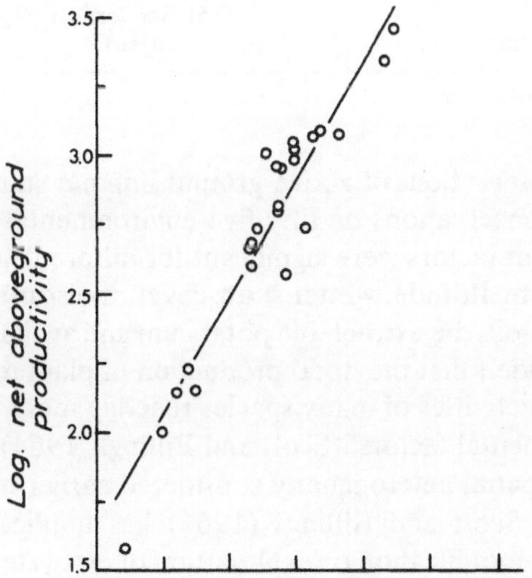

FIG. 5.7 RELATIONSHIP BETWEEN NET ANNUAL ABOVE-GROUND PRODUCTIVITY AND ACTUAL EVAPOTRANSPIRATION, FOR A RANGE OF HABITATS.
The line is described by the formula:

$$\mathrm{Log}_{10}\, NAAP = 1{\cdot}66 \quad \mathrm{Log}_{10}\, AE - 1{\cdot}66$$

where *NAAP* is net annual above-ground productivity (gm of dry matter/m^2/yr), and *AE* is the actual evapotranspiration for the year (mm). (After Rosenzweig, 1968.)

in mm) for those environments from data on latitude, monthly temperature variations and precipitation patterns, Rosenzweig (1968) was able to establish a close correlation between the two variables NAAP and AE (Figure 5.7), which could then be used in predicting productivity in unmeasured environments.

Each of these studies sought to evaluate associations between variations in ecological and environmental variables at the large (generally global) scale. They have been complemented by a number of studies of these sorts of associations at more local scales. These include studies of specific ecosystems such as the marine (Mann, 1969), arctic tundra (Bliss et al., 1973) and desert (Noy-Meier, 1973) ecosystems. A variety of approaches have been adopted in these studies, the most common being to examine the pattern and strength of correlations between environmental factors and vegetation structure and productivity at different sites. This can be illustrated by the study of an alpine tundra ecosystem in Wyoming by Scott and Billings (1964), who attempted to consider the importance of all the environmental variables which previous studies had shown to be of importance (Table 5.5). They correlated observations of above-ground summer standing crop variations for individual species with observations on fifty-five environmental factors, and found (not surprisingly) that different factors were significant for different species. The factors most frequently significant were altitude, winter snow cover, moisture regime, soil movement, the percentage of clay in soil, the extractable potassium and available water in the sub-soil. Scott and Billings concluded that the 'total production of plant material by alpine tundra was the sum of the productivities of many species reacting more or less independently to many different environmental factors' (Scott and Billings, 1964).

Table 5.5
SUMMARY OF ENVIRONMENTAL VARIABLES IMPORTANT IN CONTROLLING PRODUCTIVITY VARIATIONS IN ALPINE TUNDRA ECOSYSTEMS

(1) Mean annual temperature	(9) Soil depth
(2) Summer air and soil temperature	(10) Soil organic matter
(3) Depth and duration of snow cover	(11) Soil aeration
(4) Length of growing season	(12) Partial pressure of oxygen and carbon dioxide
(5) Available soil moisture	(13) Summer cloudiness
(6) Wind	(14) Animals
(7) Wind erosion	(15) Soil fertility (especially the availability of nitrogen)
(8) Degree of soil development	

Source: Scott and Billings (1964)

Like the problem of spatial heterogeneity considered earlier in this chapter (see pages 145–7), this finding of Scott and Billings (1964) has implication for environmental management and for the modification or exploitation of ecosystems. If different environmental factors have differing significance to ecological productivity even within the same ecosystem, it suggests that a *holistic approach* to nature reserve design and management, and to general environmental conservation and ecological resource protection, is required.

5.4 REGIONAL ECOLOGICAL SURVEYS

Spatial heterogeneity of ecological characteristics is an inherent feature of all types of ecosystems, and ecological surveys are required to establish the nature and magnitude of these variations at all scales. Goodier and Grimes (1970) outline three main needs for ecological surveys. One is *ecological research*, because data on the pattern and distribution of ecological changes through space are required in the selection of research areas for detailed study, and for the study of relationships between plants and animals and their habitat. A second need is in *resource management*, where such information is required for planning the integration of wildlife habitat conservation with other forms of land use. Finally data on spatial variations are required for the *management of nature reserves*. If an inventory of ecological resources is available (ideally in map form), reserve management programmes can take account of the spatial distribution of vegetation components in relation to the management of public access, the allocation of areas for research, and similar land-use decisions. Whilst the *biome approach*, which considers both plants and animals simultaneously, would undoubtedly provide sounder data on the entire range of ecological resources within a habitat or area, most ecological surveys focus almost exclusively on vegetation distribution. This is partly for ease of survey (plants are static and animals are mobile), and partly a reflection of the frequent demand for maps of the distribution of vegetation *per se* within an area. Bunce and Shaw (1973) point out, however, that vegetation can be used as an indicator of 'whole ecosystems' for a number of reasons. Vegetation is a sensitive and integrated measure of the entire ecosystem complex, and furthermore vegetation is the primary producer on which nearly all other organisms depend (either directly or indirectly). In addition, suitable measures of macro-vegetation are relatively easy to obtain. Vegetation also represents an integration of all environmental factors involved in ecosystem functioning, and it is (on the whole) relatively permanent and thus no time dimension is required in the data collection stages of analysis.

Vegetation mapping has thus been carried out in many countries and at many scales and, as Matthews points out, vegetation maps 'are essential as a record of the existing vegetation of the country and should form the basis for future land utilization' (Matthews, 1939). The history of vegetation mapping in Britain mirrors the general trends in the development of the field. Mapping had begun by the late nineteenth century on a localized basis, and by interested amateur botanists. Systemmatic mapping began early in the present century, and attention was concentrated initially in northern Britain. In 1906 Hardy produced a coloured vegetation map of the Highlands of Scotland, at a scale of 1:633 600. Initial progress in regional mapping was rapid, and so between 1900 and 1912 some ten primary map surveys of vegetation associations within the British Isles had been published in colour at scales of 1:126 720 and 1:63 360 (Tivy, 1954). Typical of these was Smith's (1900) mapping of vegetation in the Edinburgh district, and Lewis's (1904) study of vegetation in the catchments of the Rivers Eden, Tees, Wear and Tyne. These early mapping programmes were effectively reconnaissance surveys, which allowed the major vegetation associations to be mapped over large areas. Memoirs accompanying each vegetation map detailed the main characteristics of each vegetation type encountered, provided lists of species within each association, and suggested relationships between distributions and environmental controls; and standardization of approaches to sampling, classification and eventual publication of maps was achieved by the co-ordinating efforts of the Central Committee for the Survey and Status of British Vegetation. One of the main contributions of the Committee was a classification of English woodlands (Moss, Rankin and Tansley, 1910) (Table 5.6), although no map of the distribution of each of the woodland associations was offered at the time.

Ironically, subsequent ecologists have not followed the lead of the pioneer vegetation mappers, and Tivy has stressed that 'modern plant ecologists have tended . . . to concentrate on smaller and smaller plots with ever increasing detail and to perfect experimental techniques. Such maps as have been produced are confined to local areas and the scales and symbols used are dependent upon the author's fancy' (Tivy, 1954). Kuchler (1967) offers a different perspective on recent vegetation mapping attempts. In his view 'a vegetation map is the meeting ground of two poles; an author's systematic classification,

Table 5.6
CLASSIFICATION OF WOODLAND ASSOCIATIONS

ALDER-WILLOW SERIES: on very wet soils, consist of at least two distinct associations as yet insufficiently studied to allow satisfactory separation and characterization

OAK AND BIRCH SERIES: on non-calcareous soils
- (a) Oakwood associations: on non-peaty soils at low and moderate elevations
  - (i) Damp oakwoods – on clays, shales, loams, fine sands and moist soils generally
  - (ii) Dry oakwoods – on sandstones, grits, sands and dry soils generally
- (b) Oak-birch-heath associations: on dry, coarse sandy and dry peaty soils (low elevations)
- (c) Birchwood associations: on non-calcareous soils at high elevations (over 1000 feet)

BEECH AND ASH SERIES: on calcareous soils
- (a) Ash-oakwood association: on calcareous clays, marls, impure limestones and calcareous sandstones
- (b) Ashwood associations: on limestones
- (c) Beechwood association: on Chalk in S.E. England, with a western extension to the Oolites of the Cotswold Hills

Source: Moss, Rankin and Tansley (1910)

and nature's kaleidoscopic arrangement of plants. The degree to which these two poles meet depends on the imagination, insight and skill of the mapper' (Kuchler, 1967). One important factor which has contributed to the reluctance to continue mapping along the lines of early workers is the ongoing debate over methods of analysis of variations in vegetation between two fundamentally opposed schools of thought.

5.4a Vegetation analysis and classification

A long and wide-ranging debate in vegetation studies has centred on the best approach to adopt in describing and classifying vegetation. A wide range of opinion has been polarized into several reasonably discrete schools of thought, and the development of these has been traced by a number of vegetation geographers (such as Pears, 1968; Harrison, 1971; Shimwell, 1971; and Randall, 1978). There are three main phases in the evolution of the debate.

Tansley (1949) and the 'Dominant Species Approach'

Tansley's (1949) study of wild and semi-natural vegetation in Britain was based on the presence of dominant species. The associations were thought to represent relatively stable community types (Tansley [1949] defines a plant community as 'any collection of plants growing together which has, as a whole, a certain unity'), and they are based in part on the successional status of the vegetation community. Although the approach did provide a systemmatic inventory of the major plant communities in Britain, Harrison has pointed out that 'the informality and subjectivity of Tansley's field procedure, not only in the selection of the samples, but also in the estimation of species abundance, led to excessive generalization in the community description' (Harrison, 1971).

Braun-Blanquet (1932) and the 'Floristic Approach'

Motivated by the need for objectivity in classification, Braun-Blanquet (1932) viewed a vegetation community in two complimentary senses:

(a) *the real vegetation* – sampled as a homogeneous stand in the field, and
(b) *the community unit (nodum)* – an abstraction of the real vegetation, derived by successive comparison of similar sites on the basis of species composition.

In this approach field observation focused on recording all species present within each uniform stand of vegetation, and estimating the cover abundance (by the Domin Scale – Table 5.7) for each species. The dominant species of the field layer of vegetation could

Table 5.7
DOMIN SCALE FOR ESTIMATING COVER ABUNDANCE OF SPECIES IN VEGETATION SAMPLES

Domin Scale value	Abundance of cover
1	one or two individuals
2	sparsely distributed
3	frequent, but low cover (5%)
4	cover 5–20%
5	20–5%
6	25–33%
7	33–50%
8	50–75%
9	75–90%
10	cover complete, or almost so

Source: Ward, Jones and Manton (1972)

then be used to determine the structure of the community; and the community units within this structure (the noda) could be identified on the basis of species composition (the noda is defined by 'constant species', defined as those which are found in between 80 and 100 per cent of the samples). The approach assumes that noda are quite discrete, although it stresses that overall variation in vegetation (at the vegetation structure level) is almost continuous. Poore (1955) viewed noda as clusters of stands in multi-dimensional space, and he pointed out that it is possible to determine the environmental factors controlling vegetation variation by examining the site characteristics of each nodum. Poore and McVean (1957) extended this approach and observed how the floristic and organizational aspects of Scottish mountain vegetation vary in association with five 'factor complexes' – altitudinal zonation, oceanicity, snow cover, soil base status and moisture availability. They illustrated the nature of vegetation changes by the use of 'vegetation diagrams', in which the changes in vegetation structure and noda are plotted relative to changing exposure and altitude (Figure 5.8).

The Statistical Approach

Most developments since the middle 1950s have been numerically based, and these evolved because of the increasingly widespread availability of efficient computing facilities, and because of dissatisfaction with earlier approaches. There is currently a very wide range of numerical approaches in use – Moore and his colleagues (1970), for example, suggest that a conservative estimate would be in the order of fifty. Two very distinct schools of thought about how vegetation changes through space have evolved along separate lines. One maintains that discrete community groups can be identified within vegetation; whilst the other maintains that they cannot, and that variations between vegetational units are more or less continuous. These differences are reflected in the two basis approaches to vegetation classification.

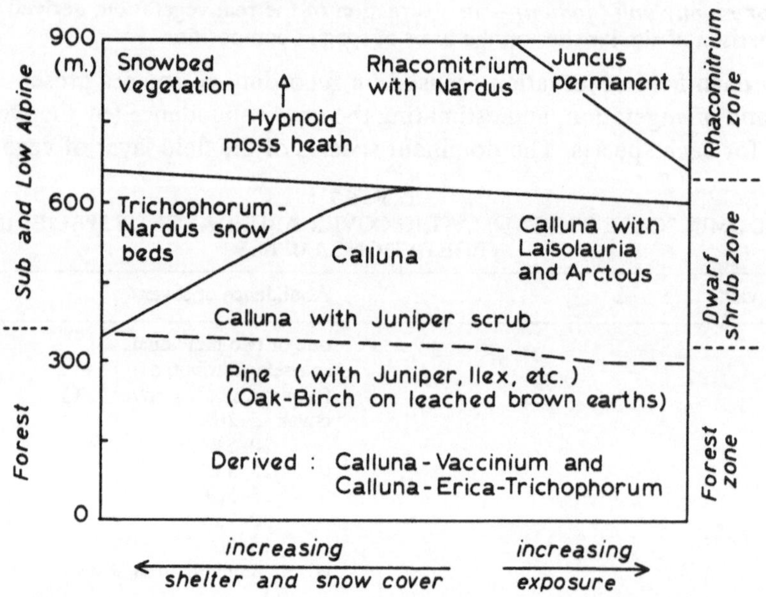

FIG. 5.8 CLASSIFICATION OF VEGETATION IN THE NORTH WEST HIGHLANDS OF SCOTLAND, ON THE BASIS OF DEGREE OF EXPOSURE AND ALTITUDE OF SITE.
(After Poore and McVean, 1957.)

Classification

Those workers who consider that vegetation can be sub-divided into discrete components (such as Tansley's 'Associations') have attempted to develop methods that group the vegetation units so that within-group variations are less than between-group variations. There are two directions from which to approach the delimitation of these units – these are either to sub-divide the overall vegetation structure into smaller components (a *divisive approach*) or, by starting with a series of individual sites, to group these together into larger units or groups (an *agglomerative approach*). The delimitation in either case can be done on the basis of one characteristic of the vegetation (that is, one species), in which case the approach is described as *monothetic*; or by simultaneously considering several characteristics (a *polythetic* approach). Most approaches to classification have been monothetic divisive ones – such as 'Normal Association Analysis' as formulated by Williams and

Lambert (1959, 1960). This aims to sub-divide a plant population so that all associations disappear (Williams and Lambert, 1959). The technique has been widely evaluated (Harrison, 1971; Pears, 1977) and the main steps in the analysis are summarized in Table 5.8. A hypothetical example is illustrated in Figure 5.9, from which the basis of the approach can be appreciated.

Association analysis allows the definition of vegetation mapping units, and the distribution of these within a study area can then be mapped. The approach provides an objective basis for classifying vegetation, and it has been widely used. For example Ward, Jones and Manton (1972) used this technique to define seven mapping units of vegetation on Dartmoor (blanket bog, grassland, grassland invaded by bracken, vaccinium moorland, valley bog, heath and grassland with gorse). Each of these units could be interpreted from

Table 5.8
SUMMARY OF ASSOCIATION ANALYSIS APPROACH TO CLASSIFYING VEGETATION

Step	Procedure
Field observation:	sample a number of randomly located quadrats of a given size in the study area
	record the number of quadrats in which each species occurs, and the number in which it is absent
	express *presence* and *absence* as frequencies
Data analysis:	compare the level of association between each pair of species (by the Chi Squared test)
	sum the Chi Squared values for each species (higher summed values denote higher levels of association)
	the species with the highest summed Chi Squared value is the highest unit of classification
	this is then sub-divided into groups – those in which the species is present and those from which it is absent
	continue this sub-division on the basis of presence and absence until the entire population has been classified with acceptable detail (the cut-off point occurs when a suitable number of 'vegetation communities' have been derived)
	interpret the 'communities' derived in the above manner

Source: Pears (1977)

colour air photographs to produce a vegetation map of the 480 km^2 of Dartmoor. Other forms of Association Analysis which have been suggested and applied include Inverse Association Analysis (Harrison, 1971), a combined method of Normal and Inverse Association Analysis (Lambert and Williams, 1962), and various agglomerative procedures (Williams, Lambert and Lance, 1966).

Ordination

Many workers have concluded that vegetation variation is continuous, and therefore a unit of vegetation cannot be classified into discrete vegetation classes. One champion of this school is Whittaker who stressed that 'vegetation is conceived as . . . a pattern of populations, variously related to one another, corresponding to the pattern of environmental gradients' (Whittaker, 1953), so that as environmental factors change, so the balance amongst plant populations changes in harmony. The thesis is thus that the natural plant cover is basically continuous along environmental gradients – and distinct breaks in the continuity only occur where there are marked environmental discontinuities. The approach of ordination has been developed to allow expression of this continuity of vegetation change. The continuity referred to, however, is continuity in composition and this is environmentally determined; it is not explicitly locationally continuous. Therefore

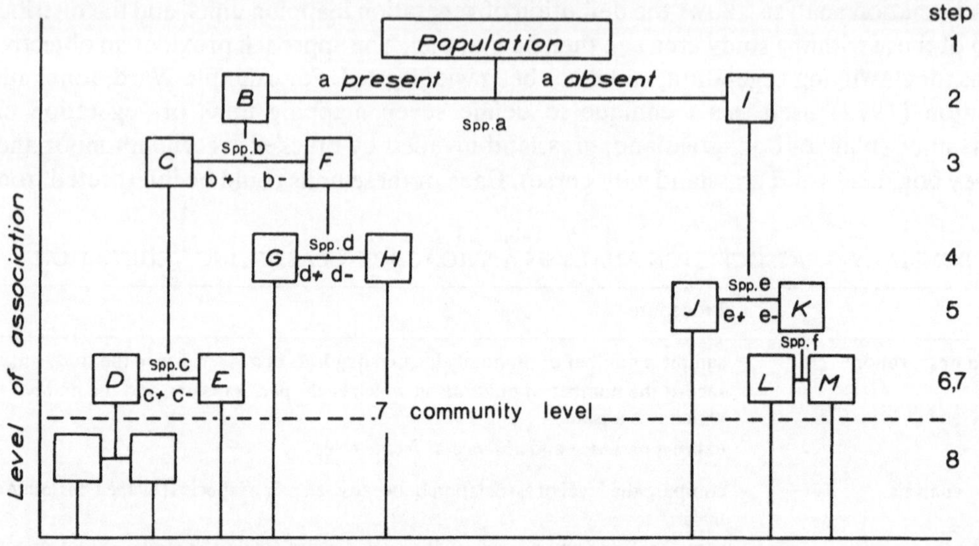

FIG. 5.9 NORMAL ASSOCIATION ANALYSIS – A MONOTHETIC DIVISIVE APPROACH TO VEGETATION CLASSIFICATION. The total plant population in a sample is sub-divided by a series of progressive steps, so that all associations eventually disappear. Sub-division proceeds on the basis of whether or not a given species is present at a given level of association. The example shown represents a seven-community classification; the communities are composed of the following species:

Community	*Species definitely present*	*Species definitely absent*
D	a, b, c	–
E	a, b	c
G	a, d	b
H	a	b, d
J	e	a, b, c, d
L	f	a, b, c, d, e
M	none of those considered	a, b, c, d, e, f

the basis of ordination is to identify the pattern of continuity. 'In ordination of stands, an attempt is made to place each stand in relation to one or more axes (in a plot) in such a way that a statement of its position relative to the axes conveys the maximum information about its composition' (Greig-Smith, 1964). Since the continuity is a reflection of environmental gradients, ordination procedures are based on *gradient analysis*. Two different but complimentary approaches to gradient analysis have been formulated (Whittaker, 1967):

(a) *Direct gradient analysis* – where vegetation samples are arranged according to a known magnitude of an environmental gradient which is selected as a basis of study.

(b) *Indirect gradient analysis* – where vegetation samples are compared with each other on the basis of differences in species composition (it is thus a gradient of variation in species composition).

Whittaker (1967) has summarized the three main concepts behind *direct gradient analysis*. One is that there exists an environmental gradient along which many characteristics of soils and climate change from moist valley to dry open slopes. Secondly, species populations are distributed – each according to its own physiological responses – along this gradient. Finally the different combinations of species along the gradient are recognized as *community types* by ecologists, and these community types are related to one another along gradients of community characteristics. Thus if samples of vegetation at equal intervals along an environmental gradient are observed, the pattern of variation of a range of factors (such as species composition, species diversity and ecological productivity) can be plotted relative to changing environmental factors (such as altitude) to demonstrate the pattern of continuous vegetational variation (Figure 5.10).

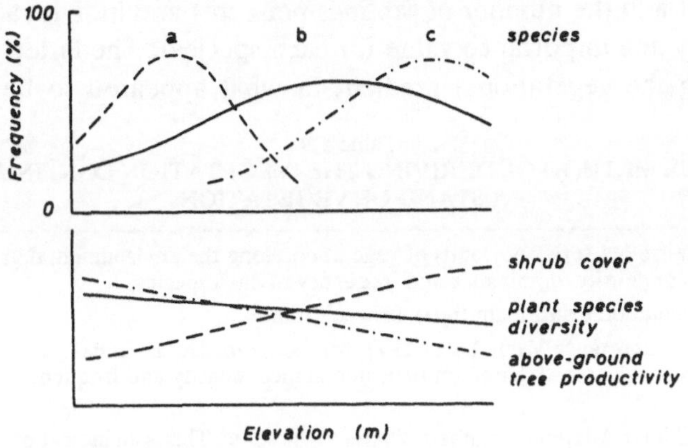

FIG. 5.10 ORDINATION – DIRECT GRADIENT ANALYSIS OF VEGETATION.
The figure shows the pattern of variation of species occurrence in relation to an environmental gradient (in this case elevation).

The *indirect gradient analysis* approach is based on relating vegetation variations to each other, and so some means of expressing the 'vegetation composition gradient' is required. Curtis and McIntosh (1951) formulated a valuable approach to the problem of comparing species and gradients by indirect analysis, in the form of a 'Vegetation Continuum Index' (Table 5.9). This is based on the species composition of vegetation and on successional status, and if the index is derived for a sample of sites along an environmental gradient the continuum of change in vegetation composition along the gradient can be evaluated by plotting a graph with species composition (the percentage contribution which each species makes to the stand, on the basis of abundance) along the vertical axis, and the continuum index along the horizontal axis. The distribution of species along the continuum axis is generally bell-shaped and continuous; different species have different amplitudes along the continuum axis; and different species reach optimum relative abundances at different points along the continuum axis (no groups of species regularly reach a peak together) (Figure 5.11). Graphs such as Figure 5.11 can be used to demonstrate that vegetation distributions vary regularly through 'environmental space'. As an alternative, vegetation stands can be plotted with the continuum index as the horizontal axis, and a measure of environmental variation (such as soil moisture capacity or soil properties) as the vertical axis. Such plots show the pattern of environmental gradients to which vegetation is adjusted. Multi-dimensional plots, whereby vegetation variations relative to a number of environmental gradients can be considered simultaneously, were also suggested and demonstrated by Bray and Curtis (1957).

The value of the continuum index framework in analysing bird communities has also been evaluated by Bond (1957), who effectively substituted observations on bird species occurrences within 64 stands of deciduous forest in southern Wisconsin into Curtis and McIntock's formula for determining the continuity index (Table 5.9), and proved the value of identifying a variety of environmental and ecological gradients by this approach. They successfully plotted against the continuity index a range of variables describing the habitat in terms of vegetation change (such as the importance values of selected tree species), canopy and under-storey structure (in terms of the number of trees per acre, the percentage canopy cover and the number of saplings per acre) and individual bird species (the average frequency and importance value for each species). The latter were shown to be 'distributed along the vegetational gradient in what appeared to be entire or partial

Table 5.9
SUMMARY OF THE METHOD OF DERIVING THE 'VEGETATION CONTINUUM INDEX' FOR A STAND OF VEGETATION

(1) Collect field data from a series of stands of vegetation along the environmental gradient; for each stand collect measures of density, dominance and frequency of each species.

(2) Calculate the Vegetation Continuum Index for each stand:

(a) calculate the 'Importance Value Index' (IVI) for the stand. This is a measure of the relative abundance of species within the stand, based on their dominance, density and frequency. The range of values is 30–300.

(b) Assign the 'Climax Adaption Number' (CAN) to the stand. This is an indication of the relative position of the stand along the succession from pioneer to climax stages. The range of values is 1–10.

(c) Calculate the Vegetation Continuum Index (VCI) for each stand, from the relationship

$$\text{VCI} = \text{IVI} \times \text{CAN}$$

The range of values of VCI is from 300–3000

Source: Curtis and McIntosh (1951)

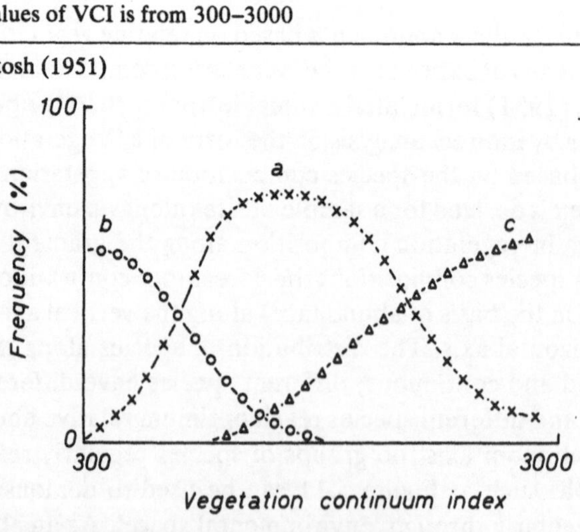

FIG. 5.11 ORDINATION – INDIRECT GRADIENT ANALYSIS OF VEGETATION.
The figure shows the pattern of variation of species occurrence in relation to the 'Vegetation Continuum Index' values for the component stands. See Table 5.9 for explanation.

gaussian [i.e. bell-shaped] curves. The curves of no two species exactly coincided' (Bond, 1957). In essence both the direct and the indirect forms of gradient analysis are based on the same concept, which is that the continuous nature of vegetation changes precludes meaningful attempts to identify discrete vegetation units which can be mapped.

Attention has also been devoted to increasing understanding of both the application and the interpretation of ordination and environmental gradient analysis. An example of the former is Bunce's (1968) study of vegetation at 58 sites in the Carneddau mountain group of North Wales, which used an ordination approach for correlating species distributions with an environmental gradient of sites from wet eutrophic (lower cliffs and in gulleys) to dry oligotrophic (upper cliffs and buttresses) conditions. Bunce was forced to conclude that 'although some species have similar distributions, no distinct groups of stands were separated from the cluster. The vegetation is therefore continuous, and divisions of stands into groups would be arbitrary' (Bunce, 1968). An example of increasing the possibilities

Table 5.10
SUMMARY OF GENERALIZATIONS CONCERNING CHARACTERISTICS OF ENVIRONMENTAL GRADIENTS

(a) Species importances along a gradient generally form curves approximating normal curves
(b) The modal optima of minor species are scattered along environmental gradients apparently at random
(c) The modal optima of major species are dispersed in a more uniform pattern that reduces competition between species
(d) These 'normal distributions' may be modified by competition
(e) In field data the tails of the normal curves will be cut off at the point beyond which the species is too rare to be found with any regularity within sample plots of a given size
(f) Environmental gradients differ in
  (i) richness in species in particular communities or samples (*Alpha diversity*)
  (ii) degree of compositional change or species turnover (*Beta diversity*)
  (iii) total species richness (*Gamma diversity*)

Source: Gauch and Whittaker (1972)

for interpreting ordination results is the work of Gauch and Whittaker (1972), who have examined the assumptions behind environmental gradients, or as they prefer, community gradients (*coenclines*) on the basis of a series of computer-based simulations. They chose simulations in favour of field data because with the latter the regularity of vegetation response to changes in environment is often obscured by local environmental factors, by sampling errors and by the complex inter-play of many environmental gradients which are difficult to isolate and interpret. Their simulations led them to draw up a series of generalizations concerning environmental gradients (Table 5.10).

A valuable contribution to increased understanding of the controls of vegetation distributions along environmental gradients was Terborgh's (1970) consideration of the mechanisms of distributional limitation in 410 species of forest birds in Peru. The study was designed to test which of three mutually exclusive 'models' (Table 5.11) could be used to account for forest bird distribution along an environmental gradient in the Eastern Andes. The field data suggest that the three mechanisms of distributional limitation differ appreciably in importance with the Peruvian birds. Less than 20 per cent of the limits for individual species could be accounted for by 'habitat discontinuities'; about a third were associated with 'competitive exclusion'; and about a half were a function of gradual environmental changes along the topographic gradient. Figure 5.12 summarizes the main characteristics of each of the three models.

In summary, there are a wide range of approaches to the problem of classifying and describing vegetation. A number of attempts have been made to evaluate several different approaches when applied to one standard body of test data. For example Moore and his colleagues (1970) compared the Braun-Blanquet approach with similarity co-efficients, cluster analysis, association and inverse analyses, and ordination when applied to field data from salt-marsh, sand-dune and heathland habitats around Dublin. They concluded that

Table 5.11
SUMMARY OF THREE MODELS OF DISTRIBUTIONAL LIMITATION ALONG ENVIRONMENTAL GRADIENTS, SUGGESTED BY TERBORGH (1970)

Model	Overall controls	Character of environmental gradient
1	Gradual environmental variations	The abundance of different species is a function of their ecological amplitudes on the environmental gradient. There are *no* exclusive reactions between competing species, no real discontinuities in habitat. Normal curves, optima fall at random along the gradient
2	Competitive exclusion	If ecological requirements of closely related species are similar, this leads to unstable co-existence; populations are forced to occupy mutually exclusive domains. This leads to sharp zones of contact and truncated distributions which do not correspond with habitat changes (repulsion interaction). Mean amplitude of species decreases as the number of species increases (amplitude compression)
3	Habitat discontinuities (Ecotones)	Massive rate of change of species at the discontinuities; population density curves assume square wave forms; the total density of individuals is nearly constant across the ecotone

Source: Terborgh (1970)

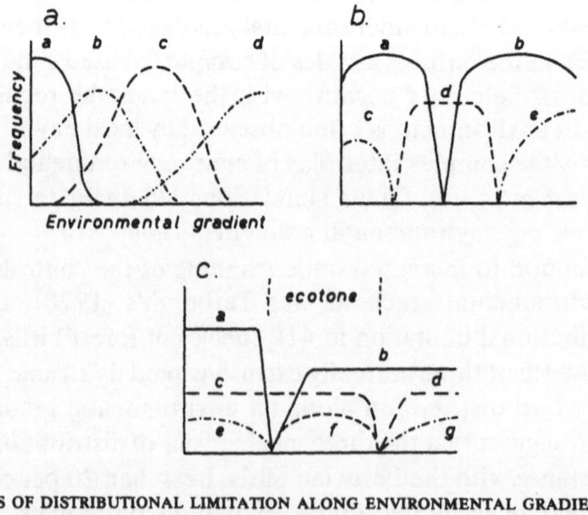

FIG. 5.12 THREE MODELS OF DISTRIBUTIONAL LIMITATION ALONG ENVIRONMENTAL GRADIENTS.
The models are based on gradual environmental variation (a), competitive exclusion (b) and habitat discontinuities (c). See Table 5.11 for explanation of the basis of each model. (After Terborgh, 1970.)

'there is, at present, no single *best* method of phytosociological analysis, and there probably never will be. A method must be judged in relation to the purpose of the investigation' (Moore et al., 1970). Their main reason for the evaluation was as a pilot study for a projected survey of the vegetation of the whole of Ireland (a total area of 68,000 km^2); and on this basis the Braun-Blanquet scheme was considered the best because it optimizes the balance between time (and hence money) input and understanding of the structural complexity of vegetation and its relationships with environmental factors. Kellman (1975) has stressed that the value of each of the various techniques of vegetation analysis and classification is a function of simplicity in field observation and recording, the level of background information available to the investigator and the scale of application (and scale of resolution required).

5.4b Vegetation mapping

Despite these methodological debates over methods of classifying and describing vegetation, vegetation mapping has advanced along several lines within recent decades. One has been an increasing concern to map the structure (the spatial pattern of growth forms) of a plant community – such as the height and density of vegetation forms – rather than simply the species composition of the vegetation. Perera maintains that 'it is the presence and particular proportion of each growth form which largely gives a plant community its unique and unmistakeable character' (Perera, 1975). Such approaches to mapping, as illustrated by the work of Dansereau (1951), are thus based on systems of notational symbols used to record the structural characteristics of the types observed. Shimwell (1971) and others have evaluated these forms of approach, and Perera (1975) has mapped the structure of vegetation in Sri Lanka by this method.

Advances have also been made in mapping vegetation distributions within small areas. For example Newbould (1960) based an interpretation of the present vegetation of Cranesmoor, a valley bog in the New Forest of southern England, on three complementary types of information on vegetation distributions. These were an overall vegetation map which recognized four basic vegetation types (dry heath and its variants; wet heath; schoenetum; and general bog communities); species maps showing the distributions of selected species (based on subjective estimates of the abundance of the species within a grid of 15m squares laid over the whole study area); and quadrat data (the presence or absence of each species within quadrats located at 50 randomly selected sites within 13 representative zones of the vegetation). These three approaches have been widely adopted in detailed studies of vegetation distributions over small areas.

Although such studies reveal considerable detail about vegetation distribution within small areas they are generally time-consuming and specialized, and Tivy (1954) has stressed the need for rapid reconnaissance techniques of vegetation analysis for large areas, to provide basic inventories of the overall large-scale pattern of vegetation changes. She outlined an approach based on a four-step procedure, applied to three sample areas in the Southern Uplands of Scotland. The four steps were to first divide the whole area to be surveyed into topographic units (watersheds); secondly to cover each of these units by foot and locate all major changes in vegetation on base maps at a scale of 1:63,360; thirdly to classify, define and delimit the vegetation associations to be mapped at the finer scale (such as 1:25 000) on the basis of the presence and combination of dominant species; and finally to compile the finished map on the basis of the field map and field notes, supplemented by information from interpretation of vertical air photographs.

The availability of detailed air photograph coverage for many areas is greatly increasing the ease with which large-scale regional ecological surveys can be carried out, and mapping based on aerial photo-ecology (Howard, 1970) – with suitable ground verification procedures – is both quicker and cheaper overall than the field-based approaches of earlier workers (such as Lewis, 1904). Goodier and Grimes (1970) point out that the most common use of air photographs in resource surveys is in vegetation mapping, and they illustrate the ease with which such maps can be constructed. Maps of this sort are of basic importance in both land management and land categorization.

Air photo mapping of vegetation can only be done accurately by trained observers, and often there is a need for approaches which can be used by non-specialists. Such approaches are required in many rapid reconnaissance surveys, for example. Bunce and Shaw (1973)

Table 5.12
STANDARDIZED ECOLOGICAL SURVEY PROCEDURE OUTLINED BY BUNCE AND SHAW (1973)

SAMPLING:	(a)	Select sites for survey, and sample each site at 16 randomly located points
	(b)	Collect field data for quadrats with a basic size of 200 m^2. Within this locate four plots of sizes 100, 50, 25 and 4 m^2, to give information on homogeneity of the samples
RECORDING:	(a)	Herbaceous vegetation – note the presence or absence of vascular plants within each of the successive plot sizes; estimate the cover of each species, to the nearest 5% (later converted to the Domin Scale)
	(b)	Tree and shrub vegetation – measure the diameter of each tree at breast height (1·3 m), as this correlates with other properties such as total dry weight, leaf area and tree age; estimate height of the tree; with shrubs measure the diameter of all major stems
	(c)	Habitat records – standardized format of recording features such as (i) type of vegetation management (ii) regeneration species present (iii) dead vegetation (iv) habitat factors such as rock exposures (v) general vegetation structure
	(d)	Soil records – dig a shallow soil pit in the centre of the plot; record characteristics of the surface layers of the soil; examine and record lower layers of the soil

aimed to produce an objective user-orientated method of mapping and classifying habitats (in particular woodland habitats), which could be widely applied yet in which subjectivity of individual observers could be minimized. They concluded that the choice of methods would need to fulfill each of four broad criteria:

(1) It must carry information relevant to the particular problem at hand (it must, therefore, be user-orientated).
(2) It must carry the necessary information in as few measures as possible (it must be 'efficient').
(3) The data must be easy to collect, even at the expense of some loss of information (it must optimize between completeness and ease).
(4) Compatible information must be available for *all* members of the study population (it must be a standardized scheme).

To fulfil these criteria Bunce and Shaw (1973) developed a standardized ecological survey scheme (Table 5.12) which was used by a body of observers (after initial training in the theory and practice of the method) in a large-scale survey of semi-natural woodlands in Britain during the summer of 1971.

The reason why Bunce and Shaw (1973) collected observations from a nested hierarchy of four quadrat sizes was to evaluate the type and detail of the data which could be collected at different sampling resolutions. If the species composition of two samples of the same vegetation but of different sizes were compared, the larger sample would probably have a greater number of species represented, simply because the probability of sampling species within a vegetation unit increases in proportion to the relative size of the unit area sampled. Kellman thus comments that 'the question of what constitutes an adequate sample thus becomes that of deciding whether information gained per unit sample area declines beyond some threshold sample size, making it progressively less worthwhile to extend sampling further' (Kellman, 1975). Experimental determination of the 'optimum' vegetation sampling scheme can be made by examining how many new species are

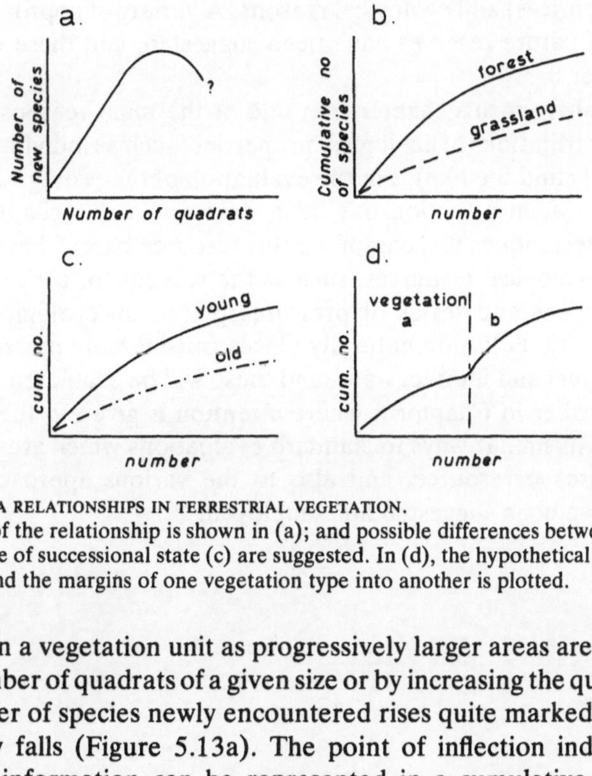

FIG. 5.13 SPECIES/AREA RELATIONSHIPS IN TERRESTRIAL VEGETATION.
The generalized form of the relationship is shown in (a); and possible differences between grassland and forest (b), and the significance of successional state (c) are suggested. In (d), the hypothetical pattern of change as the sample proceeds beyond the margins of one vegetation type into another is plotted.

encountered within a vegetation unit as progressively larger areas are sampled (either by increasing the number of quadrats of a given size or by increasing the quadrat size). In most samples the number of species newly encountered rises quite markedly at the beginning, then progressively falls (Figure 5.13a). The point of inflection indicates an optimum sample size. The information can be represented in a cumulative manner – plotting cumulative number of species sampled against sample size (Figure 5.13b) to produce a *species-area graph*. Such curves might be of value in highlighting the differences in species diversity between vegetation types, such as between forest and grassland (Figure 5.13b). The same basis might also be used to illustrate changes in species diversity through time, such as along a successional sequence within one type of vegetation (Figure 5.13c). If the vegetation sampling proceeds beyond the boundaries of a given vegetation type, a composite species-area graph might be produced (Figure 5.13d) which could be of value in identifying the 'optimum' location on the ground at which to demarcate the vegetation boundary.

Hopkins (1955) has evaluated the nature of the species-area curves for eight different vegetation types in Britain (heath, grassland, beech wood, blanket bog, salt-marsh, shingle ridge, open moor and pine wood). The forms of the curves were shown to be very similar, with a rapid initial rise in the number of species with area followed by a decrease in the rate of rise with increasing area. It was found that a species–log. area curve could also be plotted for each vegetation type. These showed a slight increase in the number of species with increasing area for small areas, and the rate of increase became higher with larger areas, and it eventually stabilized as a straight line semi-log. relation between area and number of species.

The association between area of sample and number of species recorded is of considerable value in planning and designing the size and shape of nature reserves, where it is generally necessary to establish an optimum size rather than a maximum size for both economic (high land prices) and ecological reasons. A variety of approaches to identifying the optimum size of nature reserves have been suggested, and these will be considered more fully in Chapter 8.

It will be apparent from this chapter that one of the main reasons for mapping and evaluating spatial distributions of ecological properties (such as individual species, species diversity, productivity and the like) is in the evaluation of the ecological resource base of an area or country. Much attention has been devoted in this chapter to considering approaches to, and techniques for, compiling this resource base. Chapter 6 will focus on the wider issues of ecological resources, such as the reasons for concern over ecological resources, and the types and scales of pressures placed on ecological resources from various types of activity. Pollution naturally places considerable pressures on ecological resources, in both direct and indirect ways, and these will be evaluated in Chapter 7. This theme is pursued further in Chapter 8 where attention is given to the need to evaluate ecological resources in similar ways to standard evaluations which are regularly made of other types of landscape resource, and also to the various approaches to ecological evaluation which have been suggested and employed.

6
Ecological Resources

> Preserving diversity in a world of rapidly shrinking land resources will require a prompt and universal response based on an appropriate application of ecological knowledge. Every nation should possess an inventory of its biological endowment (Terborgh, 1974).

The stability and survival of natural ecosystems in many environments are presently being threatened by exploitation of ecological resources, reduction if not wholesale removal of ecological habitats, and environmental pollution. Pollution and its ecological effects will be considered in the next chapter (Chapter 7). The aim of the present chapter is to consider the notion of ecological resources, particularly in terms of interest in the resource base and pressures on it. Approaches towards evaluating and conserving the ecological resource base will be considered in Chapter 8.

Ecological resources, as considered in this chapter and indeed throughout this book, refer to all plant and animal resources in terms of individuals, species, communities, habitats and ecosystems, other than those managed specifically for financial gain (such as commercial forestry operations and agriculture in general). These latter are forms of ecological resources, but attention here will focus specifically on 'wildlife' ecological resources. This is not to say that all natural species, or that natural ecosystems per se are more valuable in all cases than managed ecosystems – indeed it would be hard in many areas (such as most areas within Britain) to find ecosystems which have not been affected to some degree or another by some form of human impact at some point in time.

Geographical interest in ecological resources can be traced back at least to the period of interest in regional identification at the turn of the present century, because plant and animal distributions are a basic component of the character and identity of different regions. More recently interest has centred on problems such as those of identifying and reaching agricultural potentials of different regions of the world (Visher, 1955), and on the wide realm of plant and animal geography and the factors controlling distributions at local, regional and global scales (Kellman, 1975). A recent application, illustrated by the study by Mather (1974) of spatial variations in the physical productivities of major land uses (hill farming, forestry and red deer husbandry) in the Northern Highlands of Scotland, has been in evaluating the use of productivity assessments as a basis for land zoning and for optimizing the allocation of land resources between competing demands. At the global scale Eyre (1978) has evaluated the 'real wealth of nations' partly on the basis of existing ecological resources (particularly in terms of primary productivity). Geographical interest is also being focused on approaches to evaluating ecological resources so as to provide direct inputs into the planning decision-making machinery, and on identifying the needs for and appropriate approaches to the conservation of ecological resources (Warren and Goldsmith, 1974). Taylor (1974) has stressed the need for measuring the ecological resource base *before* it can be properly managed, and he offered an evaluation of the existing resource surveys available for Great Britain. The latter appear to be open to improvement because although resource surveys at the national scale, in the form of land-use surveys, were completed in the 1930s (Stamp, 1962) and 1960s (Coleman, 1961); generalized vegetation surveys are available for both Scotland (McVean and Ratcliffe, 1962) and Wales (Taylor, 1968); the Nature Conservancy Council has recently published a Review of the ecological resources and conservation value of National Nature Reserves and ecologically important sites in Britain (Ratcliffe, 1977b); and land-use and vegetation maps are now available for many local areas and for sites currently threatened with, or in the process of, development; there is still a need for a 'comprehensive, disinterested survey of land resources in Britain' (Taylor, 1974) based on actual (not assumed or historic) land use and contemporary management practices. Such a resource survey would be of value in drawing up a budget of different types and qualities of land use in Britain. Such a resource inventory is essential if suitable and effective resource management policies are to be formulated. The survey would also provide a much needed 'base-line' against which changes in the size and composition of the ecological resource base through time could be evaluated. These changes are often widespread and have lasting impacts on the environment and on ecosystem stability. The two main pressures are the direct use of, or indirect impacts on, ecological resources associated with development or changes in land-use policy (considered in this chapter), and the direct and indirect effects of pollution (see Chapter 7). Both of these can be viewed as stresses that transform community structure in ecosystems, and that reverse the usual sequence of ecological succession in terms of species dominance and diversity, as well as overall ecosystem stability and productivity (Regier and Cowell, 1972).

6.1 REASONS FOR INTEREST IN ECOLOGICAL RESOURCES

Romain Gary observed that 'we need all the dogs, all the cats, and all the birds, and all the elephants we can find . . . we need all the friendship we can find around us' (Gary, 1958), but interest in ecological resources is more fundamental and less self-centred than this. According to Moore (1969), conservation of wildlife rests on one or both of two value judgements. One is that we conserve wildlife because we believe it exists in its own right, and it has intrinsic value. The other is that we do so because we believe wildlife is of value to man, who is himself valuable. More recently Scheffer (1976) has traced two public concerns about the management of wildlife resources – a growing respect for life (the 'anti-kill' concern) and a growing respect for nature and naturalness (the environment/ecology concern). The *aesthetic reasons* can be complemented by a series of *social benefits* of wildlife conservation and ecosystem management, though these are difficult to evaluate and measure unambiguously (Westman, 1977). The third type of reason is the *scientific value* of natural and undisturbed ecosystems. Dasmann sees concern for wildlife as an insurance policy against disrupting the life-support capacity of the planet earth; 'it is because we need to know how the biosphere functions, and keep it functioning while we learn, that the protection of natural communities and wild species becomes so important' (Dasmann, 1973). A fourth main source of interest in wildlife is that stemming from a *recreational/educational need* to preserve natural landscapes and ecosystems. Usher, Taylor and Darlington (1970) have noted that 15 per cent of the visitors to two Naturalists' Trust nature reserves in Yorkshire went primarily for reasons of 'natural history'; Usher and Miller (1975) have commented on increasing interest in 'conservation features' (such as nature trails and hides) at Bridestones Nature Reserve in the North York Moors National Park; and Palmer, Robinson and Thomas (1977) have evaluated the significance of 'the countryside image', particularly in the light of perceived recreational opportunities. The importance of wildlife to countryside visitors was assessed by means of a questionnaire survey in Dalby Forest in the North York Moors National Park by Everett (1977), and he calculated that interviewees' interest in wildlife was made up of an appreciation of the aesthetic, emotive and psychological elements of wildlife, as well as a purely active pursuit in nature study. The association of 'nature' with peace and tranquility – conventionally the inspiration of poets and painters – was clearly apparent in the interviewees' minds (Table 6.1).

Table 6.1
PERCEIVED ASSOCIATION BETWEEN HUMAN BENEFIT AND VISITS TO NATURAL AREAS

The data represent the percentage of respondents choosing different answers to the question 'Could you choose the three most important of these benefits you feel that you might obtain by visiting this area?' The area was the Dalby Forest in the North York Moors National Park

Benefit	*% replying*
Relaxation	67·9
Closeness to nature	51·1
'get away from it all'	47·8
Exercise	31·3
Peace of mind	30·3
Education	27·2
Health	9·9
'something to talk about'	8·8
Variety of life	8·4
Curiosity	7·9
Invigorating	7·5
Excitement	1·9

Source: Everett (1977)

Helliwell (1969) has attempted to rationalize these interests in ecological resources into seven categories of benefit which could then be used as a basis for evaluation of wildlife resources (see Chapter 8). These are:

(a) Direct returns: the direct material and financial returns from hunting, shooting, fishing, berry picking, etc.

(b) Genetic reserve: the maintenance of a reserve of genetic material for the breeding of new varieties of crop-producing plants and animals.

(c) Ecological balance: the maintenance of natural populations of animals and plants as a 'buffer' against unnaturally large increases in pest species.

(d) Educational value:	the education of children and adults in a direct and interesting way into the manner in which the biological world functions.
(e) Research:	facilities for research into biological problems, and for the training of research workers.
(f) Natural history interest:	the basis of an interesting hobby for the amateur naturalist, and the subject matter for photographers, artists, poets, etc.
(g) Local character:	the basis, in some places, for the 'character' of the locality.

6.2 PRESSURE ON ECOLOGICAL RESOURCES – THE GLOBAL SCALE

Recognition of the dramatic, far reaching and often irreversible impacts of human activities on ecosystem stability and survival is not new. Indeed, as far back as 1864 George Perkins Marsh feared that 'man is everywhere a disturbing agent. Wherever he plants his foot, the harmonies of nature are turned to discords. The proportions and accommodations which insured the stability of existing arrangements are over-thrown' (Marsh, 1864). What is new, however, is awareness of the global scale of the problem, of the urgency with which solutions are required, and of the enormous problem of ensuring that the ecological costs of any planned development are taken into account as well as the social and economic costs when alternative schemes are being evaluated. Perhaps the area where pressure on resources is most forcefully highlighted is in terms of endangered, threatened and extinct species of plants and animals (see pages 116–19). Moyal (1976) has pointed out that the Edelweiss plant (*Leontopodium alpinus*) is no longer common on dry rocky Alpine slopes high above the timber-line because of over-collecting. In Britain alone there are currently some 321 species of flowering plant which are regarded as being either rare (161 species), vulnerable (95), endangered (46) or extinct (19) (Table 6.2).

The Council of Europe (1977) has compiled a list of about 1400 rare or threatened species of plants within Europe, of which over 100 are in danger of extinction; and the

Table 6.2
HABITATS OF SPECIES THAT ARE NATIONALLY RARE OR THREATENED IN BRITAIN

Habitat	Extinct	Endangered	Vulnerable	Rare	Total
Montane	2 (3)	0 (0)	10 (17)	48 (80)	60 (19)
Woodland, scrub and hedge	1 (3)	5 (14)	14 (40)	15 (43)	35 (11)
Arable	1 (4)	14 (62)	7 (30)	1 (4)	23 (7)
Man-made open habitats, road-sides, quarries, etc.	4 (16)	5 (19)	7 (27)	10 (38)	26 (8)
Lowland pasture, open grassland, natural open habitat	1 (1)	7 (10)	28 (40)	35 (49)	71 (22)
Heath	0 (0)	3 (18)	3 (18)	11 (64)	17 (5)
Wetlands	6 (13)	7 (15)	16 (35)	17 (37)	46 (14)
Aquatic	1 (8)	0 (0)	2 (17)	9 (75)	12 (4)
Maritime	3 (10)	5 (16)	8 (26)	15 (48)	31 (10)
TOTAL	19 (6)	46 (14)	95 (30)	161 (50)	321 (100)

Note: Figures in brackets refer to percentage of totals for that class.
Source: Perring and Farrell (1977)

International Union for the Conservation of Nature and Natural Resources has established a 'Threatened Plants Committee' to consider means of conserving and protecting over 20 000 species of flowering plants whose survival is now threatened (Lucas and Synge, 1977). The task is an urgent one, because extinctions are frequent. In October 1977, for example, the World Wildlife Fund announced that another mammal – the Caribbean Monk Seal (*Monachus tropicalis*) – must now be regarded as extinct because it has not been seen for 25 years.

Not all pressures on ecological resources have such terminal impacts, however. For example in many developed countries processes such as urbanization, development and changes in agricultural practices are changing (generally reducing) the availability of habitats and niches within ecosystems, and thus fragmenting the composition and structure of the ecosystems. Such changes often lead to the disappearance of plant and animal species to refuge areas, and they thus promote localized extinctions (see pages 116–19). Such pressures are inevitably acute on relatively small islands where refuges are limited in number, especially if additional pressure – such as the explosion of the tourist industry on

Table 6.3
SUMMARY OF THE MAIN ECOLOGICAL IMPACTS LIKELY TO BE ASSOCIATED WITH THE DEVELOPMENT OF OIL RESERVES IN NORTHERN CANADA

A Impacts of development of the oilfield sites on North Slope:
- (i) disturbance of vegetation and soils in the uplands
- (ii) removal of gravel from streams, shorelines and ridges
- (iii) possible behavioural reactions of tundra animals to oilfield facilities and activities
- (iv) hunting and harrassment
- (v) oil pollution and waste disposal

B Impact of the Trans-Alaska pipeline:
- (i) possibility of pipeline breakage or leakage (massive oil spills)
- (ii) disruption of seasonal movement of ungulates
- (iii) severe silting of streams by continued erosion along the pipeline's right of way

C Impacts of the tanker terminal at Prince William Sound:
- (i) permanent rise in the population of Valdez (urban expansion and a range of pressures on wildlife and loss of habitat)
- (ii) oil spilled offshore (as ballast from tankers; from discharge of treated ballast at shore facilities; from infrequent but massive spills caused by tanker collisions; rupture of ships and store facilities during earthquakes; accidental or negligent mishandling of oil-loading equipment) leading to oil pollution and its ecological effects

Source: Weeden and Klein (1971)

islands like Majorca (Goldsmith, 1974b) – places greater stress on the formerly 'safe' rural habitats. In areas which have been subjected to the inroads of development over long periods, ecosystem adjustment to the human control factors might be possible, either by evolutionary adaptation of ecosystem structure and processes (see Chapter 4) or by successional adjustment (such as through secondary succession); but in cases where new and large-scale development has occurred for the first time, or where formerly localized developments have been replaced by large-scale resource exploitation or ecosystem destruction, the ecological consequences are difficult to predict in detail, but the fear is that the ecological resource base will be dramatically depleted. A recent example of the former is the development of oil reserves in northern Canada, where two critical series of impacts

(Table 6.3) will probably stem from the direct effects of exploration, development and shipping of the oil on wildlife populations and habitats, and from the growth in human populations and increased industrialization made possible by petroleum extraction (Weeden and Klein, 1971). A widely debated example of the replacement of formerly localized developments by larger-scale ones concerns the large-scale depletion of the Tropical Rain Forest habitat (Gomez-Pompa et al., 1972; Stott, 1978) which in its natural state is well adapted to the exacting equatorial climate and low soil fertility, and which is believed to maximize ecological productivity under the limiting conditions of the site. Brunig (1977) has expressed concern that interference with the structure of the tropical rain forest ecosystem may throw the system's equilibrium so much out of balance that the capacity for self-regulation and the ability to attain a new state of equilibrium may be lost.

6.3 PRESSURES ON ECOLOGICAL RESOURCES – THE BRITISH SCALE

Many of these types of direct and indirect pressures on ecological resources apply at the British as well as the global scale. Mellanby (1974) has reviewed the aspects of the changing environment of Britain which affect flora and fauna, including climatic change, suburban development, gravel and mineral exploitation, air pollution, water pollution, the production of new habitats by forestry practices, and changes in the character and impact of agriculture. Loss of habitat associated with land-use change is clearly a major source of pressure on ecological resources, and the Countryside Commission (1971) identified the need for a detailed and factual survey of landscape change in Britain to enable more authoritative statements on landscape conservation, and to aid in the evaluation of regional policies of grant-aid for conservation. Although the Commission itself carried out three pilot surveys between 1967 and 1969, the main 'Changing Countryside Project' was abandoned in 1969 because of high running costs, large investments of time required, and shortcomings of their chosen survey method. No subsequent nationwide countryside change survey has been attempted, and there is currently no centralized data base for evaluating changes in habitats. Each major form of habitat change tends to be evaluated in individual studies, and this makes it difficult to consider simultaneously a series of habitat changes within similar areas.

Dower (1964) identifies eight main functions of open land in Britain. These are agricul-

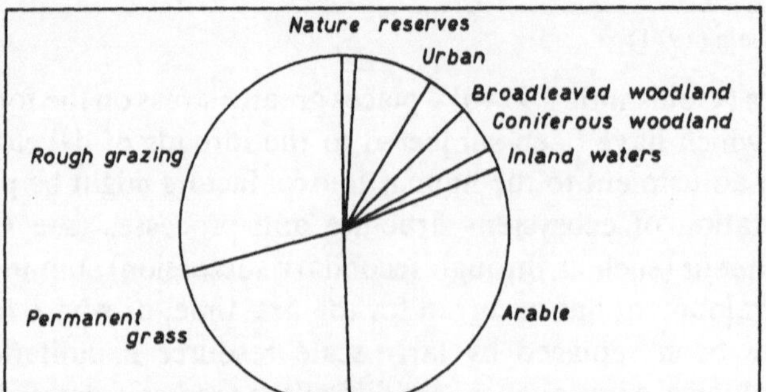

FIG. 6.1 THE PATTERN OF LAND USE IN GREAT BRITAIN (1977).
(After Nature Conservancy Council, 1977a.)

ture; forestry; water supply and minerals; recreation; landscape; employment and residence; urban limitation (green belts); and the urban reserve itself. Dower (1964) pointed out that in the future the balance between these conflicting demands might well change. The approximate distribution of land between various uses in 1977 in Great Britain is shown in Figure 6.1. Clearly a large proportion (over 75 per cent) of the land is given over to various agricultural uses (arable, permanent grass and rough grazing), and approximately equal proportions are occupied by urban areas and by woodland (coniferous plus broadleaved). Nature reserves occupy less than 1 per cent of the land area. The character and magnitude of the various pressures on ecological resources which arise from the various land uses can be evaluated by considering some of the more important land uses.

6.3a Urbanization

The expansion of urban land into formerly rural environments, especially during the present century, has caused the removal of numerous ecological habitats and the disruption of large numbers and wide varieties of ecosystems. Best (1976) has traced a rapid extension of urban land in Britain between 1920 and 1939, and a slowing down of the rate of growth after the introduction of effective planning control in 1947. By 1961 urban areas

Table 6.4
SUMMARY OF URBAN AREAS IN GREAT BRITAIN (1961)

Type of urban area	Area (10^3 ha)	Percentage of total urban area
London (Administrative County)	30	2
over 100000 population (excl. London):	312	19
10000–100000:	539	32
below 10000 (excl. isolated dwellings):	322	19
isolated dwellings:	159	9
transport land*:	326	19
TOTAL URBAN AREA:	1688	100%

Note: *includes roads, railways and civil airfields outside settlements.
Source: Best (1976)

of various sizes and forms occupied over 1·6 million ha (Table 6.4), and Best (1976) predicts the likelihood that by the year AD 2000 urban land will occupy about 14 per cent of the land surface of Great Britain. It now seems likely that urban growth has occurred mainly on above-quality farmland, and Swinnerton (1976) has demonstrated that in many areas the better quality farmland is very vulnerable to future urban growth, especially where there is a shortage of alternative land for development. Although the process of urbanization produces immediate ecological effects via the loss of species adapted to the existing natural habitats, Davis (1976) has pointed out that except in the most highly disturbed areas, natural recolonization of plant species takes place and so a new 'man-made' equilibrium form of ecosystem with birds, mammals, invertebrates and plants characteristic of parks, gardens and wasteland can be established. This in some ways compensates for the loss of the previously 'natural' ecosystems. Indeed, if ecological diversity is used as a measure of the ecological value of a habitat, many urban and suburban ecosystems may be of greater value than some natural ecosystems because the diversity of habitats within urban areas (because of differing management and maintenance strategies) could exceed the diversity of many relatively homogeneous natural ecosystems. These benefits are, however, often more than outweighed by the adverse environmental effects of pollution (of air, water and soil) associated with urban areas and with industrial activities which tend to concentrate around urban areas. There are also a number of environmental problems (of amenity, safety and reclamation potential, in particular) relating to derelict land, defined by Wallwork as 'land so damaged by industrial or other development that it is incapable of beneficial use without treatment' (Wallwork, 1976). Again, however, certain types of dereliction offer valuable habitats for some species, and they provide the first stages for secondary succession (see Chapter 4) to recolonize previously worked areas.

6.3b Mineral exploitation

Although large areas of the landscape are worked for mineral resources, Ratcliffe (1974) has stressed that there are both adverse and beneficial effects of this form of land use from the points of view of nature conservation, and pressures on ecosystems. The two main adverse effects are the direct destruction of habitats (such as that which occurs with the quarrying of important limestone areas; and with the mining of ironstone), and the production of chemical conditions which are unfavourable to terrestrial and aquatic plants and animals (such as the effects of lead mining in central Wales, particularly around the Rivers Ystwyth and Rheidol, during the last century). The main benefits relate to the creation of new or more varied habitats – including lakes and marshes (such as by the flooding of worked-out gravel pits, and of salt-mine subsidences); cliffs, screes and waste ground (in quarries and mine tips); and along old access routes and on walls and buildings. In essence the balance of conservation interest between the former 'natural' and the subsequent 'man-made' ecosystems will depend on the rarity of both the habitats and the individual species present in the two types of system. The situation can readily be envisaged whereby the 'man-made' post-exploitation ecosystem could have a higher conservation value (greater ecological diversity) than the original pre-exploitation ecosystems in the same area.

6.3c Agriculture

Farmers in England and Wales currently affect about 80 per cent of the land surface directly, although the area of agricultural land has contracted especially within the last decade (in harmony with increasing urbanization). Coppock (1970) has estimated that probably four million acres of lowland agriculture has been transferred to other uses since 1850, with over 50 per cent of this occurring since the 1920s. Ecologically this contraction of area is of less immediate significance than the fundamental changes in farming practices in recent decades (Cornwallis, 1969). Thus rotational systems have been replaced in lowland Britain by farming with an emphasis on arable cropping at the expense of livestock, with cereals (especially barley, often grown in monoculture) dominating. There has been an allied reduction in permanent grassland, coupled with drainage of wetland areas and extensive hedgerow removal. Moore (1977b) recognizes three distinct phases in this recent period of agricultural change:

(a) the inter-war period; characterized by scrub removal over thousands of acres, the turning over of many fields to grass leys, and the implementation of field drainage schemes,

(b) the 1950s and 1960s; when extensive hedge removal became common on most farms under arable or mixed farming regimes, and the use of pesticides in agriculture increased almost exponentially and their ecological effects were witnessed on both farmland (both 'target' species and other in the food webs of these were affected) and elsewhere (including nature reserves, because the spatial mobility of the pesticides was enhanced by movement along the food webs to a wide range of predatory species such as peregrine falcons),

(c) the present; characterized by (i) increased food production – agricultural food productivity is increasing at an estimated 2·5 per cent per annum, with a marked emphasis on sugar beet (through intensive cultivation and the use of toxic pesticides) and on dairy farming (through improved pasture by ploughing and reseeding, improved drainage, use of fertilizers and herbicides) (ii) economic pressures on farmers who find it increasingly difficult to justify (if only on economic grounds) the maintenance of non-productive land on the farm – even though it is these habitats which offer the best support for wildlife because most grassland is already improved and all arable land is treated with pesticides.

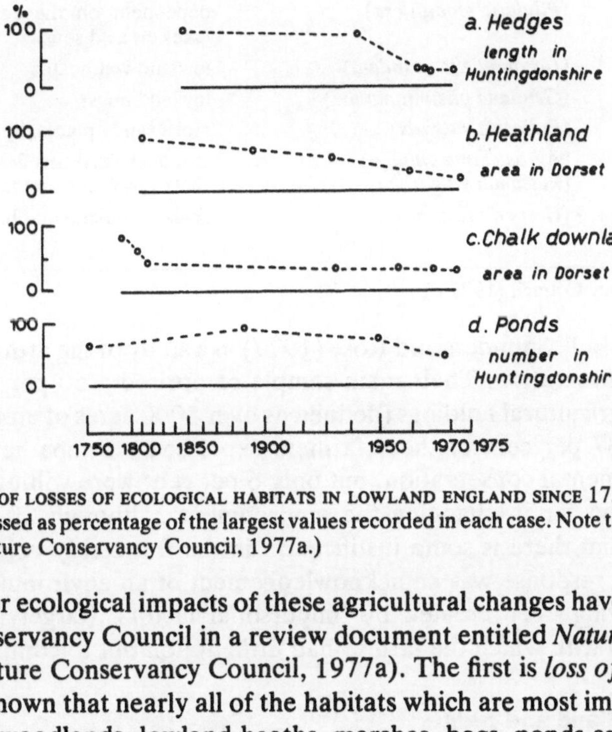

FIG. 6.2 EXAMPLES OF LOSSES OF ECOLOGICAL HABITATS IN LOWLAND ENGLAND SINCE 1750. All values are expressed as percentage of the largest values recorded in each case. Note the logarithmic time-scale adopted. (After Nature Conservancy Council, 1977a.)

The two major ecological impacts of these agricultural changes have been evaluated by the Nature Conservancy Council in a review document entitled *Nature Conservation and Agriculture* (Nature Conservancy Council, 1977a). The first is *loss of habitat*. Studies by the NCC have shown that nearly all of the habitats which are most important to wildlife – such as ancient woodlands, lowland heaths, marshes, bogs, ponds and unimproved pastures and meadows – have decreased in area in the present century. These have been complemented by significant reductions in the area of upland moorland and hedges (Figure 6.2). The second major ecological impact has been *reductions in the range of species* related to habitat loss. Moore (1977a) has commented on the contraction in the ranges of numerous plant and animal species recorded by the Biological Records Centre (see pages 142–3). For the 278 rarest English plants the decline in the number of 10 km squares in which they occurred between 1930 and 1960 was 30 per cent, and some of these (such as the Pasque Flower – *Pulsatilla vulgaris* – which has suffered the impact of the ploughing up and renovating of limestone pastures) have shown dramatic reductions in their geographical distributions to such a point that some species can now be regarded as 'endangered' (on the verge of extinction). Table 6.5 summarizes the reductions in the range of some species which can be related to habitat loss. In the light of these habitat losses the Nature Conservancy Council (1977a) has recommended that the farming community refrain from reclaiming the last remnants of good wildlife habitat still existing on their farms (areas such as small woods, old hedges and ponds), and it has stressed the need to encourage farmers to get their extra production and profits by farming the cropped land more intensively. The viability of these recommendations has been to some extent

Table 6.5
REDUCTIONS IN THE RANGE OF SOME SPECIES OF PLANT AND BUTTERFLY IN RELATION TO HABITAT LOSS

Species		Changing habitat factors
Pillwort	(*Pilularia globulifera*)	dependent on the edges of ponds and lakes on acid soils
Marsh Clubmoss	(*Lycopodiella inundata*)	lowland wet heaths
Marsh Gentian	(*Gentiana pneumonauthe*)	lowland bogs
Small Fleabane	(*Pulicaria vulgaris*)	moist sandy places, by ponds/streams
Corncockle	(*Agrostemma githago*)	unmodernized cornfields
Pasque Flower	(*Pulsatilla vulgaris*)	chalk downland and limestone pastures
Silver-spotted skipper butterfly	(*Hesperia comma*)	chalk downland and limestone pastures

Source: Nature Conservancy Council (1977 a)

evaluated by Newby, Bell, Saunders and Rose (1977) in a study of the attitudes of farmers in East Anglia to conservation. Their main sample comprised a 50 per cent sample of farmers of all large agricultural holdings (defined as over 1000 acres of crops and grass) in East Anglia. Some 87 per cent of these farmers expressed a sympathetic view to the problems of environmental conservation, but only 6 per cent were willing to accept that farmers had 'gone too far in changing the environment', although 20 per cent were prepared to accept that there is some justification in the frequently voiced criticism of farmers. The general response was an acknowledgement of an environmental problem, but one seen as having been created by 'impersonal factors' (largely economic and technological factors) with which the farmer had little option but to comply.

6.3d Changes in woodland and hedges

The total area of non-commercial woodland and of hedgerows in Britain has declined markedly within the present century, for five main reasons. One is that many of the trees in the landscape are very old, and many are dying simply through old age. With age a tree's resistance to disease, drought and other environmental pressures is reduced also. Many trees in Britain date back to the period of widespread plantings in the main Enclosure Movement (between about 1700 and 1870). A second factor is the modern trend in land management which affects trees (Nature Conservancy Council, 1974), especially those on farmland. Such factors as deep ploughing and improved land drainage (which both lead to lowering of the local water table), increased exposure for remaining trees after the removal of adjacent protective cover, and the prevention of natural regeneration amongst trees by the grazing of stock have all placed pressure on remaining tree resources. To these can be added the use of mechanical hedge trimmers and similar mechanical hedge-management techniques which prevent the survival of saplings and the natural regeneration of hedgerow species. A third factor is the extrinsic removal of hedgerows because of the need to enlarge field sizes to accommodate large modern farm machinery, and the recent emphasis on arable farming methods with minimal stock and zero-grazing techniques (Council for the Protection of Rural England, 1971). Fourthly there are increased incentives for farmers to clear deciduous woodland for conversion to agricultural use which are

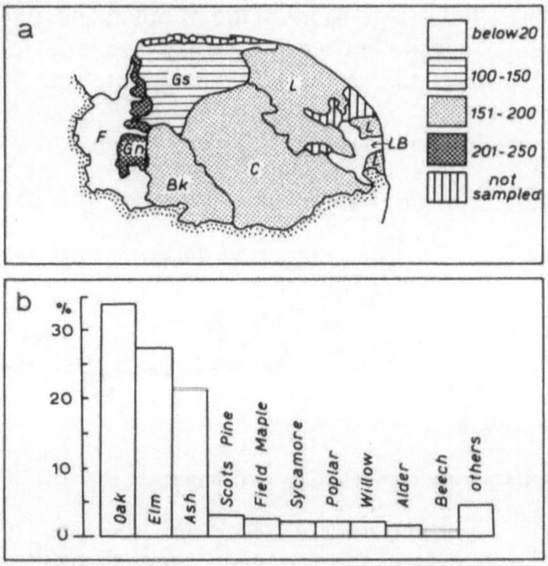

FIG. 6.3 SOME RESULTS FROM THE 'FARMLAND TREE SURVEY IN NORFOLK' (1977).
(a) Average number of trees per km² in different landscape types.
Key to landscape types:

Bk	Breckland	C	Central	F	Fen
Gn	Greensand	Gs	Goodsands	LB	Lower Broads

(b) Relative proportions of the tree species within the county. (After Hardy and Mathews, 1977.)

occasioned by increased land values and by the need to increase agricultural output (and hence provide additional income for re-investment in the farms). A further factor has been the extensive outbreaks of tree diseases in recent years (Edward, 1977), the most catastrophic being that affecting Dutch elms. Other diseases have affected beech, ash, weeping willows, Norway spruce and London plane trees in recent years.

The extent of these losses of trees and hedgerows in one area is highlighted in the results of a recent 'Farmland tree survey in Norfolk' organized by Norfolk County Council (Hardy and Mathews, 1977). The field survey of the status of the populations of different tree species within the county revealed a distinct lack of variety in hedgerow tree species. Three species (oak, elm and ash) accounted for over 80 per cent of the trees sampled

(Figure 6.3b), and many of the existing trees were found to be over-mature and likely to die. A quarter of all the trees were either stag-headed or over-mature. Some 8000 trees were estimated to be lost annually from hedgerows, although the losses were uneven between areas and between tree species (oak, for example, was affected worse than other species). The survey estimated that an annual planting programme of around 40 000 trees would be required in Norfolk alone to halt the decline in tree numbers.

Extensive woodland clearance has been recorded for West Cambridgeshire by Harding (1975). Some 17 per cent of the woodland present in 1946 had been cleared by 1973, and the clearance had proceeded in two main periods. One was immediately post-war and the other has occurred since the early 1960s mainly for agricultural use (particularly arable cultivation). Just over 1 per cent of the clearance was associated with some form of development.

Hedgerows can be considered as extensions of the woodland habitat in that they provide corridors for the migration of species between areas, and reservoirs for the preservation of many woodland plants. However, as Caborn (1971) has pointed out, there are a number of

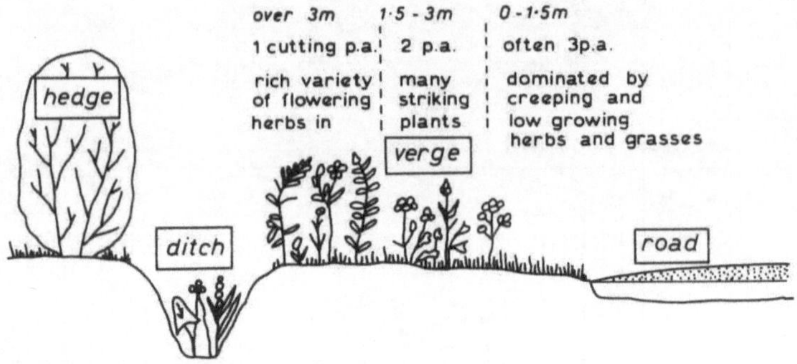

FIG. 6.4 RANGE OF MICRO-HABITATS COMMONLY FOUND ON ROADSIDE VERGES. (After Perring, 1967.)

agricultural benefits to be had from the removal of unwanted internal hedges. These include the extra workable land so obtained; savings on hedge maintenance; increased machine operating efficiency; and savings on wasted overlap of distributed seed, fertilizers, etc. On this basis an annual rate of hedgerow loss in Britain as a whole of about 7000 km between 1946 and 1970 is perhaps not surprising. On the other hand there are substantial conservation benefits to be derived from preserving hedgerows. Caborn (1971) has calculated that – assuming an average width of 2 m for the 950 000 km or so of hedgerows left in Britain – hedgerows represent an area of potential habitats for birds, insects and small mammals (as well as plants) which is greater than the area presently given aside to National Nature Reserves.

The widespread removal of internal hedges in fields has placed increasing value on the remaining hedges as wildlife habitats – in fact the last major refuges for many of the species whose former agricultural habitats have since been removed. Roadside verges, especially those bordered by hedgerows and ditches, offer valuable habitats for wildlife, and Way (1970) has stressed that about a third of the national length of hedges in Britain are now associated with roads. Perring (1967) has noted the richness of the flora and fauna of roadside verges, in that of the 1700 or so flowering plants and ferns which are either native to, or naturalized in, this country, over 700 occur on verges, and 50 are more common on verges than elsewhere. Also some 20 British mammals (out of a total of 50), 6 out of 6 reptiles, 40 out of 200 birds, 26 out of 60 butterflies and 8 out of 17 humble bees breed along roadways. This richness relates in part to the fact that many roadside verges act as refuges to species in the wake of agricultural changes, and in part to the diversity of zones and habitats within many verges because of the management practices employed. Thus the frequency and timing of cutting of roadside verges affects the successional development and diversity of micro-habitats within verge areas (Figure 6.4). Verge and hedgebank management generally involves the use of chemical herbicides and cutting machines, neither of which are favourable to wildlife; and although the value of roadside habitats for conservation has been questioned because of the direct dangers from traffic movement, Way notes that

> there are no indications that kills of individual animals on roads are in any way significant when related to the total population of the species in the area. It can be argued that it is better to have two animals killed on the road and eight others surviving in the habitat along the road, than no habitat and no animals (Way, 1970).

6.3e Recreation and amenity pressure

There are a range of ecological problems relating to the effects of visitors in the countryside, the use of certain areas for recreation and amenity, and the management of amenity habitats. Goldsmith (1974c) has pointed out that the ecological effects depend largely on the sensitivity of the vegetation, which is in turn a function of the position in the successional sequence (Figure 6.5), the proportion of the ground covered by tolerant

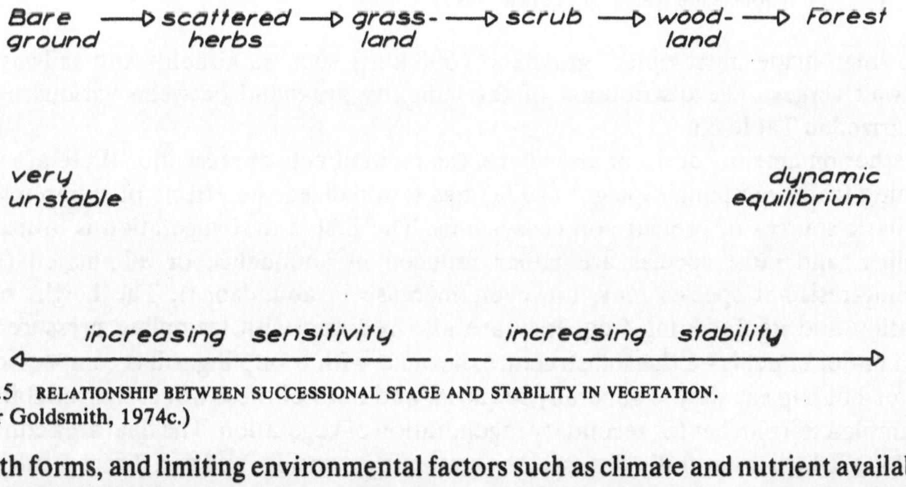

FIG. 6.5 RELATIONSHIP BETWEEN SUCCESSIONAL STAGE AND STABILITY IN VEGETATION. (After Goldsmith, 1974c.)

growth forms, and limiting environmental factors such as climate and nutrient availability. Inevitably the habitats most favourable for recreation and amenity use tend to be those devoid of shrubs and trees (for ease of access, safety, etc.), and yet these successional stages are sensitive to ecological pressure (Figure 6.5).

There are basically two types of habitat which are adversely affected by recreation and amenity use – those areas designated or created specifically for that purpose (such as amenity grassland), and the countryside in general. The Natural Environmental Research

Council (1977) has recently surveyed all amenity grassland (defined as 'all grass with recreational, functional or aesthetic value and of which agricultural productivity is not the primary aim') in Great Britain, and they classified the 8500 km^2 into four basic groups of use. One was 'semi-natural' (4100 km^2), such as National Trust land, nature reserves and common land. The second group was 'intensively managed' (1100 km^2) grassland such as school playing-fields and golf fairways. Thirdly there was 'man-made or trampled' grassland (2700 km^2) typical of urban parks and domestic lawns; and these were complemented

Table 6.6
SUMMARY OF MAIN TYPES OF AMENITY GRASSLAND IN BRITAIN

Types of grassland	Area in 1976 (km^2)
Bowling and golf greens:	20
Urban parks, sports grounds, etc.:	2020
Ornamental lawns:	20
Golf fairways:	350
Domestic lawns:	900
Road verges (District Council):	250
Outfields and airfields (Armed Services):	450
Road verges and similar (County Council):	1310
Caravan sites:	60
National Trust land:	920
Civil airfields:	110
Motorway verges:	60
Country Parks and Nature Reserves:	690
Common land:	530
Railway embankments:	200
Golf rough:	500

Source: Natural Environmental Research Council (1977)

by the 'man-made untrampled' grassland (600 km^2) such as airfields and railway and motorway verges. The distribution of this amenity grassland between various uses is summarized in Table 6.6.

Whether on amenity areas or elsewhere, the main effects of recreational use are those stemming from trampling. Speight (1973) has rationalized the effects of trampling into three basic sources of pressure on ecosystems. The first is that vegetation is bruised by trampling, and most species are either reduced in abundance or eliminated (some trampling-resistant species may, however, increase in abundance). The height of the vegetation and its flowering frequency are also reduced with trampling pressure. The second major effect is on the soil structure, because with trampling soil is compacted and its water-holding capacity is generally reduced, and both of these affect the suitability of the trampled soil patches for secondary regeneration of vegetation. The final area concerns animal life, which is naturally disturbed by trampling invasion, and as a direct result some species decline in number or move elsewhere. The relationships between these effects are illustrated in Figure 6.6, together with suggestions by Liddle (1975) of the actual ecological processes involved in trampling. For site management and ecosystem conservation purposes it is generally useful to be able to express the ability of a site or habitat to withstand trampling and recreational pressure, and attempts have been made to derive quantitative indices of site sensitivity by controlled experiments. Thus Cieslinski and

Wagar (1970) proposed an 'Index of Site Durability', derived from measuring the effects of controlled trampling (simulated with a corrugated concrete roller) at different intensities, on the amount of vegetation remaining within a habitat. Values of the index for different sites could then be related to site factors such as the soil clay content, pH, tree cover, slope, aspect and altitude. Alternatively Liddle (1973) derived an 'Index of Habitat Vulnerability', again based on controlled intensity walking on a previously unworn area of the habitat to be evaluated. This index is based on the number of passages the vegetation

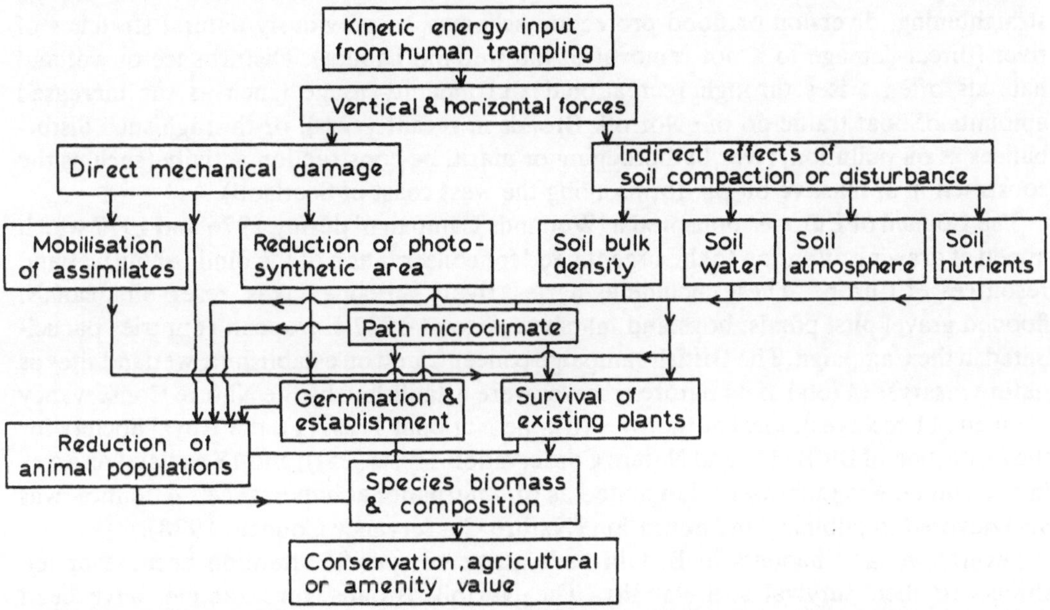

FIG. 6.6 MODEL OF SOME OF THE ECOLOGICAL EFFECTS OF TRAMPLING ON SOIL AND VEGETATION.
The relationships are considered to be causal, but not all of the same nature. (After Liddle, 1975.)

can withstand before it is reduced to 50 per cent of its original cover or biomass. Clearly the time distribution of the trampling pressure is critical for successful site management. For example Slater and Agnew (1977) have reported that over 60 per cent of the damage to the surface of a raised peat bog on the west coast of Wales (Cors Fochno) was caused during a few weeks of intense use over the Easter period, when 40 per cent of the annual total of 1890 visitors were on the site. Although the ecological damage can be inflicted within short periods of time, ecological recovery through natural secondary succession can be a lengthy process. Bell and Bliss (1973) noted from trampling experiments on alpine communities in Olympia National Park in the United States that plant cover and plant production can be rapidly reduced in even one season of light use, yet they suggest that it could take up to 1000 years for natural regeneration of the habitat.

6.3f Pressures on wetland resources

Recent years have witnessed a growing concern for the need to conserve and manage wetland resources because of the pressures on the stability and survival of many wetland areas. These pressures include *pollution*, *loss of habitat* and *habitat disturbance*. Pollution of wetland habitats derives from agricultural (fertilizer and pesticide sources in particular), industrial and domestic sources. Many environmental pollutants are soluble and so they can be dispersed throughout entire drainage systems downstream from the points of origin or emission. Thus lakes are particularly susceptible to pollution problems (see Chapter 7). Loss of habitat has arisen because many of the recent agricultural changes affect aquatic habits either through drainage of agricultural land (which lowers local water tables and can lead to the disappearance of shallow lake and pond habitats), through the intentional reclamation of pond and depression-storage areas for agricultural uses, and through straightening, diversion or flood protection schemes on previously natural stretches of river (direct damage to if not removal of the wetland habitat). Disturbance of wetland habitats often arises through recreational and amenity usage (such as the increased amounts of boat traffic on the Norfolk Broads in recent years), or through such disturbances as oil pollution, river bed dredging or maritime construction activity (such as the construction of massive oil-platforms along the west coast of Scotland).

The Council of Europe sponsored a 'Wetlands Campaign' during 1976 and 1977 which aimed at drawing attention to the urgent need for conservation of the diminishing wetland resources of Europe. These include estuaries, rivers, streams, lakes, reservoirs, canals, flooded gravel pits, ponds, bogs and marshes. A total of 22 European countries participated in the Campaign. The British campaign concentrated on establishing wetland sites as nature reserves (a total of 44 nature reserves were established by the Nature Conservancy Council (11 reserves), local authorities (6), the National Trust (1), the Royal Society for the Protection of Birds (11) and Nature Conservation Trusts (25)); and 13 wetland reserves in the United Kingdom were designated as of 'international importance'. Attention was also devoted to publicity and education (Nature Conservancy Council, 1978).

Several wetland habitats in Britain have attracted specific attention because of the threats to their survival and stability. The Norfolk Broads, for example, have been seriously affected by agricultural activities and recreational use, to such an extent that co-ordinated protection measures similar to those currently available to National Parks have been requested (Elkington, 1977). George (1976) has evaluated the main ecological pressures on the Broads. Agricultural pressures derive from the leaching of fertilizers applied to farmland which leads to eutrophication of the water bodies; from small-scale conversion of marshland to arable use; and from improved land drainage which produces a direct reduction in the size of wetland habitat. Tourism and recreation pressures appear to be not serious across the Broads as a whole, although in some favoured tourist locations the pressures on ecology can be very marked. However only 16 of the 42 Broads are available for public navigation – the rest are too shallow or separated from the main rivers by unnavigable canals, and these can thus act as refuges for birds which are intolerant of disturbance. Since the early 1950s the Broads' former fame for diversity of wildlife and submerged aquatic plant communities has declined (especially the latter), and now only 4 of the Broads have more than remnants of the original sub-aquatic macrophyte flora and its associated invertebrate fauna. Moss (1977) has concluded that the increased turbidity in the waters of the Broads, which brought about the loss of macrophytes, is largely a function of increased nutrient loadings from human activities in the catchment area of the Broads. Boat disturbance does not appear to contribute significantly to the sustained turbidity and its ecological consequences.

A second example of threatened wetlands is the Somerset Levels which comprise a series of alluvial and peat marshes bordering the Bristol Channel, and which offer the last large area of wetland habitat in lowland Britain (Harvey, 1977). There are currently two main threats to the survival in their natural state of the Levels – *the peat-winning industry* and *improved agricultural drainage*. The peat-winning industry produces huge open excavations and effectively destroys the wetland habitat rapidly and irreversibly. Some 1500 acres are currently being worked, and in 1977 consent was available for peat cutting in an additional 1200 acres. In addition, over 1200 acres of the Levels have been improved for agriculture with the expansion of dairying and grassland farming. The three main agricultural impacts have been a general lowering of the water table, intensification of animal husbandry and widespread use of herbicides and fertilizers. Water table lowering threatens to eradicate all true wetland plants and to diminish the value of the area as a migration centre for wading birds which feed in winter on soil invertebrates which lie near to the soil surface because of the high water tables. On the whole, intensive animal husbandry tends to support few cultivated species and persistent 'weeds', so this leads to loss of ecological diversity and reduction in the number and variety of micro-habitats available within the local ecosystems. Herbicides and fertilizers serve in many cases to damage natural meadow vegetation, and they generally pollute waterways within the system. The Nature Conservancy Council (1977b) – sensitive to these threats – has reviewed the use and conservation of the Somerset Levels, and evaluated five main options for future management of the resource:

OPTION 1 *Limit wildlife conservation*: continue to manage existing nature reserves, but do not establish new ones.

OPTION 2 *Control agricultural improvement and peat extraction*: not entirely realistic in an area which relies heavily on both of these activities for employment and income.

OPTION 3 *Offer protection for 'key areas'*: protect 'key areas' in an attempt to preserve the best areas of different habitats.

OPTION 4 *Improvement of derelict land*: reclaim worked-out peat cuttings for conservation use.

OPTION 5 *A land-use strategy*: a rational strategy agreed upon by all land owners and managers, and all interested parties, which aims to reconcile the conflicts of land use within the resource.

The Nature Conservancy Council has widely advocated the need for a land-use strategy in recent years – in the contexts of recent agricultural changes and their impact on conservation, of the Somerset Levels dilemma, and elsewhere – and the implications and implementation of such an integrated approach to land-use planning will be considered in Chapter 8.

This concern for wetland habitats in their entirety has been complemented by concern for individual wetland species. There is evidence, for example, of a substantial reduction in the numbers and range of otters in Britain since the early nineteenth century, and of a further decline since the early 1960s. Although there are a wide range of pressures on otter populations which could be responsible for the recent decline (Table 6.7), the recently established Joint Otter Group (with representation from the Nature Conservancy Council, the Society for the Promotion of Nature Conservation, the Institute of Terrestrial Ecology and the Mammals Society) was forced to conclude that 'there is no conclusive evidence to indicate the direction or extent of trends at present, nor is it possible to isolate any

Table 6.7
PRESSURES ON THE STABILITY AND SURVIVAL OF OTTER POPULATIONS IN BRITAIN DURING THE 1970s

Human disturbance:	of water and water banks, for recreation, exercise, etc.
Hunting:	direct effects = killing; indirect effects = disturbance. Killing is now a minor factor (5 otter kills were reported for 1976 compared with *c*. 200 per annum before the 1960s)
Riparian clearance:	reduction or loss of stream-bank vegetation cover
Pollution:	lethal and sub-lethal effects of organochlorine insecticides, industrial PCBs and heavy metals
Disease:	unknown but possible
Increased road casualties	
Severe winters:	such as the winter of 1962–3
Increased competition:	from increased mink populations for winter food
Impact of fisheries:	such as otter control for fishery protection
Trapping and killing for pelts	

Source: NCC/SPNC Otter Group (1977)

particular environmental factor which is demonstrably dominant in determining present day otter numbers' (NCC/SPNC Otter Group, 1977).

6.4 Beneficial man-made environments

Although each of these types of activity places pressure on ecological resources through trampling, loss or reduction of habitat, modification of habitat, direct persecution of species and the like, there are a number of situations in which 'man-made' environments can be beneficial to wildlife in general, and to specific species in particular. Not all developments are thus bad from an ecological point of view. Davis (1976) has outlined some ways in which wildlife can adapt to and exploit man-made environments, because – except in the most highly disturbed areas – recolonization and secondary succession is possible, and new forms of dynamic equilibrium can be established within the man-made or man-controlled ecosystems. Indeed some industrial developments offer a wide range of habitats to wildlife (Table 6.8) and Kelcey (1975) has outlined some of the ecological values of such industrial habitats. First, many linear habitats (such as canals and roadways) serve as wildlife corridors in much the same way as hedgerows do; and Twigg (1959), for example, has described the various habitats for plants and animals along a number of disused sections of the Shropshire Union Canal. Secondly, many of the discrete sites are well protected (ironically often better so than nature reserves) and so they can provide safe areas for rare and unusual (as well as common) species to survive and reproduce close to if not actually in urban areas. Thus many flooded gravel pits near London now provide habitats for wading birds. Thirdly, individual sites and linear habitats form centres from which wildlife can colonize new habitats, or to which biota can retreat. Messenger (1968) has described the railway flora of Rutland, which was ecologically quite distinct because here the plant communities were protected from the destructive influences of modern agricultural practices, and because they were subjected to a range of 'unique' controlling influences conditioned by railway practices. His data show that railways were one of the richest botanical areas in the county, with 372 species of flowering plants found on railway properties (with a total area of 4 km^2) compared to an average of 260 species (the range was from 174 to 391) found in samples from a 4 km^2 grid across the entire county. Finally

Kelcey (1975) points out that extractive industries often provide suitable conditions and opportunities for many species of plants to colonize, as well as breeding sites for birds which nest in areas devoid of vegetation. The potential value of allowing natural processes of vegetation succession and animal colonization to 'restore' areas affected by mining, quarrying, building and dumping has been considered by Hutchinson (1974). He concluded that although these natural processes will eventually 'restore' derelict and industrially-damaged land, in countries such as Britain with land shortages, damaged areas cannot be left for sufficient periods of time to heal normally. He thus stresses the need to ensure suitable conditions under which the natural processes of succession will most effectively and speedily repair the damage to the ecosystems.

Table 6.8
MAIN WILDLIFE HABITATS ASSOCIATED WITH INDUSTRIAL DEVELOPMENT

DISCRETE SITES	
(a) Extractive	(i) Clay pits – wet
	– dry
	– spoil-heaps
	(ii) Gravel pits – wet
	– dry
	(iii) Sand pits – wet
	– dry
	(iv) Chalk quarries
	(v) Other limestone quarries
	(vi) Mineral workings
	(vii) Quarries in igneous rocks
	(viii) Quarries in sedimentary rocks (other than chalk and limestone)
(b) Industrial *per se*	(i) Industrial waste
	(ii) Flashes (or subsidence lagoons)
	(iii) Mill races
	(iv) Grounds of industrial complexes
(c) Miscellaneous	(i) Sewage works
	(ii) Sludge lagoons
LINEAR SYSTEMS	
(a) Canals (used and disused) including the verges, bridges, locks and lock gates	
(b) Railways (used and disused) including the verges, ballast, bridges, walls, platforms, etc.	
(c) Roads	(i) Ancient trackways
	(ii) Classified and unclassified road-verges
	(iii) Motorways

Source: Kelcey (1975)

In this chapter attention has focused primarily on the reasons for interest in ecological resources, and the various ways in which pressures are currently being placed on remaining ecological resources at several scales of study. Clearly the direct and indirect pressures which result from changing land use and habitat modification or removal are complemented by the pressures which result from environmental pollution. These can similarly have both direct and indirect effects; and pollution is of basic concern because of the mobility of most environmental pollutants (via the water cycle and the chemical elements cycle of the Biosphere – see Chapter 2). The effects of pollution on ecosystems and on ecological resources in general will be considered in Chapter 7.

7
Pollution

The significant aspect of human action is man's total impact on ecological systems, not the particular contributions that arise from specific pollutants (MIT, 1970).

Although pollution is not an entirely recent area of concern, within the last two decades there has been a marked growth of interest in all aspects of environmental pollution. General reviews of the scale and sources of pollution (such as those by Barr, 1971; Ward and Dubos, 1972; Bugler, 1972) have been complemented by studies of the relationships between decision making and pollution control (Porteous, Attenborough and Pollitt, 1977); and the recent 'energy debate' has focused attention on the environmental effects of both established and novel sources of energy (Chapman, 1975; Patterson, 1976; Foley,

Table 7.1
SOME POLLUTION JOURNALS CURRENTLY BEING PUBLISHED

Air Pollution Abstracts
Air Pollution Control Assoc. Jnl
Air & Water Pollution Report
Atmospheric Environment
Biological Conservation
Clean Air
Critical Reviews of Environmental Control
Effluent & Water Treatment Journal
Environmental Health
Environmental Pollution
Environmental Pollution Management
Environmental Research
Environmental Science & Technology
International Biodeterioration Bulletin
International Journal of Environmental Studies
International Pollution Control Magazine
Journal of Environmental Planning & Pollution Control
Journal of Environmental Management
Journal of the Water Pollution Control Federation
Marine Pollution Bulletin
Pesticide Science
Pollution Abstracts
Pollution Monitor
Science of the Total Environment
Water, Air & Soil Pollution
Water & Pollution Control
Water Pollution Control

1976) which include particulate and heat pollution. A guide to the strength and diversity of this interest lies in the variety of journals currently devoted to different aspects of environmental pollution (Table 7.1); and one measure of the variable history of general public concern over pollution issues is the number of letters to the *New York Times* on the subject (Figure 7.1).

There are two aspects of recent findings concerning the scale and nature of environmental pollution that have aroused particular concern. One is the realization that pollution is now a problem of global dimensions. This has been convincingly demonstrated, for example, by the 1972 expedition of the Scripps Institute of Oceanography (University of California) to the central North Pacific Ocean. Within an 8·2 hour period of observation at a location over a thousand kilometres away from the nearest major civilization (Hawaii)

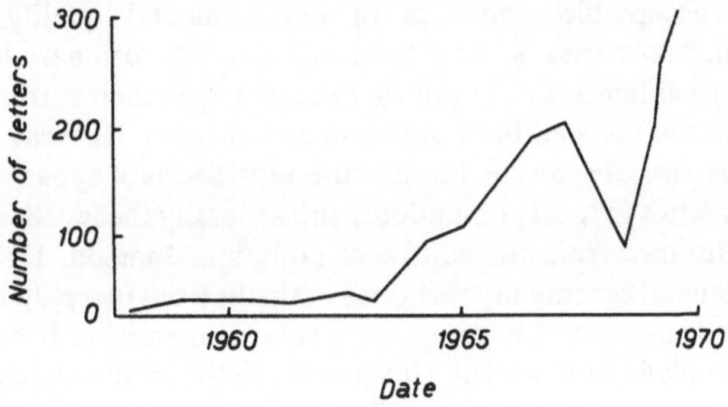

FIG. 7.1 CHANGING STRENGTH OF PUBLIC CONCERN OVER ENVIRONMENTAL POLLUTION.
The number of letters to the *New York Times* regarding pollution. (After Munton and Brady, 1970.)

outside major shipping lanes, a total of 53 man-made objects were sighted bobbing on the ocean. These included 28 fragments of, or whole, plastic bottles, a coffee can and a rubber sandal (red). Venrick and his colleagues (1973) predicted, on the basis of these observations, that there are probably around 35 million plastic bottles, and nearly 6 million red sandals, currently afloat in the North Pacific Ocean. The second aspect for concern is growing awareness that pollution is a 'vicious circle' situation, because beyond a certain critical level (or range) pollution reduces the assimilative capacity of ecosystems, and it thus reduces their ability to withstand further pollution. Ironically, the scale of some forms of pollution-creating activities which is judged to be optimal in economic terms (on the basis of cost/benefit analysis) is sometimes larger than the scale which might be considered 'ecologically optimal', and so environmental deterioration can readily and rapidly arise. In cases where the economically optimal scales of activity exceed those levels which are critical to ecological stability, this 'vicious circle' of pollution–environmental degradation–reduced assimilative capacity–further environmental degradation can quickly generate. Subsequent stabilization of the environmental degradation, let alone recovery, is often difficult and generally costly.

7.1 THE CONCEPT AND NATURE OF POLLUTION

7.1a Perspectives on pollution

The problem of pollution can be viewed in a variety of ways. One is that *perception* of pollution has improved in recent years, and awareness of the scale and repercussions of environmental pollution has heightened. A range of factors might be responsible for this, including better education and information programmes, and improved technological advances which have increased the accuracy and scope for measuring low levels of pollution. Unwin and Holtby (1974), for example, found high levels of awareness of air pollution control measures amongst residents interviewed in Manchester (England). These factors of awareness and technology have been complemented, however, by increasing concern over acceptable standards of environmental quality, particularly in industrially-developed countries. Kneese has concluded that 'ordinary folk have come to expect standards of cleanliness, safety and wholesomeness in their surroundings that were the exclusive province of the well-born or rich in earlier times' (Kneese, 1977). Pollution can also be viewed as a *social problem*, inasmuch as most forms of apparent pollution cause annoyance or disturbance in local populations, and generally the levels of disturbance are closely related to distance from the source of pollution. Jonsson, Deane and Sanders (1975) found that general community reactions to smells from two pulp mills at Eureka in northern California decreased with increasing distance down-wind from the mills. Some 74 per cent of the people interviewed by Unwin and Holtby in Manchester felt that smoke control had made the area a better place to live in. Pollution can also be thought of as an *economic problem*, because both pollution and the environmental disruption it causes are seen by economists as *externalities*. Hardin (1968) has pointed out that many human activities tend to internalize benefits but externalize a significant proportion of the costs involved. He also stressed the propensity to engage in such activities because all of the benefits accrue to the individual who initiates the action, whilst the adverse consequences are broadly shared. Finally, pollution can be viewed as an *ecological problem*, because many pollutants in the environment (in the air, in water and on land) are present in quantities which prove to be lethal for many species of plants and animals, and sub-lethal doses of many pollutants can have widespread and lasting effects on the physiology of organisms and on the biogeochemical cycles within ecosystems. Sub-lethal concentrations may exist undetected for long periods of time, and yet they can affect species and individuals in a variety of ways – such as through growth efficiency, reproductive success and general competitive ability. These ecological effects include health impairment and fatalities in humans either from exposure to environmental pollution or through occupational exposure to harmful materials and compounds.

Kneese (1977) has isolated two factors which have stimulated the emergence of pollution as a centre of major concern in economics, and these are equally applicable to the rise of general interest in pollution. First is the massive increase in industrial pollution and energy conversion in recent decades, which lead to very large flows of materials and energy from concentrated states in nature to degraded and diluted states in the environment. These have led in particular to changes in the physical, chemical and biological status of the atmosphere and the hydrosphere (both of which are of basic importance to the overall stability of the biosphere – see Chapter 2). The second factor is the increasing introduction of exotic materials into the environment. These are materials to which natural systems cannot adapt (or at least they cannot adapt quickly), or adaption may occur in some species but not in others. The balance between species can thus be affected, which can in turn trigger off latent instability in the ecosystems affected by the materials.

7.1b Definitions of pollution

Because of the differing viewpoints on pollution, there are a variety of ways in which pollution can be defined. Thus, for example, the economist might adopt Cottrell's (1978) definition of pollution as 'the consumption of environmental quality', whilst the scientist might favour the definition offered by the Natural Environmental Research Council (1976) that it is the 'release of substances and energy as waste products of human activities which results in changes, usually harmful, within the natural environment'. Lee and Wood (1972) have isolated three criteria to be used in defining pollution. One is that it must result from a particular form of human activity – the disposal of wastes. A second is that it occurs where the disposal of wastes leads to damage (whether direct or indirect, and over the short or long terms). They maintain that it is also restricted to those circumstances where the effect of the damage is met by third parties (it is not self-inflicted, or inflicted on employees). Lee and Wood (1972) also point out that although it is convenient to classify pollutants according to the environmental medium through which they are principally diffused (air, water – fresh and marine – and land), these media are all inter-related and so the conventional format of classification is rather arbitrary.

Lord Kennett has more specifically defined pollution as 'the presence at large of substances, or energy patterns, which have been involuntarily produced, have outlived their purpose, have escaped by accident, or have unforeseen effects, in quantities which harm his [man's] health or do offend him' (quoted by Robinson, 1973). A similar, though ecologically more relevant, definition is offered by Dasmann, who views pollution as 'the accumulation of substances, or forms of energy, in the environment in quantities, or at rates of flow, which exceed the capacity of ecosystems to either neutralize or disperse them to harmless levels. Pollutants are not necessarily harmful in themselves' (Dasmann, 1975). He quotes the example of human faeces, which can be regarded either as a pollutant or as a useful fertilizer. If distributed properly on agricultural soils they enter into biogeochemical cycles and enrich crop yields. However, when accumulated in areas where large numbers of people live, they become obnoxious pollutants and a source of disease. It is implicit in Dasmann's (1975) definition that a critical level exists above which the substance is regarded as a pollutant, and below which it is not. This theme is pursued by Robinson (1973) who distinguishes between a *contaminant* ('the material cause of any deviation, local or general, from the mean geochemical composition'), and a *pollutant* (where 'the quantity of the contaminant is sufficient to affect man or other organisms in an adverse manner'). A different perspective on this same problem is given by the Massachusetts Institute of Technology in the *Study of Critical Environmental Problems*. It concludes (MIT 1970) that

> *Residuals*, or *waste*, are generated in all stages of the production and consumption of goods or services. *Residuals* become *pollutants* or an environmental problem of some kind and in some

degree when they have harmful effects in the atmosphere, the oceans or the terrestrial environment. *Harmful effects* are effects that are harmful to man, or to animals, plants or inanimate objects or conditions that are important to man. Their importance to man may be biological, economic, religious, moral, aesthetic or intellectual.

7.1c Sources and types of pollution

In the Massachusetts Institute of Technology (1970) viewpoint, pollution represents a harmful accumulation of residuals from the production and consumption of goods and services. The study concluded that 'man produces more than a million kinds of products both as waste and as useful products that eventually end up as wastes' (MIT, 1970), and it specified a series of *key pollutants* whose global effects are such as to make it particularly important to bring them under satisfactory control (Table 7.2). Of the million or so

Table 7.2
KEY POLLUTANTS IDENTIFIED BY THE MIT *STUDY OF CRITICAL ENVIRONMENTAL PROBLEMS* (1970)

carbon dioxide; particulate matter; sulphur dioxide; oxides of nitrogen; toxic heavy metals (lead, mercury, arsenic, chromium, cadmium, nickel, manganese, copper, zinc); oil; chlorinated hydrocarbons (especially DDT and polychlorinated biphenyls (PCB)); other hydrocarbons; radionuclides; heat; nutrients

products, many are known to be lethal, or highly toxic, to organisms; and many others are known to be relatively harmless. In between lie a wide range of materials with varying degrees of toxicity. Dasmann (1975) has expressed concern that the long-term effects of low-level doses of various toxic materials are not known, and that for many chemicals now being produced and released into the biosphere (such as the polychlorinated biphenyls) neither the short- nor the longer-term environmental effects are known with any degree of certainty.

As an example of the wide variety of pollutants now being issued, the United States National Academy of Sciences National Research Council has identified eight broad types of pollutant entering watercourses (Table 7.3) which include heat from power plants, sediment from land erosion and radioactive materials. Kaminski (1976) has identified a variety of sources of air pollutants in Europe, and these include large heating installations, coal and oil power plants; domestic fuel systems; refuse collectors; metallurgical works (such as steel and iron works); the chemical industry; the petrol industry; coking plants; street traffic and air traffic.

Table 7.3
SUMMARY OF MAIN TYPES OF POLLUTANTS ENTERING WATERCOURSES IN THE UNITED STATES

(1) Domestic sewage and other oxygen-demanding wastes
(2) Infectious agents
(3) Plant nutrients
(4) Organic chemicals which are highly toxic at very low concentrations (such as insecticides, pesticides, detergents and petro-chemicals)
(5) Minerals and chemicals (including chemical residues, salts, acids, silts and sludges)
(6) Sediments from land erosion
(7) Radioactive substances
(8) Heat from power and industrial plants

Source: Myslinski and Ginsburg (1977)

With a multiplicity of sources and types of pollutants, it is logical to expect that environmental pollution is a problem of large-scale and wide-ranging spatial dimensions. To take the example of air pollution, few areas of England and Wales are completely free over long periods of time from air pollution stemming principally from urban areas. Indeed, Barnes (1976) has demonstrated that emissions of pollutants can produce measurable concentrations of smoke and sulphur dioxide in country districts over 100 km away; and Meetham (1950) estimated that about 0.8 million tons of smoke and 3.0 million tons of sulphur dioxide are deposited annually over Britain. There are inevitably marked contrasts in air pollution levels between urban and rural areas, and Robson (1977) concluded that although nitrogen dioxide pollution from transportation is much higher (both per capita and in total) in urban areas, air pollution in general over city areas is lower than rural areas because of the increased dispersion afforded by the aerodynamically 'rough' city surface. Allied to an increasing output of gaseous pollutants from industrial areas, there is commonly a slow drift of gaseous pollutants between urban and rural areas, and this can be manifest in the spatial distribution of precipitation reaction (acid rain). Moss (1975) has pointed out that often there is quite a marked change in pH at about the general position of the rural/urban fringe, which marks the change from precipitation influenced by many urban and industrial emission products to one influenced by the slow drift and conversion of atmospheric gases and their subsequent removal. The precipitation chemistry at a rural (unpolluted) site near Lake Windermere in North West England has been examined by Peirson, Cawse, Salmon and Cambray (1973), who found evidence of the presence of over 30 trace elements, the highest concentrations being in chlorine, sodium and calcium. A *washout factor*, defined as the ratio between the concentration of an element in the rainwater and the concentration in air, was derived to characterize the extent to which trace elements are washed out of the atmosphere. Values for sodium were 2900, for chlorine 2300 and lead 450.

To these problems of urban air pollution and acid precipitation can be added other air pollution problems such as the atmospheric effects of high-flying aircraft such as Concorde (Harries, 1973); and the effects on the atmosphere of chloro-fluorocarbons (the propellant solvents for aerosol dispensers) (Lovelock, Maggs and Wade, 1973). The latter are represented mainly by two compounds (CCl_2F_2 and CCl_3F), both of which are very stable chemically and only slightly soluble in water. Fears have been expressed that they might persist and accumulate within the atmosphere. One of the compounds (CCl_3F) is entirely anthropogenic in origin, and yet the gas was found in air and sea samples wherever and whenever it was sought on the voyage of the *RRS Shackleton* on a journey from the United Kingdom to Antarctica, and back, during 1971–2.

7.2 ENVIRONMENTAL EFFECTS OF POLLUTION

Although, as considered earlier, there are a number of perspectives on the nature of pollution problems, Woodwell (1970) has summarized these into two complimentary viewpoints.

> Practical people – toxicologists, engineers, health physicists, public health officials, intensive users of the environment – consider pollution primarily as a direct hazard to man. Others, no less

concerned for human welfare, but with less pressing public responsibilities, recognize that toxicity to humans is but one aspect of the pollution problem, the other being a threat to the maintenance of a biosphere suitable for life as we know it. The first viewpoint leads to emphasis of human food chains; the second leads to emphasis on human welfare insofar as it depends on the integrity of the diverse ecosystems of the earth . . . (Woodwell, 1970).

7.2a Human health effects

Examples of the associations between environmental pollution and human health and welfare include the deterioration of health related to increased duration and magnitude of exposure to sulphur dioxide (Figure 7.2), and the inclusion of 'health effect descriptions' in the 'Pollutant Standards Index' defined by the United States Federal Inter-Agency Task Force on Air Quality Indicators (Table 7.4). Knowledge about such associations is of

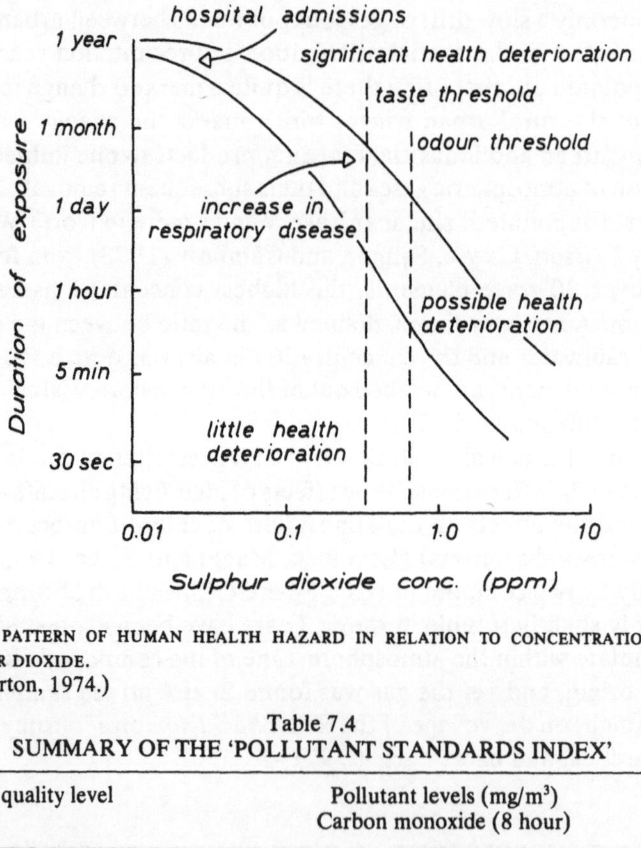

FIG. 7.2 CHANGING PATTERN OF HUMAN HEALTH HAZARD IN RELATION TO CONCENTRATION AND DURATION OF EXPOSURE TO SULPHUR DIOXIDE.
(After Berry and Horton, 1974.)

Table 7.4
SUMMARY OF THE 'POLLUTANT STANDARDS INDEX'

Index value	Air quality level	Pollutant levels (mg/m^3) Carbon monoxide (8 hour)	Health effect
500	Significant harm	57·5	
400	Emergency	46·0	Hazardous
300	Warning	34·0	Very unhealthful
200	Alert	17·0	Unhealthful
100	National Ambient Air Quality Standard (NAAQS)	10·0	Moderate
50	50% of NAAQS	5·0	Good

Source: Crossland (1978)

central concern to human health and welfare, because of the increasing recognition that the amount of morbidity and mortality for a number of specific diseases (such as bronchitis, lung cancer and stomach cancer) and pneumonia can be related to air pollution (Lave and Seskin, 1970); and that occupational exposure to some biologically persistent pollutants (such as farmers using fungicides or chlorinated pesticides) can be reflected in the concentrations of these pollutants in blood samples and body organs (Lunde and Bjørseth, 1977). This human perspective on pollution problems appears to be in some senses a problem of awareness and perception, however, because Crowe (1968) found that few respondents in his study of public response to air pollution defined air pollution in health-related terms. Most considered it more of a nuisance, defining air pollution in terms of bad odours and/or reduced visibility.

7.2b Ecological effects

Clearly, as Woodwell (1970) pointed out, the human welfare perspective is only one side of the pollution problem. It is convenient to consider health impairment as a direct effect of pollution; but, as Munn, Phillips and Sanderson (1977) suggest, there are other important direct effects, and yet other equally important indirect effects on the environment (Table 7.5). These latter include the various ecological effects on the structure and behaviour of

Table 7.5
SUMMARY OF THE MAJOR DIRECT AND INDIRECT EFFECTS OF AIR POLLUTION

A. DIRECT EFFECTS
- (i) impairment of health
- (ii) damage to vegetation
- (iii) damage to animals
- (iv) damage to materials
- (v) socio-economic effects (such as increased laundry bills, etc.)
- (vi) aesthetic effects

B. INDIRECT EFFECTS
- (i) Changes in ecosystem structure and behaviour:
  - (a) elimination of sensitive species, leading to reduced species diversity
  - (b) selective removal of large over-storey plants in favour of smaller types
  - (c) reduced standing crop of organic matter, thus also reduced amounts of nutrient elements stored in the ecosystem
  - (d) enhancement of activity of insect pests and diseases with increased mortality
- (ii) Changes in solar radiation
- (iii) Acidification
- (iv) Salinization
- (v) Climatic change

Source: Munn, Phillips and Sanderson (1977)

ecosystems – such as reduced species diversity, vegetation change, reduced standing crop and increased disease and pest activity. As evaluated in Chapter 3, the ecosystem provides a valuable conceptual framework for considering the stability and equilibrium relationships between plants and animals. Some of the specific benefits of an ecosystem approach to pollution studies, as identified by Moss (1976), include the ability to integrate all known environmental components and pathways; provision of a framework for measuring nutrient movement through the biosphere; opportunity for establishing data on ambient levels for many environmental pollutants; and a framework for delimiting the spatial aspects of pollution. Regier and Cowell (1972) have forwarded the view that pollution is a form of stress that deforms ecosystem structure, and reverses the usual sequence of ecological succession with respect to the dominance of various species, diversity, stability and production. They also make a valuable distinction between two types of pollutant – the presence of materials that serve essentially as plant nutrients, and the presence of materials that are directly toxic or deleterious to the organisms involved. An example of the former type of pollution is where nutrients enter a lake system in streamflow, and opportunist species of phytoplankton become important. Some of these (such as the blue-green algae) are poisonous to larger animals, and they lead to reduced diversity and stability through mortality of predatory species in the lake system. Blooms of these opportunists can 'burst out' at rates far in excess of the population increases of zooplankton herbivores, but not the decomposers (protozoa and bacteria). The combined effects of oxygen depletion over a diurnal range (because of phytoplankton respiration) and decomposer activity is reduced availability of oxygen to extents which are detrimental to larger animals. This forces a reversion of the natural succession back to more primitive states, in which the ecosystem has low diversity, low stability-persistence and low stability-resilience (see pages 128–33). Poisonous pollutants generally act discontinuously through time, except in such cases as where poisons near to outlet points are always sufficiently concentrated as to be lethal, and so the effects of pollution of ecosystem stability will be related to the frequency of occurrence of the poisons. With elements which are emitted perhaps once per week it is possible for organisms with short life-spans to flourish briefly. This can lead to large-scale population fluctuations. Emissions once or twice per month can create a successional sequence with relatively few long-lived organisms but stable populations overall. With infrequent emissions (perhaps once a year) in an aquatic ecosystem, for example, succession might reach the stage of medium-sized fish (especially if the migration of fish into the area is relatively free, and the environment is a warm one) (Regier and Cowell, 1972).

It is convenient to consider several aspects of the relationship between the quantity of poisonous pollutants present, and ecological response. One important relationship is that between the susceptibility of an organism and the concentration of a poisonous pollutant. Bliss (1935) has evaluated this relationship in the form of a 'Dosage Mortality Curve', which illustrates the 'normal frequency distribution of the variation among the individuals of a population in their susceptibility to a toxic agent' (Bliss, 1935). The generalized form of the curve is shown in Figure 7.3, and it is clearly similar in concept to the tolerance response curves considered in Chapter 3. The curve (Figure 7.3) shows that a range of concentrations of a poisonous pollutant may have lethal effects on the organisms involved (whether or not 'target species' or 'innocent by-standers' are involved is not relevant here), although there is an optimal concentration with maximum impact. Studies of environmental radioactivity and radioactive pollution have suggested two further concepts which are of value. If a 'primary standard', or 'Acceptable Dose Limit', can be defined for a pollutant (defined as a dose which has either no or acceptable short/long-term effects on human health or on plant and animal species into which the pollutant comes in contact), then the 'Environmental Capacity' of that pollutant can be defined as 'the rate of introduction of a pollutant which, at equilibrium, will result in a rate of exposure of the target per unit time equal to that defined by the primary standard' (Preston, 1977).

Clearly, from an environmental management point of view, it is essential that the Acceptable Dose Limits and Environmental Capacities of all of the major environmental pollutants specified in Table 7.2 are established and defined, and that pollution control programmes are formulated around these data-bases. Problems arise, however, because although an Acceptable Dose Limit can be defined for one species and for one pollutant, it is likely that the Limits will vary between species for a given pollutant because of the variability of response of different organisms to the same environmental factors (see Chapter 3). Inevitably, perhaps, pollution control programmes will tend to be based on human-orientated Acceptable Dose Limit values. Lack of complete understanding of the effects of sub-lethal concentrations of many particulate poisonous pollutants also leads to difficulties, because the long-term effects of prolonged exposure to relatively low (sub-lethal) concentrations of many pollutants have not, as yet, been fully identified or evaluated.

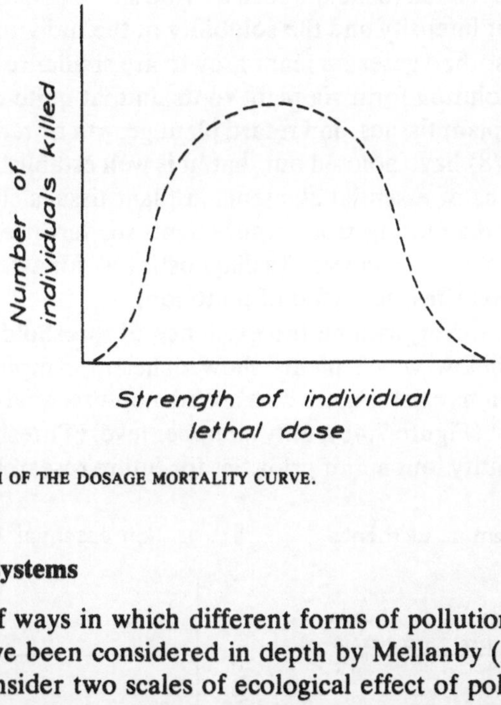

FIG. 7.3 GENERALIZED FORM OF THE DOSAGE MORTALITY CURVE.
(After Bliss, 1935.)

7.2c Pollution and ecosystems

There are a number of ways in which different forms of pollution can affect plants and animals, and these have been considered in depth by Mellanby (1967) and others. It is convenient here to consider two scales of ecological effect of pollution in general – the effects on individual organisms and effects on ecosystems. The latter naturally provide an integration of all of the individual effects on organisms present within the ecosystem, but the systems viewpoint also allows identification of aggregate effects, and the ecosystem scale of study is convenient for long-term environmental management.

Individual organisms

A number of effects of pollution on individual organisms relate to air pollution by industrial gas emissions. In many areas sulphur dioxide (SO_2) is released into the atmosphere where it is oxidized and hydrated, and it returns to the ground with precipitation in the forms H_2SO_3 or H_2SO_4. This acidification of rainfall can affect all species with which the acid rainwater makes contact. Thus, for example, Grodzinska (1977) showed a good correlation between acidification of tree bark and air pollution by SO_2 in the Cracow Industrial Region of South Poland. Many species of plants can respond to air pollution by physiological changes; and Sharma (1977) has identified a number of ways in which plants can change their leaf structure so as to decrease the amounts of poisonous gases getting into the leaf tissues. Plant protection against pollution arises through changes in stomatal frequency, stomatal size, trichome length, type and frequency, and in the subsidiary cell complex. Despite these physiological protective devices, plants can and do absorb varying quantities of atmospheric pollutants; and, indeed, the role of plants as 'sinks' for atmospheric pollution is critical in maintaining the balance of both local ecosystems and the global scale biosphere. Hill (1971) has attempted to evaluate the role of vegetation as a sink for many gaseous air pollutants under controlled laboratory conditions, by measuring the removal of gases from the atmosphere by an alfalfa canopy. Rates of absorption by plants were shown to be related to factors such as wind speed above the plant, height of the vegetation canopy, light intensity and the solubility of the individual pollutants in water. The effects of these absorbed gases on plant growth are similar to the effects of pollutant elements absorbed in solution form via plant roots, in that quite commonly the material will accumulate within plant tissues, and retard plant growth or reduce plant productivity. Davis and Beckett (1978) have pointed out that 'it is well established that plant growth is related to concentrations of essential elements in plant tissues closely enough for plant analyses to be used for diagnosing crop requirements for fertilizers'; and they have thus evaluated the use of plant tissue analysis as a diagnostic tool for assessing the seriousness of metal toxicity to crops on sites suspected of pollution.

The method does, however, assume the existence of threshold concentrations of elements in plant tissue below which plants show deficient symptoms, and above which optimum growth conditions exist (Figure 7.4a). With non-essential elements a threshold of deficiency does not exist (Figure 7.4b), only an upper level (Threshold of Toxicity). Davis and Beckett (1978) identify four major priorities for future research on the uptake of toxic

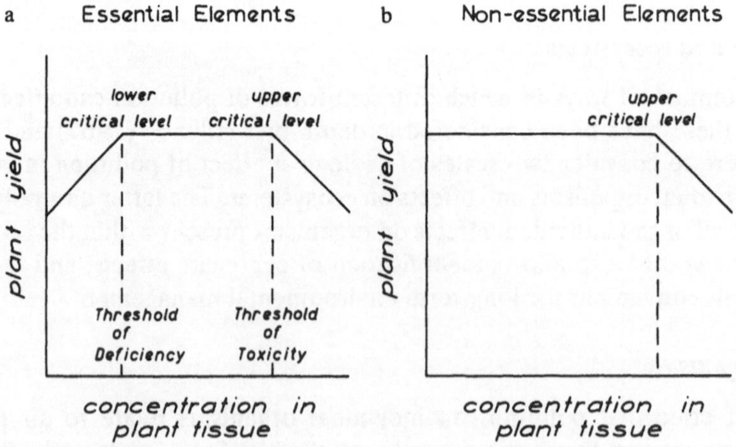

FIG. 7.4 RELATIONSHIPS BETWEEN PLANT YIELD AND CONCENTRATION OF ELEMENTS IN PLANT TISSUES.
The relationships are shown for essential (a) and non-essential (b) elements. (After Davis and Beckett, 1978.)

elements by plants. These are confirmation that the upper critical levels are constant under varying environmental conditions; establishment of the upper critical concentrations for all relevant potentially toxic elements; comparison of upper critical concentrations in different plant species; and investigation of the possible effects on plant growth of high concentrations of more than one potentially toxic element in plant tissues.

Ecosystems

The effects of pollution of ecosystem structure and stability have been considered in a number of reviews, including those by Woodwell (1970), Patrick (1972) and Regier and Cowell (1972). Because similar pollutants can affect different species in different ways, ecosystem response will depend largely on its structure and species composition. Species composition in an ecosystem may change because of pollution, because in some species reproductive success is inhibited by pollution, whereas other species may tolerate or even benefit from pollution (if the pollutant was formerly a 'limiting factor' – see Chapter 3). The age structure of the ecological populations in the ecosystem may also change, because the effects of pollution may be differential between different age groups with varying tolerance levels and abilities to recover. Thus the impacts of pollution may be more marked amongst very young and very old members of a population. Higher levels of pollution might trigger off more dramatic changes in ecosystem structure, because those species that are specialists (often quite sensitive to small-scale environmental changes) will tend to disappear, and although total population size may remain the same if opportunist or generalist species increase in numbers as a compensation (through loss of competitive stress for food, habitats, etc.), the number and diversity of species present in the ecosystem will generally decrease. Species which are tolerant of pollution can greatly increase in numbers as a result of increasing nutrient levels in the system, and/or if pollution reduces pressure from predators. Further increases in pollution can lead to severe reductions in the numbers of species present, and thus a simplification of the number of food pathways through the community (see Chapter 3). This in turn would increase the susceptibility of the ecosystem to further change, because of reduced stability within the system. The presence of highly toxic pollutants may reduce the entire reproductive capacity of the ecosystem, and lead to an overall decrease in biomass. Table 7.6 summarizes the main ecological effects of pollution on ecosystems.

Perhaps the most dramatic type of example of ecosystem response to pollution concerns *eutrophication* of aquatic ecosystems. Patrick (1972) has pointed out that pollution affects the efficiency of transfer of nutrients and/or energy in aquatic systems, and often an increase in certain species of blue-green or green algae leads to accumulation of massive standing crops of these algae. Eutrophication occurs when nutrients are imported into a lake from the upstream drainage basin. The nutrients can be from domestic, industrial or agricultural sources; and frequently the most acute problems arise when all three sources are providing nutrients in large quantities simultaneously (such as in the Great Lakes in Canada, and in Lough Neagh in Northern Ireland). Bartsch (1970) defines eutrophication as 'the nutrient enrichment of waters which frequently results in an array of symptomatic changes, amongst which increased production of algae and other aquatic plants, deterioration of fisheries, deterioration of water quality, and other responses, are found objectionable and impair water use'. The three least welcomed ecological symptoms identified by Bartsch (1970) are the massive blooms of phytoplankton, large-scale growth of attached algae, and unwanted rooted macrophytes; and Rawson (1956) has summarized the main differences in plankton conditions between natural (oligotrophic) and eutrophic lakes

Table 7.6
SUMMARY OF MAIN ECOLOGICAL EFFECTS OF POLLUTION ON ECOSYSTEMS

Before pollution	After pollution
complex arrangements of specialist species	generalist species common
forest ecosystem	hardy shrubs and herbs
phytoplankton in open ocean	algae of sewage plants
diversity of species	monotony of species
integrated nutrient cycles	loose nutrient cycles; terrestrial cycles depleted, aquatic ones overloaded
stable ecosystem	unstable ecosystem

Source: Woodwell (1970)

(Table 7.7). Odum (1969) has observed that the net effect of the various impacts of eutrophication on ecosystem structure and functioning is to push back the ecosystem in successional terms, to a younger stage. The corollary, as stressed by Odum, is that lakes can and do progress to more oligotrophic conditions when nutrient input from upstream slows or ceases; but this self-repair through successional development can only happen if eutrophication has not proceeded to such a point that the regenerative capacity of the lake ecosystem has been reduced below a critical lower functioning level.

Table 7.7
MAIN DIFFERENCES IN THE PLANKTON CONDITIONS OF OLIGOTROPHIC AND EUTROPHIC LAKES

	Oligotrophic	Eutrophic
Quantity	poor	rich
Variety	many species	few species
Distribution	to greater depth	only shallow distribution
Diurnal migration	extensive	limited
Water-blooms	very rare	frequent

Source: Rawson (1956)

In summary, therefore, environmental changes through pollution can lead to changes in the composition of an ecological community. A pollutant might eliminate many species from an ecosystem, and those that remain could become abundant through decreased competition. The introduction of toxic substances into the environment often leads to reductions in species numbers and increases in the number of individuals of the same species. More serious toxic elements can result in complete reductions of species numbers in most groups; and even more severe toxic effects can lead to death of all organisms. This suggests, therefore, that the presence or absence of certain species might be valuable in providing a guide to environmental conditions, especially as a rapid reconnaissance means of detecting pollution in an ecosystem.

7.3 POLLUTION MONITORING

The basic approaches for monitoring environmental pollution are the same for both water and air pollution. Williams and Dussart (1976) classify water-quality assessment procedures into three broad groups – chemical, bacteriological and biological ones. The *chemical methods* focus either on the hardness or the chemical status of the water sample, or on its oxygen status. Hardness and chemical content are direct measures of the amount of chemical pollutants present in the aquatic ecosystem. The significance of oxygen is perhaps not so immediately apparent. Dissolved oxygen is present in water as a result of photosynthesis by aquatic species, or diffusion from the atmosphere; but it can be depleted by organic pollution, especially in sluggish streams and standing water. Mabey (1974) has highlighted the two main sources of organic pollution in water bodies. These are in the form of complicated chemicals containing carbon, which are important parts of living matter (such as fats, proteins, etc.), and synthetic substances such as detergents and insecticides. Natural streams have a relatively high capacity to cope with natural waste material provided to them, because this material is broken down into simple substances by bacteria and fungi in the water. These oganisms, however, need oxygen, and they use it up in the process of conversion. Normally the oxygen used in waste disposal is replaced from one or both of two sources – seepage through the surface of the water body, and oxygen given off in respiration by submerged green plants during daylight. With heavy pollution and large amounts of organic waste in the water, oxygen can be used up faster than it is replaced. When this happens the breakdown of waste matter in the water slows down, leading to an accumulation of wastes, and encouraging the immigration into the ecosystem of bacteria which can work without oxygen. These bacteria break down the waste material, but in doing so they release foul-smelling gases (such as methane, ammonia and hydrogen sulphide). Oxygen levels in the water are thus both important controls, and valuable indicators, of the chemical status of the equatic ecosystem. Two indices of aquatic oxygen are widely used. The *Dissolved Oxygen* level (DO) is a measure of the free oxygen content of the water, and it is measured in mg/l. The oxygen saturation levels in natural waters is normally about 10–11 mg/1 (at a normal temperature of about 10°C). A Dissolved Oxygen level of about 8 mg/1 (80 per cent saturation) can support most natural life; below 5 mg/1 the water is seriously polluted and aquatic life is significantly affected (Sutton, 1975). *Biochemical Oxygen Demand* (BOD), measured in mg/1 also, is a measure of the demand for oxygen (from aquatic fauna and decomposers) in a water body, and it is normally derived by measuring oxygen consumption in sealed samples of water incubated for 5 days at 20°C. Both the chemical content and the oxygen levels approaches to chemical determination of pollution have been extensively used, but Price (1978) has isolated a series of inherent limitations behind the approaches, at least when applied to aquatic ecosystems (Table 7.8).

Bacteriological methods of pollution monitoring are based on the association between density of coliform bacteria and degree of organic pollution. Health and sanitation studies often use bacteriological methods of pollution assay. Thus, for example, Regnier and Park (1972) have considered the evidence on faecal pollution of beaches in Britain, and they conclude that the lack of generally accepted bacteriological standards for bathing waters in the United Kingdom can be attributed to two main difficulties in the formulation of reliable standards. One is that there is as yet no epidemeological evidence that disease is transmitted by sea bathing; but the main problem is that many factors affect the numbers of coliform bacteria recovered on successive sampling days for individual sites, so that generalizations are difficult to arrive at.

Table 7.8
SUMMARY OF THE MAIN LIMITATIONS BEHIND CHEMICAL METHODS OF MONITORING AQUATIC POLLUTION

(1)	It is not practical to monitor all determinants which may be relevant to protection of fisheries and other sections of fauna, or to various other uses which the river system might serve
(2)	Routine analytical methods may not be sufficiently sensitive to measure reliably low concentrations of pollutants which may be significant in ecological terms
(3)	The significance of different concentrations of many substances is not adequately understood
(4)	Combinations of pollutants normally encountered in the field may produce effects which are different from those observed for individual constituents
(5)	Chemical monitoring procedures based of regular snap samples may fail to show occasional, but significant, deteriorations in water quality

Source: Price (1978)

Biological assessment of pollution is based either on the presence or abundance of 'indicator species', or on the overall diversity of species within the ecosystem. There are a range of approaches to biological assessment – indeed, Williams and Dussart (1976) outline 17 different methods in general use. The *indicator species* approach has been widely used, and Wielgolaski (1975a) has stressed the three major advantages of this:

(a) the approach is relatively inexpensive; it can provide information for a large number of sites (it is not just restricted to localities with fixed or expensive monitoring stations),

(b) it can be used to distinguish the accumulative effects of pollution (in that it integrates ambient 'background' levels with various transient episodes of pollution),

(c) it can show the synergistic effects of several pollutants acting simultaneously.

Wielgolaski (1975a) does stress, however, the caution necessary in interpreting the occurrence or absence of single species. The soundest evidence of changed environmental conditions stems from observing the response of groups of species simultaneously. Perhaps the main benefit of the indicator species approach is its simplicity, because it is based largely on identification of types of organism, rather than individual species. Indeed, the *ACE/Sunday Times Clean Air and Water Surveys* (Mabey, 1974) were implemented in the field by 12- to 14-year olds, and the indicator species approach (Table 7.9) for both aquatic and atmospheric environments proved to be valuable in collecting field data from which a general study of the spatial patterns of environmental pollution in Britain was possible.

Table 7.9
INDICATOR SPECIES OF WATER AND AIR POLLUTION USED IN THE *ACE/SUNDAY TIMES CLEAN AIR AND WATER SURVEYS*

Zone	Pollution state	Indicator species
WATER POLLUTION		
A	clean water only	stone-fly nymph; mayfly nymph
B	tolerate slight pollution	caddis-fly larva; freshwater shrimp
C	quite bad pollution	water louse; bloodworm
D	tolerates bad pollution	sludgeworm; rat-tailed maggot
AIR POLLUTION		
0	badly polluted	*Pleurococus* lichens on trees
1	quite bad pollution	crusty *Lecanora* species at base of trees, on walls, concrete and paving stones
2	relatively little pollution	*Xanthoria* on alkaline stone
3	little pollution	Leafy lichens (*Parmelia*) on walls, *Lecanora* and *Pleurococus* on trees
4	relatively clean air	Leafy lichens (*Parmelia*) on trees
5	air almost free from pollution	shrubby lichens (such as *Ramalia* species) on trees
6	very clean air	shrubby (*Usnea*) lichens widespread

Source: Mabey (1974).

7.3a Aquatic indicator species

An interesting example of the use of biological assessment of pollution is the study by Landner, Nilsson and Rosenberg (1977) of the biological composition of fifteen different regions along the Baltic coast of Sweden, affected mainly by organic waste pollutants from the pulp and paper industry. They identified three broad 'pollution zones' on the basis of species present and absent:

ZONE 1: a heavily polluted zone – restricted in extent, all macrofauna had been wiped out and overall biomass in the system was low.

ZONE 2: a transitional zone – characterized by reduced bottom fauna, only very tolerant species were found, and these only in small numbers.

ZONE 3: a slightly polluted zone – characterized by a mass occurrence of opportunist species and some mobile animals, and considerable bio-stimulation of some resistent species.

Natural fish populations appear to be important general indicators of water quality, especially when used in the arena of public relations. Thus the Thames Water Authority (1977) has recently reported that the restoration of the tidal River Thames (previously severely polluted) in recent decades has been marked by a dramatic return of fish since about 1965. The recent reappearance of the salmon in the tidal Thames has been widely publicized. Price (1978), however, has concluded that although fish toxicity tests are widely used for statutory effluent water quality control, natural fish populations are less suitable for providing detailed scientific assessment of water quality than other biotic species. Macro-invertebrates (generally bottom-dwelling organisms (*benthos*), and consisting largely of insects, crustaceans, annelids, roundworms, flatworms and molluscs) appear to be valuable pollution indicators. Myslinski and Ginsburg (1977) identify four main values of these types of indicators:

(a) their ability to retain contaminants (such as pesticides, radioactive materials and trace elements),

(b) their relatively large size makes identification less of a problem than with smaller species,

(c) their limited mobility restricts the organisms to a specific environment,

(d) their life-span of months/years means that they provide evidence of responses to conditions that have prevailed over reasonably long periods of time.

The presence of certain pollutants in the aquatic ecosystem brings about a decrease in the concentration of dissolved oxygen in the water, and this (coupled with the collection of sludge deposits) affects the types of macro-invertebrates which can survive in the habitat.

Table 7.10
SOME POLLUTION-TOLERANT AQUATIC MACRO-INVERTEBRATES

Mosquito (*Culex pipiens*) Rat-tailed maggot (*Tubifera*)	} both live in thick sludge, in conditions of low dissolved oxygen
Damsel fly	
Dragon fly	
Diving beetle (*Dytiscus*)	adapted to a low oxygen environment
Certain midge larvae and red-blooded chironomids (*Chironomus riparius*)	
Some crustaceans (e.g. *Asellus*)	associated with water recovering from pollution
Aquatic earthworms (*Tubifex* and *Limnodrilus*)	abundant in polluted environments where food is abundant, and competitors intolerant of low oxygen concentrations are absent
Leeches (e.g. *Trocheta subvirdis*, *Erpobdella octoculata*)	
Molluscs (e.g. *Physa integra*)	

Source: Myslinski and Ginsburg (1977)

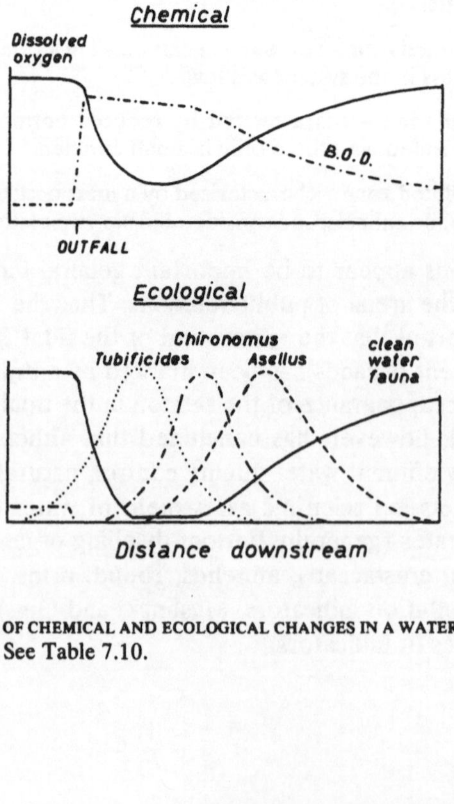

FIG. 7.5 GENERALIZED PATTERN OF CHEMICAL AND ECOLOGICAL CHANGES IN A WATERCOURSE, DOWNSTREAM FROM AN OUTFALL OF ORGANIC EFFLUENT. See Table 7.10.

Pollution-tolerant species that indicate degeneration of environmental quality (Table 7.10) dwell in areas of low oxygen content (they have on the whole evolved specialized structures for obtaining oxygen). Those which are intolerant of pollution indicate clean water conditions, and they are absent in reaches with low levels of dissolved oxygen. Because of this, there are commonly marked spatial patterns of distribution of animal populations along watercourses, related to the presence of organic pollutants and to the varying tolerances of different macro-invertebrates to survive in such habitats (Figure 7.5).

An alternative to the use of macro-invertebrates as water pollution indicators is the use of algae or phytoplankton. There are special problems inherent in using phytoplankton as indicators, however, and these have been evaluated by Rawson (1956). The main problem is that although there are a large number of algal species which are characteristic of eutrophic lake systems, there are relatively few which are distinctly characteristic of the oligotrophic state.

Most studies of water pollution using the indicator species approach are based on samples taken from within the water body. Matson, Horner and Buck (1978) have advocated the use of samples taken from river sediments, because of the higher recovery rate there for most individual organisms. Their study in the Shetucket River Basin in Connecticut revealed that for all of the micro-organisms they considered, higher concentrations were found in sediments than in the water itself. Indeed, they found sufficient supplies of biological indicators in the sediments to create a potential health hazard under existing standard indicator densities.

A variety of characteristics of indicator species have been employed in the biological assessment of pollution. Four principal ones have been used.

Density of indicator species

Wright and Tidd (1933) used the total density of aquatic earthworms as a measure of degree of pollution. Densities of below 1000 per m^2 were regarded as evidence of 'negligible pollution'; between 1000 and 5000 per m^2 as evidence of 'mild pollution'; and 'severe pollution' was characterized by densities in excess of 5000 per m^2.

Relative abundance

As an alternative, Goodnight and Whitley (1960) used the abundance of aquatic earthworms relative to the abundance of other benthic organisms. A ratio of over 80 per cent was taken as indicative of a 'highly polluted' condition; 60 to 80 per cent was regarded as 'doubtful'; and below 60 per cent was taken as indicative of a clean condition.

Presence/absence of indicator organisms

Beck (1955) suggested an approach based on the presence or absence of different types of species of known tolerance to pollution. He derived a 'biotic index' of the form $BI = 2$(n class 1) + (n class 2), where n is the number of macro-invertebrate species in either class 1 ('tolerant of no appreciable organic pollution') or class 2 ('tolerant of moderate organic pollution but not of conditions nearly anaerobic'). A BI value of 0 indicates a stream nearing a polluted state; values of between 1 and 6 indicate streams with moderate amounts of organic waste; and values approaching or exceeding 10 indicate very slightly polluted streams.

Ecological diversity

This approach is based on the assumption that relatively unpolluted environments are characterized by a large number of species, with no individual species present in overwhelming quantities. Thus Wilhm and Doris (1968) suggested a diversity index based on richness of species and distribution of individuals amongst the species, of the general form

$$d = \Sigma \frac{ni}{n} Log_{10} \frac{ni}{n}$$

where ni is the number of individuals per taxon, and n is the number of taxa in the sample of the community. Clean water is assumed to have a high diversity index value, because there will be a great number of species each represented by a large number of individuals. Polluted waters would have few species or one individual present. If $d = 0$, all of the organisms present belong to the same species; if d is less than 1 then there is heavy organic pollution; values of between 1 and 3 indicate moderate pollution; and a value in excess of 3 indicates clean water conditions.

7.3b Lichens as pollution indicators

One of the most illustrative approaches to biological assessment of pollution in terrestrial habitats is that based on lichen species. Considerable interest has centred around the species of lichens present in polluted areas, and on the physiological effects of particulate pollution on lichens (Hawksworth and Rose, 1976; Ferry, Baddeley and Hawksworth, 1973), and the value of lichens as pollution monitors has been evaluated in a number of areas. Most attention has focused on the relationships between sulphur dioxide levels in the atmosphere and lichen distributions because, as Hawksworth (1974a) has pointed out, sulphur dioxide limits the occurrence of particular lichen species, it reduces fertility in most species, it reduces overall lichen luxuriance, and it reduces the ability of lichens to colonize new surfaces.

An interesting early example of lichen studies was described by Fenton (1960), who was able to generalize about the distribution of lichen species growing on trees by one of the main exit roads from the city of Belfast. Crustaceous lichens were found to be the most smoke-tolerant, yet in habitats with very high levels of atmospheric pollution (such as at the city boundary) even these disappeared, with the exception of one or two very poorly developed species. A decline in atmospheric pollution away from the city was reflected in a rise in crustaceous species, until the less smoke-tolerant filicaceous lichens began to appear. A subsequent increase in non-crustaceous lichens was recorded, which was associated with a decline in crustaceous types (because of competition and over-growth); and these were eventually replaced by the fruticose lichens which can tolerate only relatively clean air. Figure 7.6 illustrates the changing lichen cover and varying lichen types along the urban-rural transect through Belfast. A later study by Fenton (1964) confirmed this general spatial pattern of lichen species.

This notion of identifying lichen zones associated with particular levels of air pollution has been refined somewhat subsequently, and several zonal scales for estimating sulphur dioxide levels in the atmosphere by using lichen indicator species have been suggested (Hawksworth and Rose, 1976). One, which identifies six air pollution zones, is summarized in Table 7.11. This was the basis of the scheme adopted in the *ACE/Sunday Times Clean Air and Water Surveys* (Mabey, 1974), which produced a generalized map of air pollution conditions throughout Great Britain. Clearly the accuracy and detail of this map (Figure 7.7) are perhaps questionable, but it does nonetheless provide a good reconnaissance approach to evaluating spatial distributions of air pollution on the large scale. For

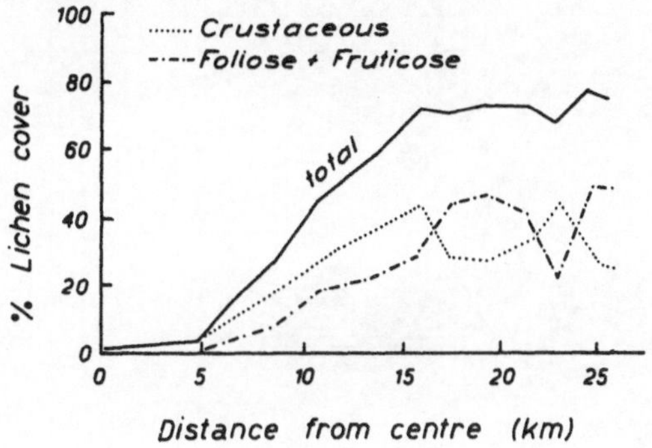

FIG. 7.6 VARIATIONS IN LICHEN COVER ALONG A TRANSECT OUT FROM THE CENTRE OF BELFAST. (After Fenton, 1960.)

Table 7.11
THE ACE ZONAL LICHEN SCALE

Zone	Lichens and mosses present	Mean winter SO_2 ($\mu g/m^3$)
0	*Pleurococcus* growing on sycamore	over 170
1	*Lecanora conizaeoides* on trees & acid stone	150–60
2	*Xanthoria parietina* on concrete, asbestos	about 125
3	*Parmelia* appears on acid stone	about 100
4	grey leafy (foliose) species appear on trees	about 70
5	shrubby (fruticose) lichens on trees	about 40–60
6	*Usnea* becomes abundant	about 35

Source: Gilbert, 1974

Table 7.12
MULTIPLE REGRESSION EQUATIONS FOR PREDICTING AIR POLLUTION LEVELS

SULPHUR DIOXIDE

$$SO_2 = 18{\cdot}72 - 3{\cdot}94\,(\text{reflectance}) - 0{\cdot}15\,(\text{height}) - 2{\cdot}38\,(\text{foliose})$$

SMOKE

$$SM = 9{\cdot}44 + 5{\cdot}29\,(\text{bare bark}) - 0{\cdot}09\,(\text{height}) - 2{\cdot}48\,(\text{reflectance}) + 2{\cdot}32\,(\text{crustose})$$

where: SO_2 is mean sulphur dioxide concentration ($\mu g/m^3$); *reflectance* is the mean log. reflectance of 10 sample trees at 1·5m height; *height* is the square root of the mean maximum height at which mosses/lichens are found on each of 10 sample trees; *foliose* is the mean % cover at 1·5 m height on the 10 sample trees; *SM* is mean atmospheric smoke concentration ($\mu g/m^3$); *bare bark* is the mean % cover at 1·5 m height over the 10 sample trees; *crustose* is as for foliose.

Source: Creed, Lees and Duckett (1973)

detailed prediction, however, this reconnaissance approach is too crude to be of widespread value. Creed, Lees and Duckett (1973) have illustrated a more refined approach, based on establishing relationships between sulphur dioxide and smoke levels, and a

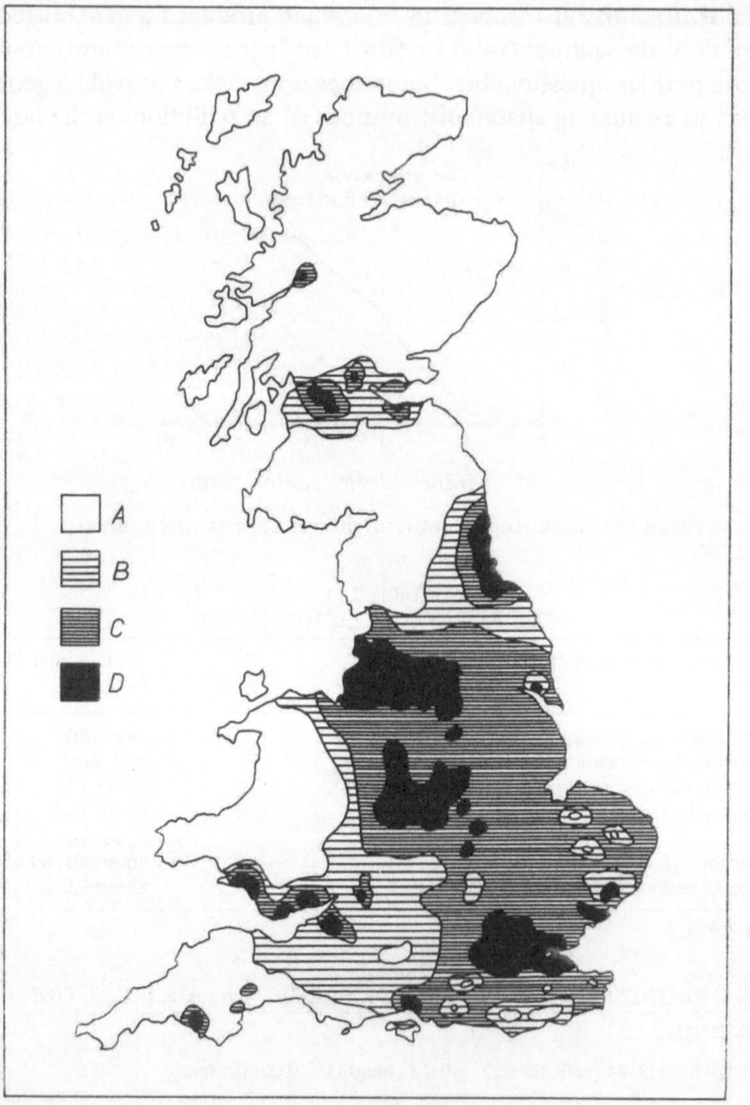

FIG. 7.7 APPROXIMATE DISTRIBUTION OF AIR POLLUTION ACROSS GREAT BRITAIN, BASED ON USE OF LICHENS AS INDICATOR SPECIES.

Four classes of pollution are mapped:

A clean air, *Usnea* lichen cover

B some pollution, crusty, leafy and some shrubby lichens

C moderate pollution, crusty lichens and leafy lichens

D heavy air pollution, crusty lichen forms only.

See Table 7.11. (After Mabey, 1974.)

variety of environmental parameters (including lichen characteristics) for a sample of 104 sites where pollution levels are known. They derived two multiple regression equations (Table 7.12) as 'a useful means of assessing long term atmospheric pollution levels'.

Although lichens have been quite widely used as indicators of atmospheric pollution (especially by sulphur dioxide), Hawksworth (1974a) has outlined other aspects of environmental change for which lichens can act as monitors. These include water pollution, the effects of agricultural chemicals, heavy metal pollution, and certain components of radioactive fallout. Regular surveys of the radionuclide concentrations present in lichens provide a valuable index of fallout from nuclear explosions; and it has been found that radionuclide concentrations in lichens in arctic and sub-arctic areas have reduced since the general cessation of wide-scale nuclear weapons testing. Hawksworth (1974a) has also suggested that lichens could be of value in monitoring the escape of radionuclides from the wastes discharged by nuclear reactors.

7.3c Bioassays and concentration factors

In recent years there has been growing interest in the ability of organisms to accumulate certain substances in their body tissues, which apparently play no part in their normal metabolism. These substances are normally taken in by digestion, and so pollutants which enter a food web or chain can readily, and quite rapidly, accumulate at the higher tropic levels in an ecosystem by passing along the trophic structure. This appears to be the case with substances such as organochlorine insecticides, polychlorinated biphenyls (PCBs) and heavy metals such as cadmium and mercury. There are two main aims behind studies of these accumulations in body tissues – to assess the ecological significance of the residues found (in terms of behavioural response in the host organisms, reproductive changes, physiological and metabolic changes, mortality, etc.), and to assess the public health risk of residues in biological materials which are consumed by man. Determinations of the accumulation of such substances have been made under both experimental and field observational conditions.

Hasselrot (1975) has described some experimental approaches – normally referred to as *Bioassays* – used by the National Swedish Environmental Protection Board. One example concerns captive fish which were kept in a river downstream from a mercury discharge. Assessment of the mercury content of the discharge was made by determining the ratio of the tissue concentration in the test fish (after one month of exposure) to the background concentrations in fish in the hatchery from which they had originated. Ernst (1977) has described similar experimental determination procedures for determining what he terms the 'bioconcentration potential' of mussels (*Mytilus edulis*) for seven chlorinated pesticides. Bioassays provide information on rates of uptake of the elements, and on differences in the overall concentration of the elements in study species. Often, however, rapid determination of the concentration effect is required; and the suitability of the bioassay approach also presupposes that the most important (or potentially most concentrated) elements have been determined before the experimental study begins. In Japan, where a wide range of species of marine organisms (both plants and animals) is consumed by man, knowledge about the accumulation of potentially harmful elements in body tissues is important from the point of view of human health and welfare. Thus Ichikawa (1961) formulated a 'concentration factor', defined as the ratio of the concentration of an element in an organism, to the concentration of that element in normal sea water, and he illustrated the wide range of values of the factor for different elements (Table 7.13). The concentration factor has subsequently been adopted in a number of studies (e.g. that by Policarpov, 1966). It is clear from Table 7.13 that some species have greater capacities than others to accumulate toxic materials in their body fat; and that these capacities vary between elements for a given species. Attention has also been focused on assessing the value of different species as pollution indicators, by determining the concentration factors of different species and different elements. This can be illustrated by the study of two tunicate species by Papadopoulou and Kanias (1977); who found that the species *Microcosmus sulcatus* could be used as a good indicator for selenium, chromium, zinc and cobalt, whereas *Ciona intensinalis* was a good indicator for iron.

Table 7.13
CONCENTRATION FACTORS FOR SOME RADIONUCLIDES IN MARINE FOOD ORGANISMS

Elements		Range of concentration factors	Organism with maximum conc. factor
Cesium	(CS^{137})	1–20	fish
Strontium	(Sr^{89}, Sr^{90})	0·1–18	algae
Iodine	(I^{131})	11–6800	algae
Iron	(Fe^{55}, Fe^{59})	400–78 000	echinoderm
Manganese	(Mn^{54})	70–50 000	molluscs
Cobalt	(Co^{57}, Co^{58}, Co^{60})	14–4000	crustacean
Zinc	(Zn^{65})	80–40 000	mollusc

Source: Ichikawa (1961)

7.3d Environmental classification and pollution

Because species of plants and animals differ in their tolerance of polluted environments, and in their behavioural, physiological and reproductive reactions to pollution, there are several ways in which ecological factors can be used in the classification of pollution states. One is the notion of using ecological indicators, discussed on pages 207–13. A second approach would be to study and evaluate the changing stages of health of a diagnostic species with increasing pollution. This is analogous to determining the tolerance response curve for that species as an index of pollution. An alternative is to view pollution state as an environmental gradient, and carry out a direct gradient analysis of the type outlined in Chapter 5.

Myslinski and Ginsburg (1977) have suggested a framework for classification which is based in part on the concept of a pollution gradient. They first identify four 'pollution tolerance states' for ecological organisms (Table 7.14); then they suggest a four-phase classification of pollution level which depends on the presence or absence of species diagnostic of each of the stages. The four pollution levels they identify are:

BALANCED: characterized by large numbers of intolerant forms, with a wide variety of species. Intolerant forms outnumber other forms.

UNBALANCED: intolerant forms are less in number than other forms, but intolerant plus moderate forms generally outnumber tolerant groups.

SEMI-POLLUTED: in which intolerant forms are infrequent if not absent; moderate and facultative forms are present in large numbers.

POLLUTED: intolerant forms are absent, and generally only tolerant species are present. Facultative forms can be present; but often (with very severe pollution) no species or organisms at all are present.

The distribution of the tolerance forms between the various pollution classes is illustrated in Figure 7.8. The analogues between this form of environmental classification procedure, and the vegetation classification schemes based on the concept of gradient analysis (outlined in Chapter 5) are clear. This seems to offer a convenient means of classifying environments according to levels of pollution, adopting an essentially ecological basis of classification.

Table 7.14
POLLUTION TOLERANCE STATES OF ECOLOGICAL ORGANISMS

INTOLERANT	organisms have life cycles which depend on a narrow range of ideal environmental conditions; species are sensitive to pollution, and they are generally replaced by more tolerant species if environmental quality is degraded
MODERATE	includes organisms that are less extremely sensitive to environmental degradation; such organisms cannot adapt to severe degradation, but they normally increase their population with slight to moderate enrichment (in the case of aquatic ecosystems, with enrichment introduced from municipal wastewater and treatment plane effluent)
FACULTATIVE	organisms which are able to survive over a wide range of environmental conditions; have higher tolerances to adverse conditions than either the intolerant or moderate organisms
TOLERANT	organisms which are able to survive over a wide range of conditions; they are generally capable of thriving in water of poor quality; often found in large numbers in areas of organic pollution

Source: Myslinski and Ginsburg (1977)

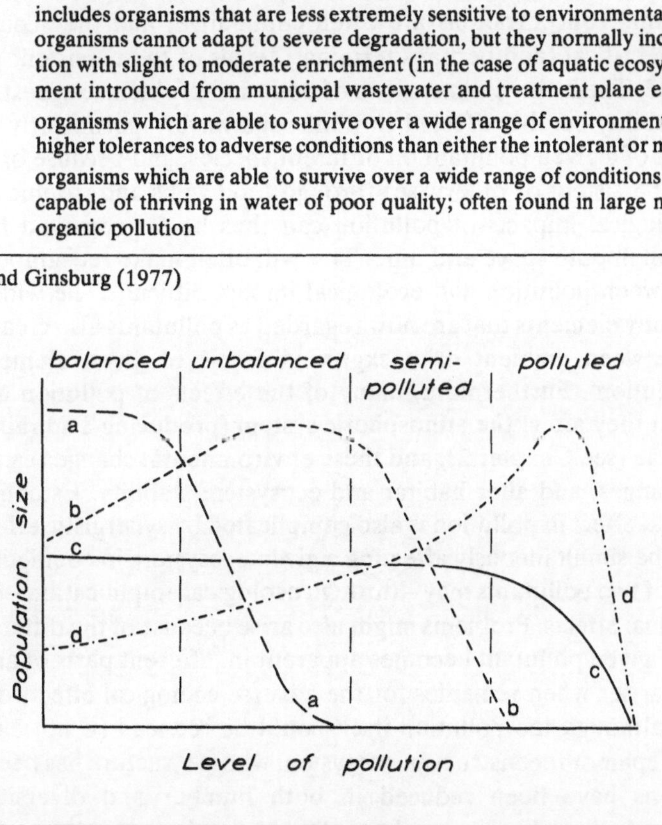

FIG. 7.8 HYPOTHETICAL PATTERN OF ECOLOGICAL CHANGES ASSOCIATED WITH VARYING LEVELS OF POLLUTION. The figure shows the changing pattern of distribution of organisms with differing pollution tolerances (see Table 7.14) as the local environment changes from a balanced to a polluted state. Key to organisms:

(a) intolerant species (b) moderate species
(c) facultative species (d) tolerant species

Based on terminology proposed by Myslinski and Ginsburg (1977).

7.4 POLLUTION – SOME GENERAL CONCLUSIONS

There are clearly a variety of ways in which environmental pollution can affect the stability and structure of ecosystems, and a number of specifically ecological problems of pollution. The distinction between two types of pollutant – those that act as plant nutrients, and those which are toxic or otherwise harmful to the organisms involved – is a valuable one, because it demonstrates that the ecological impacts of pollution can be manifest through either excessive production (such as with eutrophication) or through impairment of reproductive success, metabolic or behavioural change, or direct mortality in the species affected. The distinction also between 'target species' and non-target species is important, although most types of pollution (other than those associated with agricultural practices such as the use of organochlorine insecticides) affect exclusively non-target species. Similarly, the identification of lethal and sub-lethal ecological impacts, and specification of the sub-lethal effects, is of vital importance because although the effects of lethal concentrations or levels of exposure are immediately apparent and often quite simple to trace or demonstrate, the effects of sub-lethal concentrations are often cumulative, and they could well remain unsuspected until critical (and perhaps terminal) levels of the persistent pollutants have been accumulated. These three dimensions of pollution problems suggest that ecosystem structure and stability can be affected in numerous ways – particularly because of the differential effect of a given pollutant on different species, and because of the mobility of many elements (persistent or otherwise) through food webs and trophic structures over wide areas. Ecological impacts of pollution can thus be far removed from the initial pollution emission in both space and time. This will often make recognition of the causal relationships between pollution and ecological impact difficult. The widespread spatial distribution of many elements that are now regarded as pollutants also creates problems of distinguishing between ambient or background levels of given elements and levels regarded as 'pollution'. Furthermore, many of the effects of pollution are not directly ecological, in that they affect the atmospheric system (producing acid rain, for example) and the water cycle (see Chapter 2), and these environmental changes can in turn trigger off ecological changes, and alter habitat and ecosystem stability. Establishing the links between cause and effect in pollution is also complicated by synergistic effects, in that two pollutants might be simultaneously affecting a given ecosystem in conflicting ways, or the combined effect of two pollutants may – through ecological amplification – exceed the sum of the two individual effects. Problems might also arise because of the differential time lags as the effects of a given pollutant becomes apparent in different parts of an ecosystem. A further problem arises when remedies for the adverse ecological effects of pollution are sought, because although the pollution itself could be reduced (if not eliminated), it is often difficult to repair or reconstruct an ecosystem whose structure has been undermined, whose populations have been reduced in both number and diversity, and whose biogeochemical cycles have been severely modified from the natural state. If the pollution has had small-scale or localized effects, then the ecosystem can carry out 'self-repair' by secondary succession, evolution of new equilibrium forms of niche structure, creation of new stabilized forms of nutrient cycles, and similar changes which are essentially endogenous to the ecosystem. With badly polluted ecosystems, however, natural repair might be difficult, particularly if most organic life has been removed through pollution or its side-effects. In such cases some form of ecological 'cosmetic surgery' or reconstruction might be required. This is a costly and time-consuming eventuality, however, the costs of which have to be met by someone. Ironically, although this chapter opened with a brief consideration of the balance between economic and ecological perspectives on pollution, in final analysis economics must be considered in dealing with remedies for pollution. Some of the major economic means of reducing environmental pollution have been evaluated by Rees (1977), and these are summarized in Table 7.15.

Table 7.15

SOME IMPORTANT ECONOMIC MEANS OF REDUCING ENVIRONMENTAL POLLUTION

(1) Price all environmental resources: charge producers and/or householders for waste disposal, this would then lead to a reduction in the quantity of waste produced
(2) Establish appropriate air and water quality standards: and ensure that individual discharges are compatible with these standards
(3) Subsidize producers to reduce the level of waste discharged into the environment
(4) Ban or severely control discharges
(5) Increase the capacity of the environment to absorb waste

Source: Rees (1977)

8
The Ecological Basis of Environmental Management

> The magic of the natural world beckons and challenges, and lures the receptive soul ever onwards; but, like the Holy Grail, it is never finally found and possessed. And this is its fascination (Ratcliffe, 1976).

Many people share Ratcliffe's (1976) fascination with the 'natural world'; and the strength of opinion is now such that concerted efforts are being made to ensure that ecological factors are taken into account when planning future environmental and land-use changes. The factors underlying recent growing concern over wildlife and natural environments, as outlined in Chapter 6, are wide-ranging; but there are healthy signs that decision-makers responsible for environmental planning and management are becoming increasingly aware that the stability of the biosphere in general, and the survival and stability of individual ecosystems in particular, are criteria which need careful consideration when environmental management policies and priorities are being evaluated and implemented.

Many of the fundamental ecological issues have been considered in previous chapters. These include the overall composition and dynamics of the biosphere (Chapter 2); the form, function and equilibrium of ecosystems (Chapter 3); ecological changes in both time (Chapter 4) and space (Chapter 5); the concept of pressures on ecological resources (Chapter 6) and environmental pressures from pollution (Chapter 7). It remains in this chapter to evaluate some of the management issues of relevance to ensuring that future generations have access to natural and stable ecosystems, and a wide variety of plants and animals. It seems appropriate to concentrate on three broad themes. One is the general dimension of resource management, and in particular the scope for developing methods of evaluating ecological resources so that due consideration can be given to ecological factors in situations of environmental conflict. Secondly it is valuable to consider approaches to, and the scope for, preserving ecological resources for the future. Finally, wider aspects of the environmental management decision-making machinery can be considered, particularly in terms of approaches to assessing the likely environmental impacts of proposed landscape changes or environmental modifications.

8.1 RESOURCE MANAGEMENT

Terborgh's (1974) recommendation that every nation should possess an inventory of its biological endowment highlights the fact that hitherto ecological resources have been largely accepted as almost limitless, and that to many planners and developers the need for conservation of wildlife has been regarded as subservient to the needs for short-term economic, industrial or social gain. This 'cowboy economy' approach is regarded by many as one of the basic roots of the present environmental crisis (Chapter 1), and it devalues the importance of ecological resources *per se*. Furthermore, it fails to identify and accept the need for sensible resource management and use, or to rationalize the conflict between the different needs of conservation of nature and development of the environment.

8.1a Perspectives on resource management

Resource management in its widest sense is an inherently inter-disciplinary field, and although a growing investment of geographical interest in the field is apparent in recent years (e.g. books by O'Riordan (1971) and Simmons (1974)), geographical contributions to resource studies can be traced back at least to the early post-war period (Fairchild, 1949; White, 1949). Geographical contributions have in general been geared at regional, national and international scales of resource systems, rather than at the scale at which individual resource users operate (Coppock and Sewell, 1975).

Table 8.1
FIVE-STAGE ITERATIVE PLANNING PROCESS BASED ON SYSTEMS ANALYSIS

Stage	Action
1	Identify and agree goals and objectives
2	Initiate research leading to proper understanding of the relevant issues
3	Identify and evaluate alternative strategies for achieving the objectives
4	Select and implement a particular strategy
5	Monitor the results; possibly modify the plan in the light of changing demands and values

Source: Jeffers (1973)

Because the notion of resources is important to many disciplines, the most effective approaches to the study, evaluation and management of resources are themselves inter-disciplinary. *Resource science* has been welcomed as a new field in environmental studies; but, as Holling and Chambers (1973) point out, this has not happened without teething problems. Inevitably one of the main problems is that of communicating ideas and concepts between disciplines, and between resource scientists and decision makers. Because of this communication problem, and because many of the major problems within resource management are of global and/or long-term dimensions, Holling and Chambers (1973) suggest that it may be more expedient to seek narrow solutions for specific problems, one at a time, than to seek all embracing solutions for multi-dimensional problems. One direction which is being widely explored, and which promises to be of considerable value in resource and environmental management, is systems modelling (Bennett and Chorley, 1978), which to some extent has evolved from the growing emphasis on systems analysis in the study of the biosphere (Chapter 2) and its individual ecosystems (Chapter 3). Jeffers (1973) has suggested a five-stage planning process for both land-use planning and resource management (Table 8.1) which is 'an iterative, dynamic and continuous process', and in which systems analysis and modelling is a basic part of the evaluation of alternative strategies for achieving selected objectives. More recently Jeffers (1978) has reviewed the application of systems analysis within ecology in general.

Birch (1973) has outlined three main aspects of resource systems in which geographers have made valuable contributions, and these are the spatial, ecological and institutional relationships of the systems. Spatial analysis has had, as yet, limited progress in systems modelling, except at relatively high levels of aggregation, perhaps because resource systems have less clearly defined spatial structures than other systems considered by geographers. In a spatial sense, Manners (1969) has stressed the need to study how technological, economic, transport and political factors influence the size and nature of resource requirements. Perhaps the main contribution of geographers to resource management, according to Birch, is 'to ensure the appropriate spatial dimensions and accuracy in ecological analysis in order to achieve understanding at the scale of the regional system' (Birch, 1973, p. 6). Interest in the institutional relationships of resource systems has also been reviewed by Coppock and Sewell (1975), who stressed the value of constructing analytical models of the decision-making process in resource management. This geographical contribution has been complemented by the development and application of techniques of analysis such as the measurement and evaluation of the costs and benefits of resource development, and techniques which have sought to identify and measure public preferences. Decision-making processes and machinery have also been considered (e.g. Hutchison, 1969; Hamill, 1968).

The background to the present period of concern over resource management is a history of changing perceptions on resource availability and of fluctuating demand for resources. Zobler (1962) has identified three main periods within this history:

(a) 1850–1925: an era of wasteful use of natural resources, when the belief was widely held that 'nature was a limitless storehouse of raw material resources'. After the turn of the present century there were, however, some requests for the conservation of natural resources (such as by McGee, 1909);

(b) 1930–50: a transition period of rapidly accelerating demands for raw materials resources (due to widespread wind and water erosion of soils; flood damage; resource use during World War Two; world population increase; rising per capita standards of living; political fragmentation and allied emphasis on self-sufficiency in raw materials);

(c) post-1950: a period of relative shortage of resources, although by the 1950s the absolute abundance of many types of natural resources had been established.

A more detailed, and more recent, chronology is given by O'Riordan (1971), who identifies a phase since about 1960, triggered off by the threat of environmental collapse and by concern over national strategic safety. This phase has witnessed a search for co-operative environmental policies, regional economic improvement, rational resource planning, increased public awareness of environmental problems, and the rise and importance of public action and pressure groups.

Two aspects of the present phase of interest in resource management are worthy of particular consideration. One is the exponential growth in the world-wide use of many resources, particularly in the post-war period. Exponential increases in the use of fossil fuels, and in the production of industrial metal ores (Table 8.2) are producing problems of resource supply in many situations. Constraints on the growth of demand for many resources will be inevitable because of the finite nature of food and mineral resources, and because of the effects of environmental degradation which often accompany rapid growth of industrial production. Many have expressed concern over this period of exponential increase in resource use (e.g. Hanson, 1977), and most agree with the conclusion reached by Engelhardt and his colleagues (1976, p. 193) that 'mankind faces an inevitable transition from a brief interlude of exponential growth to a stable condition characterized by rates of growth so slow as to be regarded essentially as a state of no growth'. The second aspect is the changing perspective on the nature of 'resources'. Chapman (1969) has pointed out how, in economies with already high per capita levels of development, further increments of growth contain smaller requirements for material input. In such economies, reduction in the rate of growth of the required quantities of new material output (i.e. resources) is accompanied by a growing concern for the 'quality of the environment' (Jarrett, 1966). Thus some aspects of the environment not previously considered to be *resources* – such as landscape attractiveness, wilderness, and plant and animal diversity – are now perceived as resources worthy of evaluation and management.

Table 8.2
ANNUAL RATES OF EXPONENTIAL INCREASE IN PRODUCTION OF SOME IMPORTANT MINERALS

Mineral	Period	Annual rate of growth
Iron ore	1949–72	4·8%
Copper	1949–72	4·5%
Mercury	1948–58	7·7%
Mercury	1958–72	0·9%

Source: Engelhardt et al. (1976)

8.1b Terminology and classification of resources

Before considering ecological resources in detail, it is convenient to clarify some of the terminology widely used in resource management, as outlined by Chapman (1969). Economists generally draw a distinction between three main concepts:

Resource base (or Total stock): the sum total of all components of the environment that would become resources as such if they could be extracted from the environment.

Resource: the proportion of the resource base that man can make available under given social and economic circumstances, within limits set by the level of technological advancement (Figure 8.1).

Reserves: the proportion of the resource that is known (with reasonable certainty) to be available under prevailing social, economic and technological conditions.

The definition of a 'resource' as such thus depends less on the notion of an absolute asset, and more on the idea of a 'cultural appraisal'. For this reason, Zimmerman concluded that 'Resources are not, they become' (Zimmerman, 1951, p. 15); and Barnett and Morse stress that 'the notion of an absolute limit to natural resource availability is untenable when the definition of resources changes drastically and unpredictably over time' (Barnett and Morse, 1963, p. 7).

It is useful to classify resources into two broad categories:

Natural resources, which have some practical value. Dasmann (1976) classifies natural resources into inexhaustable, non-renewable, recyclable and renewable types (Table 8.3), depending on the extent to which they can be replaced by natural or man-made processes.

Non-utilitarian resources, which have social rather than practical values. For example, a population with a relatively high standard of living will generally give greater consideration to the need for clean supplies of air and water, and access to unspoiled recreation areas than a population with food shortages, widespread unemployment and lower material standards of living.

Problems of resource conservation commonly arise because many of the resources which are currently being used in industry and in economic expansion are non-renewable ones. Additional problems arise because the use of some resources destroys other resources. For example, exploitation of mineral resources in a locality concurrently reduces the quality of scenic resources.

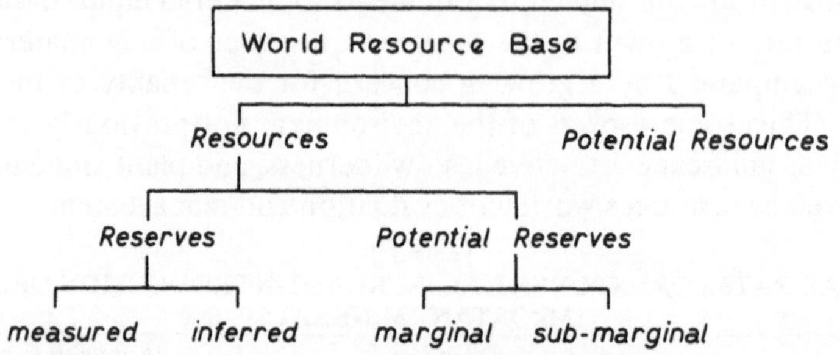

FIG. 8.1 SIMPLE CLASSIFICATION OF RESOURCE BASE, RESOURCES AND RESERVES. (After Manners, 1969.)

Table 8.3
CLASSIFICATION OF NATURAL RESOURCES

NON-RENEWABLE RESOURCES:	'are not generated or reformed in nature at rates equivalent to the rate at which we use them',
RECYCLABLE RESOURCES:	a special category of non-renewable resources, 'resources . . . which are not lost or worn out by the way we use them, and can be reprocessed and used again and again', e.g. many metals
RENEWABLE RESOURCES:	'include all living things that have the capacity for reproduction and growth. As long as the rate of use is less than their rate of regeneration, and as long as their environments are kept suitable, they will go on replacing themselves. However, living communities are not necessarily renewable, if the way in which we use them is destructive. No living species can survive if we crop it at a rate more rapid than it can reproduce, or if we destroy the habitat in which it depends',
INEXHAUSTABLE RESOURCES:	'those such as sunlight, which will continue to pour onto the earth as long as humanity will be around, whether we use it in certain ways or not.' Other examples include water resources on the world scale

Source: Dasmann (1976)

8.2 SURVEY AND EVALUATION OF ECOLOGICAL RESOURCES

The growing magnitude and diversity of pressures on ecological resources, and their effects on habitats, ecosystems and species, coupled with a growing awareness of the values of ecological resources (Chapter 6), have created a demand for appropriate consideration to be given to the impacts of planned developments and environmental changes on ecosystems and wildlife *at the planning stage*. Ecological and environmental consideration would then become an input into the planning process, to be weighed alongside social and economic considerations. This has inspired a quest for methods of evaluating the resource base of wildlife. Attempts have also been made to assess the relative values of different ecosystems, habitats and species, so that they can be ranked in order of conservation priority. Such rankings could be of considerable value in planning decision-making, because they would ensure that – given limited potential for conservation and preservation of ecosystems within an area – the ecologically most valuable sites and species would be preserved. They are also of basic value in the planning, design and management of protected environments such as nature reserves, at both the national and international scales.

8.2a Inventory of ecological resources

Ecological resource evaluations are of considerable value in planning and environmental management. The minimum requirement is to establish an inventory of ecological resources as a management data-base. In the context of geological resources, Merriam has stressed that 'data are important, and their half-life has been estimated as nearly infinite in contrast to conclusions which have a half-life of only about 3·7 years' (Merriam, 1974, p. 38). There is no reason to expect this to be any different for ecological resources. Many resource data-bases have been compiled for individual studies, and some of these have already been considered in Chapter 5. A detailed example is the survey of the flora and fauna of proposed Channel Tunnel sites near Folkestone in Kent (Brightman, 1976), which was designed to assess the plant and animal communities in the area, to evaluate ways of protecting these in the light of (what was then) imminent development, and to provide a factual basis for monitoring any changes arising from construction work. Localized studies cannot be compared and synthesized, however, unless standardized procedures are adopted, and so a variety of 'standardized resource survey' methodologies have been developed. Typical of these are Bunce and Shaw's (1973) user-orientated classification of woodland ecosystems, and Aultfather and Crozier's (1971) needle sort-card inventory system for documenting a physical inventory of the wildlife resources and appropriate management strategies of different areas. Problems of data compatibility inevitably arise if different inventory schemes need to be compared. For example, the method of vegetation classification used by the International Biological Program is based upon vegetation spacing and life forms; whereas the scheme used by UNESCO is based on structure, function and habitat criteria. Although each offers a quick and repeatable approach, and each is widely used, Goldsmith (1974a) has noted the difficulties of comparing the two schemes based on field trials in Majorca.

Grime (1974) has pointed out that most standard vegetation classification schemes do not include recent or unstable vegetation, they are often very abstract, and they rarely provide data which is intelligible to the non-specialist. Clearly this latter problem needs to be resolved before ecological resource evaluations can become a basic input into planning decision-making, and so Grime has proposed an alternative vegetation classification scheme based on triangular ordination. The characteristics of the vegetation are considered:

(a) *competition*: 'the attempt by neighbouring plants to utilize the same units of light, water, mineral nutrients or space',

(b) *stress*: usually imposed by the physical environment (shortage of light, water, mineral nutrients, etc.),

(c) *disturbance*: generally arises from the activity of grazing animals, man (through trampling, mowing, ploughing, etc.) and physical disturbance (e.g. through soil erosion).

This classification scheme allows an evaluation of individual habitat types based on meaningful criteria for resource management (Figure 8.2a), and it offers a means of comparing the sensitivity and reaction of species in different habitats (Figure 8.2b).

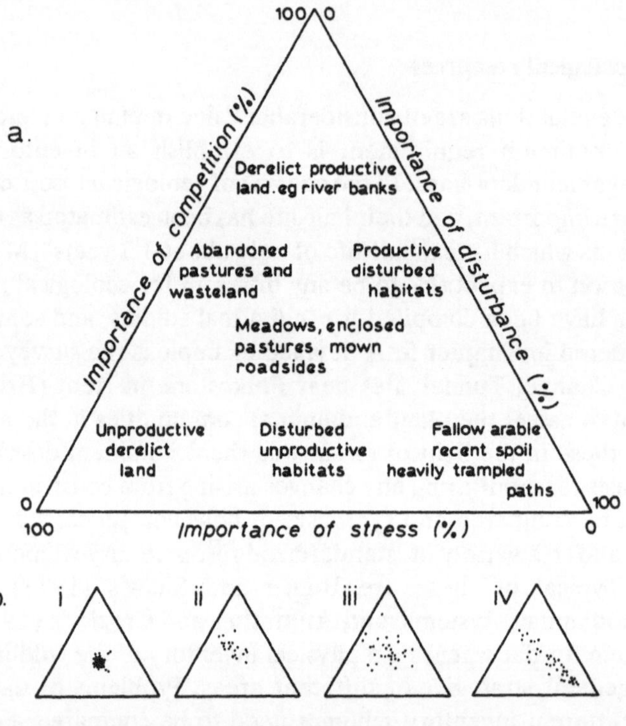

FIG. 8.2 TRIANGULAR CLASSIFICATION OF VEGETATION ON THE BASIS OF COMPETITION, DISTURBANCE AND STRESS.
(a) The generalized pattern of various types of habitats within the triaxial field.
(b) Some examples of ordination of specific samples of herbaceous vegetation from meadows (i), road verges mown frequently (ii), derelict banks of rivers, ponds and ditches (iii) and demolition sites of brick and mortar rubble (iv).
(After Grime, 1974.)

8.2b Evaluation of ecological resources

These types of resource surveys and classification schemes provide an invaluable data base for resource management, but they do not in general seek to *evaluate* the habitats or species involved. Clearly evaluations are required in formulating management and conservation strategies, to assess the suitability of an area for different uses (Olschowy, 1975; Whittaker and McCuen, 1976). These can vary considerably in their approach and content. On the one hand there are relatively simple 'listings' of evaluations of major habitats (Table 8.4), and general observations of those species of wildlife generally considered to be of greater 'value' than others (Table 8.5). These are complemented by schemes which seek to rank the ecological value of sites simply on the basis of degree of human interference (Pickering, 1977); and specific regional evaluations based on vegetational criteria which are comprehensive, relatively inexpensive, and generally valuable in resource planning and management (Hawes and Hudson, 1976). Most available resource evaluations seek to rank resources on the basis of conservation value rather than on economic terms, although the latter has been suggested and attempted by Helliwell (1973). Goldsmith advises that

Table 8.4
THE NATURE CONSERVATION VALUE OF VARIOUS HABITATS, AS PERCEIVED BY THE NATURE CONSERVANCY COUNCIL

CATEGORY 1	Most important for wildlife:	
	Primary woodlands	Lowland heaths
	High mountain tops	Unpolluted and untreated rivers, lakes, canals, permanent dykes, large marshes and bogs
	Permanent pastures and meadows untreated with fertilizers or herbicides	Coastal habitats (cliffs, dunes, salt-marsh, etc.)
CATEGORY 2	Moderate importance for wildlife:	
	Broad-leaved plantations	Mature conifer plantations
	Recently planted conifer plantations	Copses, corner plantations, etc.
		Hedges
	Moorland and rough grazing	Gravel pits, clay pits
	Farm ponds	Small marshes and bogs
	Road and railway verges	Disused quarries
	Arable land with rich weed flora	Neglected orchards
	Large gardens	Golf courses
CATEGORY 3	Little importance for wildlife:	
	Conifer plantations with no ground cover	Derelict land in towns
		Polluted water of all kinds
	Temporary water bodies	Grass leys
	Improved pasture	Playing fields
	Airports	Small gardens
	Allotments	Arable land with poor weed flora
	Horticultural crops and commercial orchards	Industrial and urban land

Source: Nature Conservancy Council (1977a)

> it is preferable to compare areas in terms of their plant and animal species richness and diversity, as well as habitat diversity, so that they can be ranked in order. It is impractical to place any kind of monetary values on either end, or any point, of the scale as they are affected by so many other extrinsic factors (Goldsmith, 1975).

Nonetheless ecological evaluations are required to complement evaluations of land-use capability (Bibby and Mackney, 1969) and landscape attractiveness (Crofts and Cooke, 1974), and to assist in the goal of many County Planning Departments to 'maintain large tracts of land, in addition to sites, of relatively high ecological interest' (Selman, 1976).

Available approaches to ecological evaluation fall into four broad categories:

(a) those which seek to evaluate individual species – generally based on degree of uniqueness,
(b) those which seek to evaluate ecological zones within an area, generally based on ecological diversity,
(c) those which seek to evaluate habitats, generally on the basis of individual species,
(d) those which seek to evaluate habitats and zones on the basis of vegetation and other criteria simultaneously.

Table 8.5
WILDLIFE SPECIES GENERALLY CONSIDERED TO BE OF GREATER CONSERVATION VALUE THAN OTHERS

(a) Species of known economic value (such as whales or herrings)
(b) Species valuable for the study of man and of behaviour (such as apes)
(c) 'Living fossils' to further the study of evolution
(d) Species which give some aesthetic pleasure (such as primroses, butterflies)
(e) Species valuable in the study of population ecology (such as colonial sea birds)
(f) Species which have already been studied intensively, and which are now already well documented
(g) Species which are evolving or extending their previous ranges into different types of habitat

Source: Moore (1969)

Species

Guidance on the species most in need of protection is required so that priorities for the acquisition of new nature reserves and other protected areas can be objectively established. Perring and Farrell (1977) have outlined the approach adopted by the Biological Records Centre of the Nature Conservancy Council, in classifying rare plant species in Great Britain on the basis of the threat to their survival. This particular study was inspired by the finding that up to two thirds of the localities of the 300 rarest species had probably been lost since recording began in Britain some three centuries ago. Their study formulated a 'Threat Number', based on subjective elements (attractiveness, remoteness and accessibility) and arbitrary categories (conservation index) (Table 8.6), and values of the Number were calculated for each of the 321 species regarded as 'rare' in Great Britain. It was then possible to arrange rare and endangered species into an order of priority for conservation; the species with the highest Threat Numbers are those most likely to

Table 8.6
METHOD OF CALCULATING THE 'THREAT VALUE' FOR INDIVIDUAL PLANT SPECIES

(1) Derive values for each of the following characteristics of the study species:

(a) The rate of decline of that species over a decade of observations:
- 0: decline less than 33%
- 1: decline between 33 and 66%
- 2: decline over 66%

(b) The number of localities of the species known to the Biological Records Centre:
- 0: over 16 sites
- 1: 10–15 sites
- 2: 6–9 sites
- 3: 3–5 sites
- 4: 1–2 sites

(c) A subjective assessment of the attractiveness of the species (a measure of the likelihood of it being picked):
- 0: not attractive
- 1: moderately attractive
- 2: highly attractive

(d) The 'Conservation Index' for that species – an arbitrary figure related to the % of the localities of that species which are in nature reserves:
0: over 66% in nature reserves
1: 33–66%
2: less than 33%
3: less than 33% and these sites are subject to exceptional threat

(e) The remoteness (relative ease with which the species could be reached by the public):
0: not easily reached
1: moderately easily reached
2: easily reached

(f) Accessibility (ease of access once the site has been reached):
scoring as in e.

(2) Calculate the 'Threat Number' for that species from the formula:

$$\text{Threat Number} = (a + b + c + d + e + f)$$

The maximum Threat Number possible is 15, and the observed range is 2–13.

Source: Perring and Farrell (1977)

Table 8.7
METHOD OF CALCULATING THE 'CONSERVATION VALUE' OF INDIVIDUAL PLANT SPECIES

STEP 1 Prepare a list of vascular plants within the study area

2 Assign a relative frequency value (P) to each species. This is based on a log scale 0–5. Category 0 = n plants (*rare* in the study area); Category 1 =10n plants; 2 = 100n plants; 3 = 1000n plants; 4 = 10000n plants; 5 = 100000 plants (*common* in the study area)

3 From the *Atlas of the British Flora* (Perring and Walters, 1962) determine the number of 10 km grid squares in the British Isles in which each species occurs. This is the 'National Value' for that species

4 From the *Atlas* determine the number of grid squares in a 'region' centred on the study area (with an area, for example, of 80 grid squares). This is the 'Regional Value' for that species

5 From an empirical table (Appendix 1 in Helliwell, 1974a) convert the % presence into a 'Rarity Value' for each species, for both the national (NRV) and regional (RRV) levels. (For example a 0·1% cover gives a real value of 398; 1% = 363; 10% = 150; 50% = 6·8; 100% = 1·0)

6 Derive a 'Conservation Value' for each species from the relationship:
$\text{Conservation Value} = F^{0.36}\,(RRV + NRV)$

Note that the empirical value of the exponent (0·36) is derived from the relationship: Relative Value = (Scarcity Index) × Area $^{0.36}$ mentioned in Helliwell (1973, Figure 1)

AN EXAMPLE: the plant *Agrostis canina*, in Snowdonia

The number in the study area is about 100000, thus F = 100000. The plant occurs in 54·8% of the grid squares in the British Isles, thus NRV = 5·15. It occurs in 60·5% of the 81 grid squares in the Snowdonia region, thus RRV = 3·8. The Conservation Value is thus:

$$100000^{0.36} \times (3{\cdot}8 + 5{\cdot}15) = \underline{564}$$

Source: Helliwell (1974a)

disappear without remedial action. Comparative data on the distribution of 53 species of mammals in Britain also now exist (Corbet, 1971), and so threat values could readily be calculated for these species. Whilst this Threat Number approach provides a valuable means of evaluating the relative extinction rates of individual species, some of the steps in calculating the index are somewhat subjective, and so the compatibility of derived values between different observers might be open to question. The weightings involved at each step also appear to be derived more for convenience than out of empirically established findings. Alternative methods of evaluating individual species might thus be required. One scheme, based on analysis of relative frequency of the plant species concerned both within a localized study area and in Britain as a whole, has been derived and illustrated by Helliwell (1974a). The method aims to establish a 'Conservation Value' for individual species, and it is summarized in Table 8.7. This scheme has the advantage that it is objective throughout; it combines national, regional and local assessments of species abundance; and the Conservation Value scores for each species can be summed to yield an overall Conservation Value for an area, habitat or ecosystem.

It is possible, however, that the ecological value of a habitat or ecosystem will be different from the simple sum of ecological values of its component species, and so alternative schemes are necessary for evaluating conservation values of ecological zones.

Ecological zones

Tubbs and Blackwood (1971) have advised that 'the conservationist must devise a means of evaluating the relative floristic and faunistic values of land, and of presenting this means in a form which is easily interpretable by planners, and at the same time is easily comparable with other information'. Thus attempts have been made to construct 'Ecological Evaluation Maps', based either on sub-dividing a study region into habitats or

Table 8.8
SUMMARY OF THE 'ECOLOGICAL EVALUATION' SCHEME PROPOSED BY TUBBS AND BLACKWOOD (1971)

(1) Sub-divide the study area into PRIMARY ECOLOGICAL ZONES:
 ZONE 1: unsown vegetation (including non-plantation woodland)
 ZONE 2: plantation woodland
 ZONE 3: agricultural land

(2) Evaluate the ECOLOGICAL VALUE of each zone, using three main concepts:
 (a) unsown or semi-natural habitats have limited distribution in lowland Britain, and are subject to pressures from reclamation and development (thus they have a high conservation value)
 (b) areas of plantation woodland often form valuable wildlife reservoirs (thus they too have a relatively high conservation value)
 (c) ecological interest in agricultural land will vary inversely with the intensity of agricultural land

 Ecological evaluations for the zones are thus:
 ZONE 1: Category I or II (the final distinction rests on subjective estimates of rarity of habitat type and presence of features of outstanding scientific importance)
 ZONE 2: Category II or III (based on subjective estimation of the value of the habitat as a wildlife reservoir)
 ZONE 3: Relative value is a function of habitat diversity;
 (a) this is related to the presence of definable features:
 (1) permanent grassland
 (2) hedgerows and hedgerow timber
 (3) boundary banks, roadside cuttings and banks, verges
 (4) park timber and orchards (other than those in commercial production)
 (5) ponds, ditches, streams and other watercourses
 (6) fragments of other unsown vegetation (including woodland) smaller than 0.5 km^2
 (b) score for the presence of each group of features:
 0 = none/virtually none in the zone
 1 = present (but not a conspicuous feature)
 2 = numerous (conspicuous feature)
 3 = abundant
 (c) Evaluation of the value for the Zone (based on sum of scores for individual features present in the Zone):

Total score	*Category*
15–18	II
11–14	III
6–10	IV
0– 5	V

topographic units, or on a regular grid square basis, and seeking to present data in a comprehensive manner for the non-specialist. Tubbs and Blackwood (1971) describe a scheme developed for Hampshire, and used in planning studies by the Planning Department of the Hampshire County Council. The scheme (Table 8.8) is based on sub-dividing an area into 'Primary Ecological Zones', and then deriving an ecological evaluation for each zone depending on general land use and habitat diversity. The final product is a 'Relative Ecological Evaluation Map' defining the boundaries and relative values of each zone. This is accompanied by a written statement which defines 'the characteristics which distinguish each zone, and including an indication of desirable conservation policies' (Tubbs and Blackwood, 1971). The technique was devised for lowland England, and clearly its application to upland areas would only be possible with modification. An alternative and more widely applicable approach was devised by the Conservation Course at University College, London (Conservation Course, 1972; Goldsmith, 1975), and this is based on objective assessment of habitat area, scarcity, number of species present and vegetation structure. These assessments are subsequently used to produce a numerical index of Ecological Value, on a grid square (km^2) basis (Table 8.9). The major problem with these zonal types of evaluation schemes is the difficulty of applying suitable weightings to the

Table 8.9
SUMMARY OF CONSERVATION COURSE METHOD OF DETERMINING ECOLOGICAL VALUES, ON A GRID SQUARE BASIS

STEP 1 Divide the whole study area into 'distinct land systems':
- SYSTEM 1: unenclosed upland (mostly moorland over 300 m)
- SYSTEM 2: enclosed cultivated land (mostly permanent pasture)
- SYSTEM 3: enclosed flat land (mostly arable land in valley bottoms)

STEP 2 Record the distribution of habitats within each land system:
- (a) arable and ley
- (b) permanent pasture
- (c) rough grazing
- (d) woodland
  - (i) deciduous and mixed
  - (ii) coniferous
  - (iii) scrub
  - (iv) orchard
- (e) hedges and hedgerows
- (f) streams
- etc.

STEP 3 Determine the following parameters for each habitat (on the basis of individual km grid squares):
- (i) Extent (E); for habitats a–d is area in ha, for linear features is length in km/km^2
- (ii) Rarity (R); $R = 100 - \%$ area per land system
- (iii) Plant Species Richness (S); the number of species of flowering plant recorded in sample plots (20 m × 20 m) in each habitat type in each land system
- (iv) Animal Species Richness (V); this correlates with vegetation stratification, thus V = the number of vertical layers in the vegetation (grassland $V = 1$, well developed woodland $V = 4$)

STEP 4 Determine the Index of Ecological Value (IEV) for each grid square, based on the formula:

$$IEV = \Sigma_{N}^{i=1} (E. \times R. \times S. V.)$$

(having standardized each parameter before the calculation)

STEP 5 Standardize the IEV values to a range of 1–20, then plot the standardized values in map form, on the grid square basis

Source: Goldsmith (1975)

component factors. For example the Tubbs and Blackwood (1971) method suggests that the conservation value of unsown vegetation is three times as high as that of category three habitats; and the scoring of habitat diversity features is devised more for simplicity than for empirically determined ecological values of the various features. The Conservation Course method applies no weightings in the calculation of the index of Ecological Value – yet clearly each of the four main parameters could be of different significance to conservation. Alternative evaluation schemes have thus been explored.

Species within Zones

Helliwell (1974a) devised a method of estimating the conservation value of individual species, so that a measure of the overall value of vegetation in an area could be derived by summing the conservation values of each species present there. The system was tested by evaluating the conservation value of four different areas of land in Britain – Snowdon, South West Scotland, the English Lake District, and Upper Teesdale. The total conservation value scores for each area, standardized to a score per unit area (ha) were: Snowdon = 743; Upper Teesdale = 378; Lake District = 129; South West Scotland = 100. This led Helliwell to conclude that 'the vegetation in the Snowdon area is worth, on average, twice as much per hectare as the vegetation in Upper Teesdale, and about seven times as much as vegetation in parts of the English Lake District and South West Scotland' (Helliwell, 1974a, p. 73). A subsequent attempt to evaluate the conservation value of vegetation along 282 km of the M1 motorway, using the same approach, failed because the linear motorway sample yielded distorted conservation values in comparison with the more circular regional samples (Helliwell, 1974b).

The Helliwell approach has proven to be valuable in distinguishing between the relative conservation values of different areas, and there are very few subjective elements in the method. But the approach is time consuming if rapid reconnaissance evaluations are required. A method for assessing the conservation value of woodland flora based on a simple count of a selective list of species, has been suggested by Peterken (1974). 'Indicator species' are selected as diagnostic of woodland conditions, and the method (Table 8.10) is based on ranking individual sites on the basis of increasing numbers of 'primary woodland species' present. The richest sites (those with most indicator species) are deemed to have the highest conservation values. Ward and Evans (1976) have developed a similar evaluative approach based on indicator species. Their method is based on assessing the conservation value of limestone pavements in Britain, on the basis of

Table 8.10
SUMMARY OF 'INDICATOR SPECIES' WOODLAND EVALUATION SCHEME SUGGESTED BY PETERKEN (1974)

STEP 1	List all species of plants present in the 10 km squares covering the sites to be assessed, using the *Atlas of the British Flora*
STEP 2	Select 'woodland species' from the list, based on general knowledge of species behaviour in relation to shade, and supplemented by local observations (the Stage 1 list)
STEP 3	Survey the woodland sites to be assessed; list all 'woodland species' present (the Stage 2 list)
STEP 4	Eliminate from the Stage 2 list all those species which occur in more than a given number of sites (e.g. in more than 3 out of a total of 20 sites). This produces List 3, which represents a first approximation of a list of 'primary woodland species' in the study area
STEP 5	Rank the study sites on the basis of increasing number of 'primary woodland species' present

floristic criteria; taking into account such factors as the numbers of different indicator species present in an area, their abundance and their rarity.

Broad based zonal evaluations

Each of the approaches outlined previously seeks to evaluate ecological resources by adopting one or both of two basic premises:

(a) that diversity of species is beneficial to ecosystem stability and conservation value, and
(b) that characteristic or indicator species within a given habitat are of greater conservation value than other species found there.

These are important to a number of the criteria for evaluating wildlife resources identified by Helliwell (1969) – such as the maintenance of the genetic reserve, ecological balance, and research and natural history interest. But other criteria cited by Helliwell (1969), such as educational value, would realistically require other characteristics of the habitats and species to be considered as well – such as access and availability. To fulfill each of Helliwell's criteria, more widely based evaluations might be required, and several approaches have been tried and tested. An illustration is provided in the assessment of the conservation value of the Yare Valley in East Anglia by Watts, Hornby, Lambley and Ismay (1975), which was based on assessing the value of individual sites specifically for research, education and amenity, as well as on ecological criteria (such as species and

Table 8.11

SUMMARY OF ECOLOGICAL EVALUATION SCHEME ADOPTED IN A STUDY OF THE YARE VALLEY, EAST ANGLIA

STEP 1 Assess the conservation value of each habitat, by scoring on the basis of 11 criteria:

Criteria	*Max. value score*
Presence of scarce species	10
Diversity of species	10
Presence of scarce habitats	5
Diversity of habitats	10
Quality of higher plants (botanical significance)	20
Quality of lower plants	5
Quality of vertebrates	15
Quality of invertebrates	10
Value for research	5
Value for education	5
Value for amenity	5
TOTAL POSSIBLE SCORE	100%

STEP 2 Calculate the conservation value for each site:

$$CV = \sum_{i=1}^{11} C_i$$

where C_i is the score for criteria i

STEP 3 Classify the sites into grades on the basis of the CV scores:

Grade	*Score* (%)	*Grade*	*Score* (%)
1	65–100	4	35–44
2	55–64	5	25–34
3	45–54	6	0–24

Source: Watts, Hornby, Lambley and Ismay (1975)

habitat diversity, rarity and quality). The main characteristics of the technique are summarized in Table 8.11. The Yare Valley scheme is based simply on a summation of component values, but this implies that all criteria are of equal importance. A 'weighted-value' approach, which also considers criteria like educational value and human impact was described by Gehlbach (1975). In this scheme, numerical values are given to each of five features of natural areas in order of their importance to preservation. The features are heritage value, educational utility, species significance, community representation and human impact.

There are thus a wide range of different approaches to evaluating the conservation value of species, habitats and areas. Each tends to stress different criteria, and to place emphasis in designating conservation value on different characteristics of the ecological resources, and so comparison between the results of applying different evaluation schemes is not really possible. This means that only limited association between different studies can be expected. Smith (1976) has stressed that it is unlikely that a suitable standard method of evaluation for planning purposes (specifically for use in Structure Plans) will be developed or adopted, in part because of the very wide range of natural habitats and environmental conditions which exist throughout Britain, and in part because different methods are evolved to suit the manpower and available ecological data-bases within individual counties or planning authorities.

8.3 PRESERVATION OF ECOLOGICAL RESOURCES

Because of the many different sources of pressure on ecological resources (Chapter 6), and in the light of growing awareness of the need to consider ecological factors alongside social and economic ones in environmental management, preservation of ecological resources is becoming increasingly vital to maintain stable ecosystems and ecological diversity for the enjoyment and benefit of future generations. Although the Conservation Movement has grown in strength and stature in Britain in the present century (Park, 1977), Moore (1969) has argued that conservationists have on the whole failed to put over an acceptable philosophy and rationale that is meaningful in the modern context; and Harry, Gale and Hendee (1969) have characterized conservation as a dominantly urban-based, upper-middle-class social movement. Whilst there are clearly socio-economic forces underlying the variable strength of interest in different environmental matters, nature conservation has tended overall to receive less committed public support than it might, perhaps because the notion of nature conservation conflicts with other uses to which many people feel rural resources should be devoted. Dower (1964) identified four chief objectives for all rural planning. These are the provision of access and facilities for open-air recreation; the evolution of farming, forestry and other rural uses and occupations; planned landscape changes; and protection of wildlife and buildings and places of architectural and/or historic interest. Social and economic pressures often mean that in many situations of rural resource conflict ecological resources are sacrificed or at least undervalued. Recently, however, the Nature Conservancy Council (1977a) has stressed the need to place conservation in its proper place alongside economic and social problems, and to appreciate the integral role of nature conservation with other rural land uses. The Council concludes that 'a positive *rural land-use strategy* is becoming increasingly necessary for all users of land. If one were to be adopted it would underline the interdependence of man and his environment, and would foster understanding of that relationship' (Nature Conservancy Council, 1977a, p. 33).

Such a rural land-use strategy has been proposed as a viable option in solving resource conflicts in the case of the Somerset Wetlands, where confrontation between wildlife and modern agriculture has arisen because of expansion of the peat-winning industry, and agricultural improvement through drainage (Table 8.12).

Table 8.12
OPTIONS FOR CONSERVING ECOLOGICAL RESOURCES IN THE CASE OF THE SOMERSET WETLANDS PROJECT

OPTION 1	Limit wildlife conservation:	continue to manage existing nature reserves, but do not establish new ones
OPTION 2	Control agricultural improvement and peat extraction	not fully realistic because of financial factors
OPTION 3	Protect 'key areas':	attempt to protect the best areas of the different types of habitat present
OPTION 4	Improvement of derelict land:	reclamation of worked-out peat cuttings
OPTION 5	Land-use strategy:	agree to reconcile the conflicts of land use

Source: Nature Conservancy Council (1977b)

Preservation of natural areas is important for two main reasons – as an insurance to keep the earth suitable for human occupancy, and because many observers feel that man has an ethical obligation not to abuse the environment around him. The 'insurance' factor has been spelled out by Dasmann, who stresses that 'it is because we need to know how the biosphere functions, and keep it functioning while we learn, that the protection of natural communities and wild species becomes so important' (Dasmann, 1973, p. 115). Mattern has rationalized the ethical obligation in the form of a *landscape consciousness*, which he views as 'an exact knowledge of the ecological interrelationships of our environment; this is very different from that vague feeling for nature which is no more than an enthusiasm for nature, expressing itself in an emotional and uncontrolled manner' (Mattern, 1966, p. 14).

The machinery for preserving wildlife and natural environments falls within the realms of *conservation*. Conservation itself evolved out of a growing concern at the turn of the century, mainly in the United States, over the use and management of resources such as water and soil. Each resource was dealt with separately in planning, and resource management was characterized by piecemeal action and unco-ordinated policy formulation and management. In 1907 Overton Price termed the notion of looking at all of the resource problems together '*conservation*'; this idea was approved by the President of the United States, and the concept has since come into general use (Pinchot, 1936). Beazley has crystallized Price's early thinking, in defining conservation as 'the establishment and observation of economically, socially and politically acceptable norms, standards, patterns or models of behaviour in the use of natural resources by a given society' (Beazley, 1967,

p. 345). Fraser Darling has stated that 'conservation has been applied ecology, or ecology in action' (Fraser Darling, 1967, p. 1009).

In broad terms, conservation has attracted support from a wide spectrum of public interest, ranging from 'deep science to simple perception' (Ratcliffe, 1976), and some of the characteristics of the new social movement of ecology have been evaluated by Means (1969). The movement includes people from all walks of life, and members are not simply sentimentalists or 'pure preservationists'. In addition, attention is being focused not only on the 'great outdoors' but on wider aesthetic and moral issues concerned with nature and the environment in general. One recent characteristic of conservation is the widespread lobbying of general public opinion and the allied increase in dissemination of information on endangered species and threatened environments, by environmental groups and organizations. Typical of these has been the recent campaign to 'Save the Whale', co-ordinated by Friends of the Earth (1978).

8.3a Nature reserves

This general concern for conservation reflects public support and emotional commitment to the preservation of ecological resources, but tangible benefits of conservation action can only really be realized through action programmes which seek to manage ecological resources directly. The programme co-ordinated by the Nature Conservancy Council, and applicable to Great Britain, has been reviewed by Ratcliffe (1977a). He identifies two main areas of concern. One is the safeguarding of the most important areas of land and water by appropriate management programmes (such as through the establishment of National Nature Reserves and the designation of Sites of Special Scientific Interest); but this is backed up by a programme for areas outside those offered direct protection. This supplementary programme is based on the provision of advise on land use and planning, education and legislation. The overall strategy is to 'promote measures which minimize environmental damage resulting from human activities, following an evaluation of intrinsic importance of the factors concerned and their vulnerability to disturbance' (Ratcliffe, 1977a). Whilst both zoological gardens (Jordan and Ormrod, 1978) and botanical gardens (Polunin, 1969) offer valuable repositories for preserving threatened plant and animal species, encourage the breeding of rare species, and act as much needed gene banks, the artificial and clinical nature of the environment and the closely monitored management programmes within these ecological detention centres provide conditions for wildlife which are far removed from natural ecosystems. Consequently much attention has been devoted to the need to protect natural ecosystems, in the real life situation.

Protection strategies commonly fall into two broad categories – simply allowing relatively natural vegetation and ecosystems to act as nature reserves by keeping land and vegetation management to a minimum, or intentionally setting aside protected areas to act as designated nature reserves, wherein wildlife management programmes are geared to established management strategies. An example of the former in Britain is the use of land surrounding National Trust properties in England and Wales as conservation zones, because these comprise one of the last big land reserves in Britain and they are entirely under National Trust control, with no major conflicting land use as in the case of Ministry of Defence land, or commercial forests (Duffey, 1978). Most conservation interest is shown in specifically designated nature reserves, however, because these are on the whole designed and managed principally for nature conservation (rather than, for example, amenity). Terborgh (1974) has outlined various types of preserved areas geared to offer protection to different types of species. Large areas are needed to preserve natural vegetation formations, animals at the top of the trophic pyramid, and species with sedentary habits and poor colonizing abilities. On the other hand, endemic species or rare habitat types can frequently be protected with relatively small investment of land, so long as appropriate habitats are identified in time. Nesting groups of colonial species can be protected with even less investment of land. Migratory species often present difficult problems because of their mobility, and so appropriate action often requires international co-operation. Clearly, therefore, different types and sizes of nature reserves will be appropriate in different situations. Because land prices are high in many countries like Britain, and competing land uses (especially agriculture) place pressure on existing nature reserve areas, it is important to establish an 'optimum size' for nature reserves on the basis of management strategies and ecological understanding.

Nature reserve size

A number of different approaches to the planning and design of nature reserves have been developed, and many of these stem ultimately from the association between the area of a sample of ecosystem (e.g. nature reserve size) and the number and range of plant and animal species which are likely to be found within it (see Chapter 5). Hooper (1971) outlined the basic ecological problem, which is to identify the exact number, extent and spacing of various habitats for the survival of all species of plants and animals. He examined data on the association between area and number of species in nature reserves of a range of sizes, and he established a relation of the general form (using notation subsequently formalized by Helliwell, 1973):

$$N = n\,A^c$$

where N is the number of species, A is area, and n represents the density of the species (number per unit area). The exponent c is a function of isolation from similar sites, and it appears to have an average value of about 0·26. The relation showed that an increase in area of ten-fold would simply double the number of species observed, and so Hooper (1971) concluded that a number of small reserves, each with different habitat conditions and each designed to protect different species, would be the best solution for conserving a maximum number of individual species. Although this approach is ecologically and economically 'efficient', it ignores the fundamental coherence of ecosystems and it concentrates on individual species. An alternative approach was suggested by Moore (1962), who was concerned about fragmentation of heathland habitats in Dorset. He suggested that the individual species within an ecosystem be classified into *key species* (ones which would lead to radical ecosystem changes if removed) and inessential ones. He further recommended that definition of the smallest viable size of *habitat* would be based on the smallest unit that would support a viable population of its weakest key species. Streeter (1974) has stressed the need to identify the smallest area that can support a viable *ecosystem*, not simply the species or habitat components. His logic is that

> the ideal size for a reserve will . . . be that which will support a viable population of all those species regarded as being characteristic of the ecosystem concerned. This will depend on the territory sizes of the species with the largest territories. These will normally, but not always, be the species occupying the end of the food chain (Streeter, 1974).

Definition of the optimum size of a nature reserve by Streeter's criteria would thus be based on the smallest area that will support a viable population of the species characteristic of the end of the food web, or the species having the largest territories in an ecosystem. In woodlands this optimum size would thus be over 100 acres (Table 8.13).

Table 8.13
TERRITORY SIZES OF SOME FOREST VERTEBRATES

Vertebrate	Territory size (acres)
Sparrowhawk	96–1280
Tawny Owl	20–30
Woodcock	15–30
Pine Marten	220–640
Red Deer	1 deer per 120 acres*
Fallow Deer	1 deer per 70 acres*

*Forestry Commission 'acceptable' density

Source: Streeter (1974)

Because nature reserves are not available in unlimited amounts, Helliwell (1973) pointed out that although a species of animal which becomes very rare will be reduced in total value as a species, this will not be in direct proportion to the reduction in numbers. For example, if a population of 10000 waterfowl were reduced by 50 per cent, its overall value would not be reduced by as much as 50 per cent. This suggested to Helliwell (1973) a relationship between potential of a species to recover from a reduction in numbers, and the value of those individuals which remain. Thus when a resource becomes scarce, the value of a unit amount is likely to increase at a greater rate than normal. Consequently it is possible to define a relationship between number of individual units and total value of a resource which is becoming scarce (Figure 8.3a); and this can be transformed into a value/area relationship (Figure 8.3b) which is of importance in determining optimum sizes of individual nature reserves. Although the overall relative value of a site rises steadily with increasing size, Helliwell points out that often the value per unit area declines, suggesting that perhaps two or more sites of a particular habitat close together may be of greater total value than one large site of the same aggregate size. From the management point of view, however, undue fragmentation of wildlife habitats should be avoided if possible, partly because of the need to protect entire functioning ecosystems and partly for ease of implementing management strategies.

Total area in nature reserves

Helliwell (1975a) has considered wider dimensions of the problem of nature reserve design practice, in considering the value of undisturbed natural areas in relation to the total area involved (Figure 8.3c). The pattern of changing values of nature reserves follows a simple logarithmic curve, because if 50 per cent of a region were designated as nature reserves, the area so defined would contain over 50 per cent of the 'ecological wealth' of the region. On the other hand, the value of land not left as nature reserves follows a straight line, because the value of agriculture and other land uses is likely to rise more or less in direct proportion to the area involved. On the basis of such analysis, Helliwell (1975b) concluded that in the region of 6 per cent of the land areas in countries like England and Wales ought to be primarily devoted to the conservation of nature. In 1977 the actual area of National Nature Reserves run by the Nature Conservancy Council (as opposed to nature reserves run by local bodies and County Naturalists Trusts) was 4 per cent (Ratcliffe, 1977a).

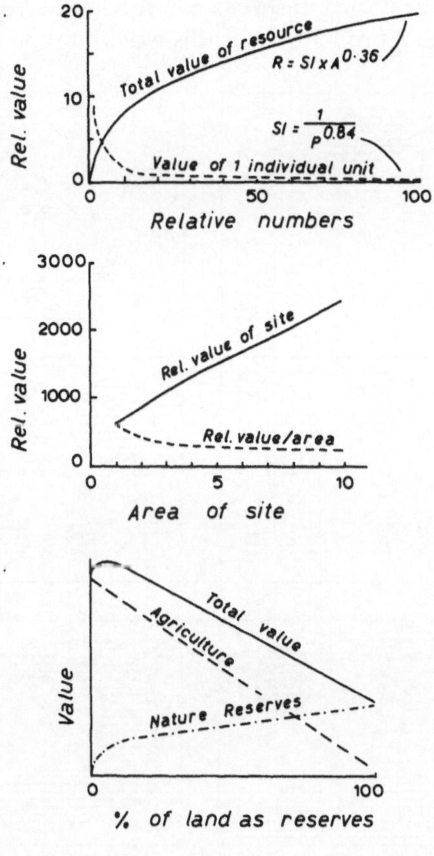

FIG. 8.3 EVALUATION OF WILDLIFE RESOURCES AND HABITATS.

(a) The relationship between number of individual units and the total value of a resource which is becoming scarce.

R = relative value, SI = scarcity index, A = area or number,
P = % of resource left. (After Helliwell, 1973.)

(b) Relationship between area of site and 'conservation value', assuming equal numbers of sites in each size class. (After Helliwell, 1973.)

(c) Relationship between amount of land as nature reserves, and its value in relation to other land uses such as agriculture. (After Helliwell, 1975a.)

Layout and design of nature reserves

Attention has also been focused on establishing optimum patterns of distribution and design of nature reserves. Helliwell (1976) has pointed out that certain potential locations for nature reserves will be relatively obvious – such as mountain tops, forests and the coastline. Within the remainder of a region or country, optimal distributions of nature reserves are required. Helliwell (1976) considered various alternative schemes for locating nature reserves within a region characterized by a series of ecological zones (Figure 8.4) (assuming unrestricted possibilities for purchasing the land!). One possibility is to establish a large nature reserve area, overlapping two of the important ecological zones (Figure 8.4a), but this would offer no protection to species in the other zones. Alternatively a number of smaller nature reserves could be scattered throughout the region to cover each of the zones (Figure 8.4b), but this would not be ideal because of the lack of

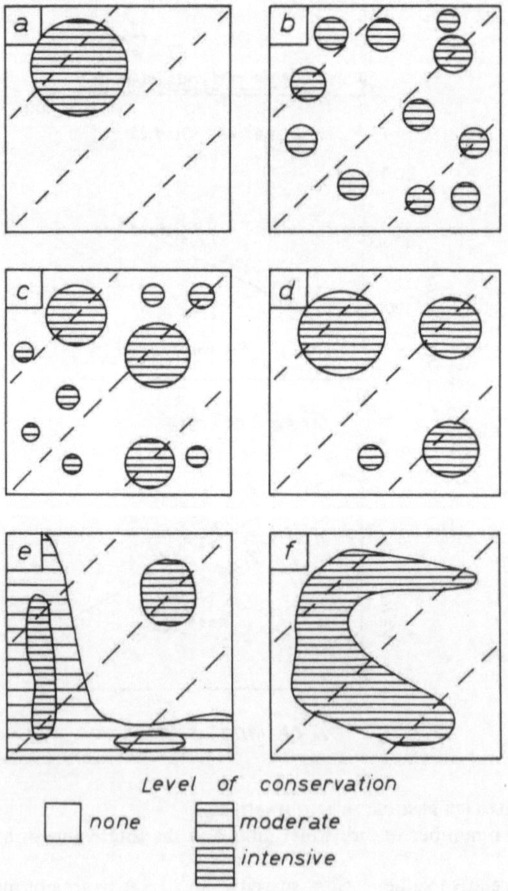

FIG. 8.4 ALTERNATIVE STRATEGIES FOR DESIGNING THE SIZE, SPACING AND LOCATION OF NATURE RESERVES WITHIN A REGION CHARACTERIZED BY FOUR ECOLOGICAL ZONES.
See text for discussion. (After Helliwell, 1976.)

areas large enough to include the overall range of species likely to be found in the region. A mixture of small and medium-sized reserves (Figure 8.4c), or fewer large-sized areas (Figure 8.4d) would perhaps be better, although there may be some adverse effects of isolation, especially in the smaller nature reserves. The isolation effects could be reduced by having areas of less intensive nature conservation linking areas of more intensive protection (Figure 8.4e); or by having a single large area which includes all ecological zones within it (Figure 8.4f). Helliwell (1976) maintains that the latter two alternatives are preferable because they involve a minimum of fragmentation together with extensive regional coverage. Management would also be simpler in larger areas. Helliwell concluded that 'patterns which combine a minimum of fragmentation with extensive regional coverage are likely to be preferred to more fragmented patterns involving the same total area of land' (Helliwell, 1976, p. 257).

Ecosystem analogues have been drawn between nature reserves and oceanic islands, and a number of aspects of island biogeographic theory have been applied in the design of nature reserves. Diamond (1975), for example, pointed out that the number of species that a reserve can hold at equilibrium will be proportional to its area and isolation; and that reserve size affects extinction rates of species within it. Diamond (1975) also formulated a series of geometric design principles for nature reserves (Table 8.14) which offer

Table 8.14
DESIGN PRINCIPLES FOR NATURE RESERVES, BASED ON ISLAND BIOGEOGRAPHIC THEORY

PRINCIPLE A	Large reserves are better than smaller ones – because they can hold more species at equilibrium, and they have lower species extinction rates
PRINCIPLE B	Reserves should be divided into as few disjunctive pieces as possible – for the same reasons as A (although separate reserves in an inhomogeneous region may each favour survival of different groups of species)
PRINCIPLE C	If reserves must be split, then they should be located as close to each other as possible – because proximity increases immigration between reserves, and it thus reduces extinction probability
PRINCIPLE D	Grouping should ideally be equidistant rather than linear – because of the increased exchange and recolonization opportunities so afforded
PRINCIPLE E	Provision of 'corridors' between reserves should be encouraged – these aid dispersal and improve the conservation function of individual sites, with little cost in land
PRINCIPLE F	Any given reserve should be as nearly circular in shape as other considerations permit, to minimize dispersal distances within the reserve

Source: Diamond (1975)

guidelines for designing the size, fragmentation, proximity, shape and patterns of distribution of reserves. Pickett and Thompson (1978), however, have questioned the relevance of island biogeographic theory to nature reserve design, because on islands immigration is important in maintaining species equilibrium, whereas in nature reserves immigration of species in the future might well not be possible (because of the disappearance of recolonization sources through habitat removal, etc.). They advocate the use of 'patch dynamics' for design purposes. They define patch dynamics as the internal dynamics generated by patterns of disturbance and subsequent patterns of succession (disturbances in the form of floods, hurricanes, etc.); and their approach is based on the notion of identifying 'minimum dynamic areas', or the smallest areas with a natural disturbance regime which maintains internal recolonization sources.

In the light of these studies concerning the design aspects of nature reserves, it is convenient to consider the existing machinery for designating National Nature Reserves in Britain, and recent re-appraisals of the nature conservation strategy in Britain.

8.3b National Nature Reserves in Britain

Although in Britain there are a large number of nature reserves of different types, established and managed by different bodies with different aims, in a formalized sense the body responsible for the designation and management of reserves in Britain is the Nature Conservancy Council. The Council was set up in 1949 as the Nature Conservancy, principally to establish a series of National Nature Reserves (NNRs), following recommendations made in *The Conservation of Nature in England and Wales* (Cmd 7122) and *Nature Reserves in Scotland* (Cmd 7814). Initially the Conservancy identified a balanced selection of 101 sites of different sizes, chosen to include both the unique and the typical; although, as Sheail (1976) has pointed out, many of the initial list of sites have not subsequently been designated as National Nature Reserves partly because they were not the best available, and partly because many were not available for acquisition. The first National Nature Reserve to be purchased was Beinn Eighe NNR in Wester Ross (10 450 acres), in 1951; and by September 1960 there were 84 National Nature Reserves totalling some 139 000 acres. Between 1949 and 1965 the Nature Conservancy had operated as an independent research council. When the Natural Environment Research Council was set up by Royal Charter in 1965, the Conservancy became a component part of NERC. Under the Nature Conservancy Council Act of 1973, however, the Council was made independent of NERC, and it was charged with fulfilling four main functions:

(1) the establishment, maintenance and management of nature reserves;
(2) the provision of advice for the Secretary of State, or any other Minister, on the development and implementation of policies for, or affecting, nature conservation in Britain;
(3) provision of advice and dissemination of knowledge about nature conservation;
(4) commissioning of support of research relevant to these functions.

One of the first major tasks carried out by the Nature Conservancy Council was to initiate a review and evaluation of all existing National Nature Reserves in Britain, and of all sites considered worthy of designation or suitable for protection. The 'Nature Conservation Review' was launched in 1966, and the final (two-volume) report on the Review was published in 1977. The three main phases of the Review programme have been outlined by Ratcliffe (1977c). Phase One saw the assembly of information on the sites considered, based on existing sources and on field surveys and checking carried out between 1967 and 1969. All sites were classified on the basis of major habitats (coastlands; woodland; lowland grassland heath and scrub; open water; peatland; upland grassland and heaths; artificial ecosystems). Figure 8.5 illustrates the large number and wide distribution of peatland sites examined during the Review. Phase Two involved chosing criteria for the evaluation of key sites. Considerable attention was devoted to the selection of appropriate criteria for evaluating each of the sites, and ten major criteria were adopted (Table 8.15). As Ratcliffe (1977c) advises, there was a tendency to highlight as important those features which are likely to disappear if no remedial action is taken. Thus natural, fragile and rare ecosystems tended to be highly valued. Phase Three was the final assessment of the sites.

Each site was evaluated on the basis of the ten criteria summarized in Table 8.15, and allocated to one of four points on a scale of conservation value with differing implications for their future management. Grade I sites are those considered to be of national or international importance, and the best examples of each type of habitat that are known.

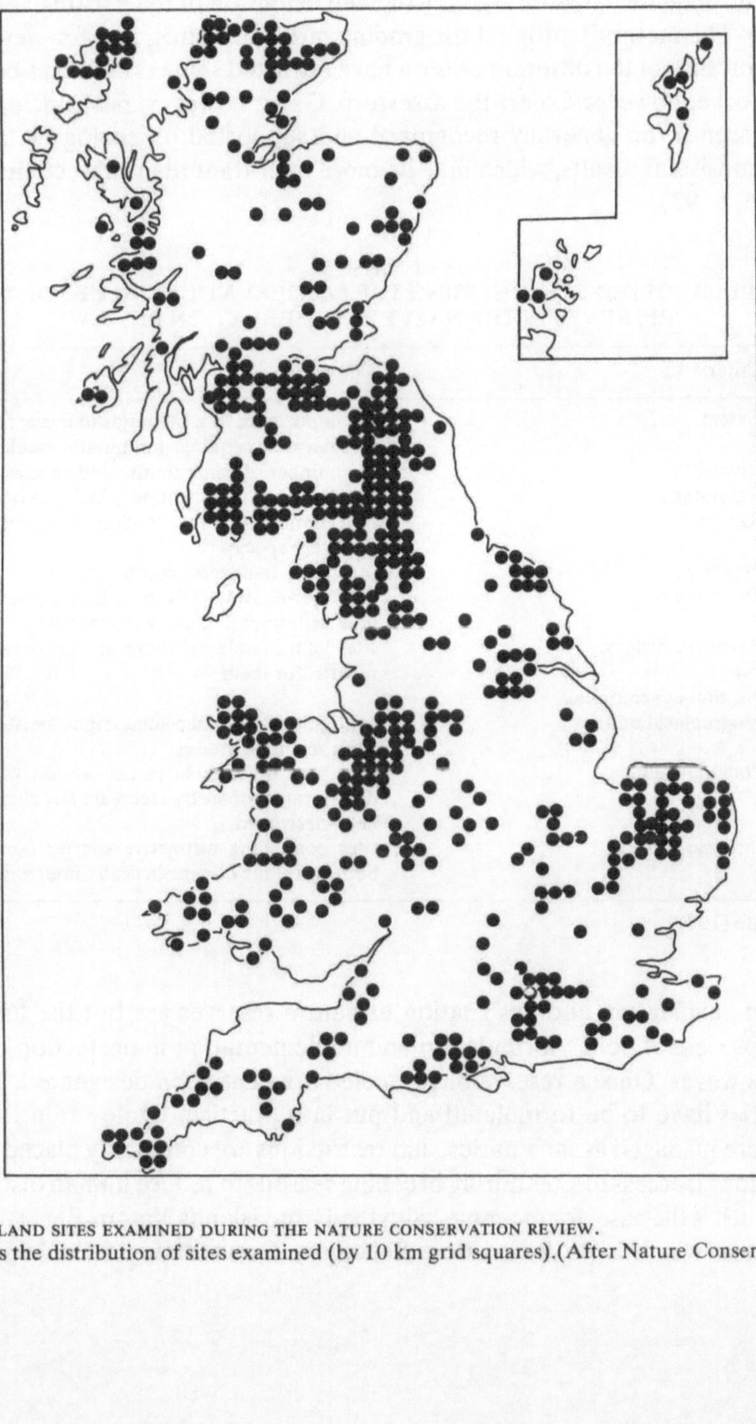

FIG. 8.5 PEATLAND SITES EXAMINED DURING THE NATURE CONSERVATION REVIEW.
The map shows the distribution of sites examined (by 10 km grid squares). (After Nature Conservancy Council, 1970.)

Grade II sites are close substitutes for Grade I, and of nearly equivalent scientific merit. Grade I and II sites are viewed as 'key sites' of National Nature Reserve quality. Grade III sites are of 'high regional importance', and Grade IV are of 'local significance'; both are deemed not to be worthy of the status of National Nature Reserves, but should be designated (if not already done so) as 'Sites of Special Scientific Interest'. Ratcliffe (1977c) stresses that the list and evaluations are not intended to be final, but that the Review seems to provide an adequate basis for urgently needed extension of the existing network of key reserve sites. The methods adopted for grading individual sites, and for weighing up the relative importance of the different criteria have attracted some criticism as being perhaps highly subjective, however expert the assessors. Gane, however, has concluded that 'the criteria are seen to be generally recognized and supported by ecologists, and produce reasonably consistent results, which may be more important than subjectivity in practice' (Gane, 1976, p. 97).

Table 8.15
CRITERIA ADOPTED FOR ASSESSING THE CONSERVATION VALUES OF NATURE RESERVES IN THE NATURE CONSERVATION REVIEW

Number	Criteria	Comments
1	Extent:	the importance of a site tends to increase with area, and safeguarding requires a minimum viable unit size
2	Diversity:	the number of communities and species present
3	Naturalness:	the degree of modification which has occurred
4	Rarity:	rare communities have tended to receive more attention than rare species
5	Fragility:	sensitivity to adverse conditions
6	Typicalness:	all major ecosystems should be represented, no matter how widespread or common they are
7	Recorded history:	sites gain in value if there are long-standing scientific records for them
8	Position in ecological/ geographical unit:	contiguity with other sites might assist in sustaining a site's scientific interest
9	Potential value:	sites may be capable of restoration where they have deteriorated, or used to recreate an ecosystem which has been destroyed
10	Intrinsic appeal:	sites containing distinctive species (such as colourful butterflies) may be intrinsically interesting

Source: Ratcliffe (1971)

The design, evaluation and designation of nature reserves are but the first steps in a continuous process of policy formulation and implementation in protection of ecological resources, however. Once a reserve or protected area has been designated, the management priorities have to be formulated and put into practice. Quite often the protected ecosystems are managed as sanctuaries, and restrictions are commonly placed on numbers of visitors allowed access to sites during breeding seasons to reduce human disturbance to a minimum. Such is the case, for example, with the Farne Islands Nature Reserve, run by the National Trust, which is protected by a Sanctuary Order dating from 1964 (Hawkey, 1976).

Restrictions on access and disturbance are but one element in most management plans, however. Priorities adopted within management strategies will inevitably reflect local circumstances and the history of use and previous management of the area. Thus a relatively accessible area which has long been associated with amenity and recreation will have different management priorities to an isolated area. An example of the former is the Malham Tarn Estate in Yorkshire, whose management policy has been described by Disney (1975). The three main priorities are:

(1) *Primary land use*: to conserve (and enhance, where possible) the *wildlife* and *scenery*.
(2) *Secondary land use*: to realize the maximum potential of the *educational* and *research* values as will not prove incompatible with the primary land use.
(3) *Tertiary land use*: to allow the fullest enjoyment of as much of the inherent *recreational* values as can be accommodated without diminishing the higher priorities.

The island of Rhum, in the Inner Hebrides, on the other hand, has the following management objectives, documented by Eggeling (1964):

(1) *Primary objective*: 'to restore vegetation covers that have been lost, and to bring the island to a higher level of biological production than at present – and one that can be sustained naturally by the environment'.
(2) *Secondary objectives*:
   (a) *Conservation objectives*: 'to conserve the flora and fauna and other features of scientific interest, and to enhance the diversity of habitats, provide shelter and increase the biological turnover by means of reafforestation'.
   (b) *Research objectives*: 'to undertake such research as necessary to attain the primary aim of management, and the conservation objectives; to carry out on the reserve such fundamental and applied research as may be of value in the development of conservation techniques for use elsewhere'.
   (c) *Other objectives*: 'to manage the reserve in the best traditions of estate management in order to ensure a stable and contented population of estate staff, wardens and scientists, co-operating as a community and a team'.

These two examples of management plans illustrate the general tendency for conservation objectives to be regarded as of primary importance, and for other objectives to be ranked by decreasing importance commensurate with meeting the primary objective. They also illustrate the specifically local orientation of most planning schemes. These site-specific management schemes are complemented by a series of general management policy

Table 8.16
GENERAL MANAGEMENT AIMS IN WOODLAND NATURE RESERVES

(1) Some woodlands should be permitted to develop naturally, without intervention by man
(2) Woodland nature reserves should contain examples of all the important semi-natural woodland types of native tree species, preferably with different stages of development represented, especially mature woodland
(3) Examples of intensively managed woodlands with a special conservation value should be included and kept under the traditional management system, especially if there is a danger of wide-scale conversion to other forms of use (such as economic forestry)
(4) Woodlands containing trees of species with restricted distributions (notably rowan, whitebeam, elm and willow) should be protected against destruction
(5) Facilities should be provided for education, for research into the biology and management of woodlands of native species, and for determining the importance of these woodlands to the conservation of natural resources

Source: Ovington (1964)

Table 8.17
SOME GENERAL GUIDELINES FOR MANAGEMENT OF NATURE RESERVES

(1)	Preservation of botanical richness of nature reserves is best assured if management is closely modelled on methods applied previously in the area
(2)	If the character and size of the reserve are suitable, internal regulation can be increased by increasing the development of ecoclines (this will be related to the degree of human interference)
(3)	Management operations should be done gradually and on a small scale
(4)	External protection of nature reserves is related to their form and size. The more nearly circular the form, and the larger the area, the safer the reserves will be against alterations threatening them from outside
(5)	If the size of nature reserves is suitable, the controller can take advantage of ecoclines which develop along the outskirts by interaction of internal and external influences
(6)	If alterations are induced from outside, the internal botanical control must be directed towards delaying the processes evoked by them

Source: Westhoff (1970)

guidelines for nature conservation in various situations. Thus for example, Ovington (1964) has specified a series of overall management aims of woodland nature reserves in Britain (Table 8.16), and he concluded that the best type of management in such habitats involves a zonal classification of the overall ecological resource, on the basis of differing degrees of disturbance. An example is afforded by the management of Yarner Wood National Nature Reserve in Devon (Figure 8.6). Westhoff (1970) has outlined a series of

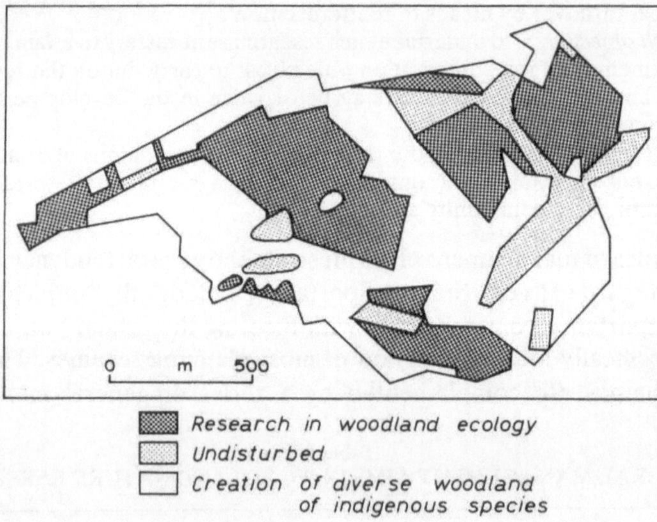

FIG. 8.6 PATTERN OF ZONAL MANAGEMENT PRIORITIES IN A WOODLAND NATURE RESERVE – YARNER WOOD NATIONAL NATURE RESERVE, DEVON.
(After Ovington, 1964.)

more general management guidelines for nature reserves in general (Table 8.17), based on ecological principles of species and environmental diversity. Because management schemes for preserving ecological resources in nature reserves, national parks and wilderness areas often identify a series of objectives, Stone (1965) has warned that in many situations objectives may be mutually incompatible. For example, two common objectives in American Parks are to preserve particular successional stages (for ecological, scenic or other reasons), and to slow down succession so as to obtain a vegetation mosaic containing most of the successional stages characteristic of a given ecosystem. Peterken (1977) has also evaluated the conflict between nature conservation under the control of the Forestry Commission, visual amenity and unrestrained timber production by cheapest methods, which commonly arises in British woodlands.

8.4 ECOLOGY AND ENVIRONMENTAL IMPACTS

The approaches to preservation of ecological resources outlined in the previous section illustrate the various forms of protection and management commonly adopted in conservation. Although these are important in protecting ecosystems and endangered and rare species and habitats, on the whole relatively small areas of the countryside are preserved in these ways, and in most of the environment there is relatively little conservation management. The Nature Conservancy Council (1977a) has formulated a 'Conservation Management Strategy' (Table 8.18) which includes both scheduled sites and the wider

Table 8.18
A 'CONSERVATION MANAGEMENT STRATEGY'

Nature of land	Area	Degree of conservation management
Scheduled sites		
National and other nature reserves	small	Conservation management primary
Sites of Special Scientific Interest		Conservation management primary, or agricultural, forestry, etc., management primary, but compatible with maintaining scientific interest of sites
The wider countryside		
Good wildlife habitats (primary woodland, old pastures, marshes, etc.)		Conservation management secondary, but important
Poor wildlife habitats (intensively farmed croplands)		Conservation management secondary and, apart from soil organisms, unimportant
'Lifeless' component of urban areas	large	Conservation requirement nil

Source: Nature Conservancy Council (1977a)

countryside, and which attempts to offer most direct protection to better quality wildlife habitats. This is understandable in the light of the social and economic constraints within which conservation operates, and because it would be impractical (if, indeed, desirable at all) to seek to protect all parts of the landscape directly. Many opponents of conservation fail to appreciate the basic difference in management strategies between overall *preservation* (a static approach which assumes no change, where management is geared to reducing or stopping all environmental change and external stimuli), and *conservation* (a dynamic approach which allows for change, and in which management is of basic importance), although few conservationists would argue that all habitats must be conserved, regardless of their ecological value.

From an environmental management point of view, however, some form of protection from adverse environmental change should be offered to non-scheduled sites, because of the overall unity and inter-relatedness of the biosphere and individual ecosystems, and because it would be ecologically and ethically unjustifiable simply to conserve scheduled areas and ignore the wider countryside environment. Within rural planning, therefore, there are increasing demands for some form of environmental accountability. Planners called upon to evaluate a proposal for a change in land use need to consider the current state of all environmental relationships in the area. Edington and Edington (1977) have pointed out the two main problems behind such planning evaluations. One is that each activity has its own site requirements in relation to natural landscape features (for example, active river floodplains are unsuitable for urban development); and the other is that different activities show different degrees of compatibility when close to each other. The planner, faced with the need to take full account of the various environmental consequences of proposed developments or landscape changes, appears to have two broad options open to him. Option one is to seek information on environmental impacts on an *ad hoc* basis from specialists in the various environmental fields, and, indeed, this is often done. The alternative option is to adopt a framework in which each potential environmental modification which might stem from a proposed development is considered before the development is approved, and such integrated evaluations are becoming increasingly possible through the use of 'Environmental Impact Assessment' techniques developed initially in the United States (Johnson, 1975).

8.4a Environmental Impact Assessment

Environmental planning in the United States has made considerable progress since the passing of the *United States National Environmental Policy Act* (NEPA) in 1969. The Act had four main aims:

(1) to declare a national policy to encourage productive and enjoyable harmony between man and environment,
(2) to promote efforts to prevent or eliminate damage to the environment and the biosphere, and stimulate the health and welfare of man,
(3) to increase understanding of ecological systems and natural resources important to the nation,
(4) to establish a Council on Environmental Quality.

One main outcome of the NEPA legislation was that all Federal Agencies were required to prepare an Environmental Impact Statement for all major developments, before approval would be given for the development. Each Statement was to consider a wide range of aspects of likely environmental impacts, such as the environmental impact of the proposed action; any adverse effects which could not be avoided if the proposal were to be implemented; alternatives to the proposed action; the relationship between local, short-term use of the environment and the maintenance and enhancement of long-term productivity; and any irreversible commitments of resources which would be involved in the proposed action if it were to be implemented. Although, under the 1969 Act, the Statements were designed to be made by Federal Agencies, many individual States have subsequently used the concept of Environmental Impact Assessment as a basis for State legislation. Considerable attention was given to the development of suitable procedures for carrying out the Assessments, and the scheme widely used is summarized in Table 8.19.

Table 8.19
SUMMARY OF AMERICAN PROCEDURES FOR ENVIRONMENTAL IMPACT ASSESSMENT

(1) Statement of objectives: definition of the objectives sought by the proposed development
(2) Technical possibilities of achieving the objectives
(3) Proposed actions and alternatives: for achieving the stated objectives
(4) Report on the character of the environment *before* action begins
(5) Principle alternative engineering proposals submitted as reports; with analysis of the monetary costs and benefits of each engineering alternative
(6) Proposed plan (engineering report) and report on the present environment are considered; this allows evaluation of the likely environmental impact of the proposal. Impacts are evaluated for each major alternative plan. Attention centres on
  (a) Magnitude of the impact (scale),
  (b) Importance of the impact (significance)
(7) Assessment of environmental impacts of each alternative plan of action
(8) Environmental Impact Statement is produced; this summarizes the whole analysis, and lists final recommendations and the relative merits of each of the main alternatives

Source: Leopold et al. (1971)

The Assessment Stage of the analysis is clearly of fundamental importance, and Leopold and his colleagues (1971) presented a matrix approach which acts as a check-list of environmental impacts, and which 'provides a format for comprehensive review to remind the investigators of the variety of interactions that might be involved. It helps the planners to identify alternatives which might lessen impact.' The 'Leopold-matrix' approach focuses attention on the likely 'first-order' (rather than secondary or indirect) effects of specific actions, and an illustration of the approach is given in Figure 8.7. Each possible impact of a specific action is first of all identified (the diagonal lines on the matrix), then evaluated. Evaluations are made of the relative *magnitude* (magnitude 10 is largest, on a scale 1 to 10; these values appear in the upper left corner of the boxes in the matrix), and *importance* (similar scale 1 to 10, shown lower right) of each impact. When alternative proposals or actions are being considered, a matrix of the form shown in Figure 8.7 is compiled for each alternative. Numerical comparison of the environmental impacts for each alternative can thus be allowed by comparing identical boxes in the two matrices. The final Environmental Impact Statement includes four reports – a justification of the proposal (with analysis of the needs for it); a description of the environment to be affected; details of the proposed action; and the completed Environmental Impact Assessment. It is important to note that it is the Federal Agency which makes the Assessment and Statement, *not* the applicant, although the applicants are commonly required to provide most of the information used in the Assessment.

Since 1969 there have been a large number and wide range of Environmental Impact Assessments in the United States. Two illustrative examples are the report by West (1976) on the conflict between the development of mineral exploitation and the preservation of natural wilderness in Alaska, which recommended the orderly development of certain areas and the preservation of the remainder of the region for wildlife habitats and recreation; and Kilburn's (1976) analysis of the likely environmental impacts of developing oil shale resources in Colorado, which considered impacts on hydrology, water quality and salinity; surface disposal of processed shale; revegetation of worked areas; ecology (ecosystem modification or removal; processes interrupted; increased human interference and trampling); and air quality.

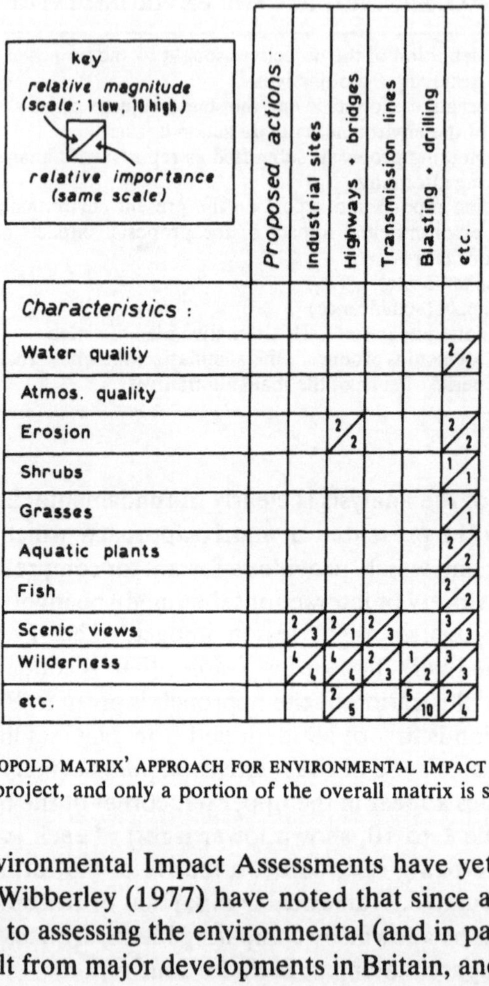

FIG. 8.7 EXAMPLE OF THE 'LEOPOLD MATRIX' APPROACH FOR ENVIRONMENTAL IMPACT ASSESSMENT.
The example is of a mining project, and only a portion of the overall matrix is shown for illustration. (After Leopold et al., 1971.)

Although formal Environmental Impact Assessments have yet to be introduced into Britain, Davidson and Wibberley (1977) have noted that since about 1965 much more attention has been paid to assessing the environmental (and in particular the ecological) damage that might result from major developments in Britain, and a number of planners have commented that there is scope for further action in reducing the environmental effects of development, and for preparing Environmental Impact Statements *before* mandatory public enquiries. There is a growing lobby of opinion that the normal processes of development control and the Public Enquiry system are not the most efficient ways of seeking to ensure that the environmental consequences of major developments are given proper consideration, and the Department of the Environment has itself considered whether a formal Environmental Impact Assessment approach could be adopted profitably in Britain. Dobry (1975), for example, has suggested a much broader perspective on the assessment process than that accepted in the United States, because he stresses the benefits of objective assessments of the effects of proposed developments on areas such as traffic and transportation, drainage, public services, employment, as well as landscape and pollution. It seems likely that in the near future Britain might adopt formalized schemes for evaluating environmental impacts, because at the present time (December 1978) the EEC is considering the possible application of Environmental Impact Assessment by member states, and the scope for co-ordinating the assessment schemes operated by individual states to ensure that procedures are not in conflict. The international nature of the problem at this level means that the EEC has to evaluate how and to what extent Impact Assessment can be implemented at the Community level, especially in view of assessing common policies and measures, and to ensure that the application of Environmental Impact Assessment procedures in member states will not become a source of competitive distortions.

8.5 CONCLUSIONS

It is clear from this chapter that increasing awareness of the need for wise use and management of ecological resources is being complemented by the development of approaches towards evaluating ecological resources, formulating appropriate management strategies based on sound ecological theory, and evaluating the likely environmental impacts (which include specifically ecological impacts) of proposed landscape changes and developments. In the light of the recognition of the need to adopt environmental management policies, which stems in part from what in Chapter 1 was termed the 'Environmental Crisis', it is comforting to think that in the 1970s steps were taken to ensure that natural ecosystems and environments will be preserved for the enjoyment and benefit of future generations. To what extent the application of ecological knowledge in planning in general, and in environmental planning in particular, will be of value in ensuring preservation of our wildlife heritage and in perpetuating the stability of the biosphere will only become apparent in the future. Ironically many conservationists fear that it has taken so long to become aware of the need for wise environmental management based on the principles of ecology, that the best we can hope for now is to prevent further disruption to natural ecosystems, and resource exploitation and environmental pollution. During the present decade there has been much speculation about future environments and about the directions in which human societies and economic systems might have to change in the face of resource depletion, environmental pollution and degradation, and a number of observers have predicted a 'Post-Industrial Society' along the lines advocated in the *Blueprint for Survival* (Ecologist, 1972) and prophecied by the *Limits to Growth* study (Meadows et al., 1972). Crystal-ball gazing is not within the realms of this book, but whatever the future holds in store the present generation of environmental managers will have failed miserably if wildlife resources diminish in quality and quantity in the future, and if future generations are not bequeathed natural habitats, functioning ecosystems, a stable biosphere and as diverse an assemblage of plant and animal species as economic, social and technological/scientific factors will allow. As Gary observed, 'we need all the dogs, all the cats, and all the birds, and all the elephants we can find, . . . we need all the friendship we can find around us' (Gary, 1958).

Bibliography

ABELSON, P. H., 1971, 'Changing attitudes towards environmental problems', *Science*, 172, p. 517.

ACKERMAN, E. A., 1959, 'Population and Natural Resources', pp. 621–48 in *The study of population; an inventory and appraisal*, Hanser, P. H. & Duncan, O.D. (eds.) (University of Chicago Press. Chicago).

AGER, D. V., 1976, 'The nature of the fossil record', *Proceedings of the Geologists' Association*, 87, pp. 131–60.

ALLEE, W. C., 1926, 'Distribution of animals in a tropical rain forest with relation to environmental factors', *Ecology*, 7, pp. 445–68.

— and SCHMIDT, V. P., 1951, *Ecological Animal Geography* (Wiley. New York).

ALLEN, J. R., 1975, *Physical Geology* (George, Allen & Unwin. London).

ARVILL, R., 1967, *Man and Environment – Crisis and the Strategy of Choice* (Penguin. Harmondsworth).

ASH, M., 1972, 'Planners and ecologists', *Town and Country Planning*, 40, pp. 219–21.

AULTFATHER, W. and CROZIER, E. S., 1971, 'A resource inventory and planning system for wildlife areas', *Journal of Wildlife Management*, 35, pp. 168–74.

BAER, J. G., 1967, 'The International Biological Programme', *Nature and Resources*, 3, pp. 1–3.

BAKKER, R. T., 1971, 'Ecology of the Brontosaurs', *Nature*, 229, pp. 172–4.

— 1972, 'Anatomical and ecological evidence of endothermy in dinosaurs', *Nature*, 238, pp. 81–5.

BALMER, F., ROTHWELL, P. and BOATMAN, D., 1975, 'Dieback of trees in North Humberside, with special reference to oak', *The Naturalist*, 935, pp. 55–8.

BARNES, R. A., 1976, 'Long-term mean concentrations of atmospheric smoke and sulphur dioxide in country areas of England and Wales', *Atmospheric Environment*, 10, pp. 619–31.

BARNETT, H. J. and MORSE, C., 1963, *Scarcity and Growth* (Johns Hopkins Press, Baltimore).

BARNETT, S. A., 1968, *The Human Species – The Biology of Man* (Pelican, Harmondsworth).

BARR, J., 1970, *The Assaults on our Senses* (Methuen, London).

— 1971 (ed.), *The Environmental Handbook – Action Guide for the UK* (Ballantine/Friends of the Earth. London).

BARTSCH, A. F., 1970, 'Accelerated eutrophication of lakes in the United States – ecological response to human activities', *Environmental Pollution*, 1, pp. 133–40.

BASSHAM, J. A., 1977, 'Increasing crop production through more controlled photosynthesis', *Science*, 197, pp. 630–8.

BEARD, J. S., 1944, 'Climax vegetation in tropical America', *Ecology*, 25, pp. 127–58.

BEAZLEY, R., 1967, 'Conservation decision-making; a rationalization', *Natural Resources Journal*, 7, pp. 345–60.

BECK, W. M., 1955, 'Suggested method for reporting biotic data', *Sewage and Industrial Wastes Engineering*, 27, pp. 1193–7.

BELL, K. A. and BLISS, L. C., 1973, 'Alpine disturbance studies – Olympic National Park, U.S.A.', *Biological Conservation*, 5, pp. 25–32.

BENNETT, C. F., 1960, 'Cultural animal geography – an inviting field of research', *Professional Geographer*, 12, pp. 12–14.

BENNETT, R. J. and CHORLEY, R. J., 1978, *Environmental Systems – Philosophy, Analysis and Control* (Methuen. London).

BERRILL, N. J., 1967, *Inherit the Earth – the story of man and his changing planet* (Fawcett. Greenwich, Connecticut).

BERRY, B. J. L. and HORTON, F. E., 1974, *Urban Environmental Management – Planning, for Pollution Control* (Prentice Hall. Englewood Cliffs, N.J.).

BEST, R. H., 1976, 'The extent and growth of urban land', *The Planner*, 62, pp. 8–11.

BEYERS, R. J., 1963, 'The metabolism of twelve aquatic laboratory micro-ecosystems', *Ecological Monographs*, 33, pp. 281–306.

BIBBY, J. S. and MACKNEY, D., 1969, 'Land use capability classification', *The Soil Survey Technical Monograph 1. Soil Survey of Great Britain* (Harpenden).

BILLINGS, W. D., 1952, 'The environmental complex in relation to plant growth and distribution', *Quarterly Review of Biology*, 27, pp. 251–65.

— 1964, *Plants and the Ecosystem* (Macmillan. London).

BIRCH, J. W., 1973, 'Geography and resource management', *Journal of Environmental Management*, 1, pp. 3–11.

BLACKMAN, F. F., 1905, 'Optima and limiting factors', *Annals of Botany*, 19, pp. 281–95.

BLISS, C. I., 1935, 'The calculation of the dosage-mortality curve', *Annals of Applied Biology*, 22, pp. 134–67.

BLISS, L. C., COURTIN, G. M., PATTIE, D. L., RIEWE, R. R., WHITFIELD, D. W. A. and WIDDEN, P., 1973, 'Arctic tundra ecosystems', *Annual Review of Ecology and Systematics*, 4, pp. 359–400.

BOND, R. R., 1957, 'Ecological distribution of breeding birds in the upland forests of southern Wisconsin', *Ecological Monographs*, 27, pp. 351–84.

BORGSTROM, G., 1969, *Too Many – a Study of the Earth's Biological Limitations* (Macmillan. New York).

BORMANN, F. H., and LIKENS, G. E., 1967, 'Nutrient cycling', *Science*, 155, pp. 424–9.

— LIKENS, G. E., and EATON, J. S., 1969, 'Biotic regulation of particulate and solution losses from a forest ecosystem', *Bioscience*, 19, pp. 600–11.

— LIKENS, G. E., SICCAMA, T. G., PIERCE, R. S., and EATON, J. S., 1974, 'The export of nutrients and recovery of stable conditions following deforestation at Hubbard Brook', *Ecological Monographs*, 44, pp. 255–77.

— SICCAMA, T. G., LIKENS, G. E., and WHITAKKER, R. H., 1970, 'The Hubbard Brook Ecosystem Study – composition and dynamics of the tree stratum', *Ecological Monographs*, 40, pp. 373–88.

BOTKIN, D. B., JANCK, J. F., and WALLIS, J. R., 1972, 'Some ecological consequences of a computer model of forest growth', *Journal of Ecology*, 68, pp. 849–72.

BOYKO, H., 1947, 'On the role of plants as quantitative climate indicators and the geo-ecological law of distribution', *Journal of Ecology*, 35, pp. 138–57.

BRADSHAW, M., 1977, *Earth – the Living Planet* (Hodder & Stoughton. London).

BRAUN-BLANQUET, J., 1932, *Plant Sociology* (McGraw Hill. New York).

BRAY, J. R., and CURTIS, J. T., 1957, 'An ordination of the upland forest communities of southern Wisconsin', *Ecological Monographs*, 27, pp. 325–49.

BRIERLEY, J. K., 1956, 'Some preliminary observations on the ecology of pit heaps', *Journal of Ecology*, 44, pp. 383–90.

BRIGHTMAN, F. H. (ed.), 1976, 'Survey of the flora and fauna of proposed Channel Tunnel sites near Folkestone, Kent, 1974', *Transactions of the Kent Field Club*, 6, pp. 5–51.

BROCK, T. D., 1967, 'The ecosystem and the steady state', *Bioscience*, 17, pp. 166–9.

BRUNIG, E. F., 1977, 'The tropical rain forest – a wasted asset or an essential biospheric resource?', *Ambio*, 6, pp. 187–91.

BUDYKO, M. I., 1971, *Climate and Life* (Hydrological Publishing House. Leningrad).

BUGLER, J., 1972, *Polluting Britain – a Report* (Penguin. Harmondsworth).

BUNCE, R. G. H., 1968, 'An ecological survey of Ysgolion Duon, a mountain cliff in Snowdonia', *Journal of Ecology*, 56, pp. 59–75.

— and Shaw, M. W., 1973, 'A standardized procedure for ecological survey', *Journal of Environmental Management*, 1, pp. 239–58.

Burdekin, D. A., and Gibbs, J. N., 1972, 'Dutch Elm Disease – recurrence and recovery in Britain', *Nature*, 240, p. 306.

Burgess, A., 1960, 'Time and size as factors in ecology', *Journal of Ecology*, 48, pp. 273–85.

Burton, I., and Kates, R. W., 1978, *The Environment as Hazard* (Oxford University Press. Oxford).

Caborn, J. M., 1971, 'The agronomic and biological significance of hedgerows', *Outlook on Agriculture*, 6, pp. 279–84.

Cain, S. A., 1939, 'The climax and its complexities', *The American Midland Naturalist*, 21, pp. 146–81.

— 1944, *Foundations of Plant Geography* (Harper & Row. New York).

Calder, N., 1967, *The Environment Game* (Panther. London).

Caldwell, L. K., 1966, 'Problems of applied ecology', *Bioscience*, 16, pp. 524–7.

Carson, R., 1962, *Silent Spring* (Houghton Mifflin. Boston, Mass.).

Castri, F. Di, and Mooney, H. A. (eds), 1973, *Mediterranean-type Ecosystems – Origin and Structure* (Chapman & Hall. London).

Chapman, J. D., 1969, 'Interactions between man and his resources', pp. 31–42 in *Resources and Man – a Study and Recommendations*, Committee on Resources and Man (ed.) (Freeman. San Francisco).

Chapman, P., 1975, *Fuel's Paradise* (Penguin. Harmondsworth).

Chorley, R. J., 1969, 'The drainage basin as the fundamental geomorphic unit', in *Water, Earth and Man*, Chorley, R. J. (ed.) (Methuen. London).

— and Kennedy, B. A., 1971, *Physical Geography – a Systems Approach* (Prentice Hall).

Churchill, E. D., and Hanson, H. C., 1958, 'The concept of climax in Arctic and Alpine vegetation', *Botanical Review*, 24, pp. 127–91.

Cieslinski, T. A., and Wagar, J. A., 1970, 'Predicting the durability of forest recreation sites in northern Utah. Preliminary Results', *United States Department of Agriculture, Forest Service Research Note* INT 117.

Clapham, W. B., 1973, *Natural Ecosystems* (Macmillan. London).

— 1976, 'An approach to quantifying the exploitability of human ecosystems', *Human Ecology*, 4, pp. 1–30.

Clarke, R., 1973, 'Technology for an alternative society', *New Scientist*, 57, pp. 66–7.

Clayton, K., 1971, 'Reality in conservation', *Geographical Magazine*, 44, pp. 83–4.

— 1976, 'Environmental sciences/studies; a decade of attempts to discover a curriculum', *Area*, 8, pp. 98–101.

Clements, F. E., 1916, *Plant Succession* (Carnegie Institute, Washington, Publication Number 242).

— 1936, 'Nature and structure of the climax', *Journal of Ecology*, 24, pp. 252–84.

— and Shelford, V. E., 1939, *Bioecology* (Wiley. New York).

Cole, L. C., 1958, 'The Ecosphere', *Scientific American*, 198, pp. 83–92.

Cole, M., 1971, 'Plants, animals and environment', *Geographical Magazine*, 44, pp. 230–1.

Coleman, A., 1961, 'The Second Land-Use Survey – Progress and Prospect', *Geographical Journal*, 127, pp. 168–86.

Colinvaux, P., 1973, *Introduction to Ecology* (Wiley. New York).

— 1976, 'Review of "Human Ecology", edited by F. Sargent', *Human Ecology*, 4, pp. 263–6.

Commoner, B., 1972, *The Closing Circle: Confronting the Environmental Crisis* (Cape. London).

Connell, J. H., 1978, 'Diversity in tropical rain forests and coral reefs', *Science*, 199, pp. 1302–10.

Conservation Course, 1972, *The Assessment of Ecological Value; a new approach and field evaluation* (University College, London).

Cooke, R. U., and Doornkamp, J. C., 1974, *Geomorphology in Environmental Management* (Clarendon. Oxford).

Cooper, W. S., 1923, 'The recent ecological history of Glacier Bay, Alaska', *Ecology*, 4, pp. 93–128, 223–46 and 355–65.

— 1926, 'The fundamentals of vegetational change', *Ecology*, 7, pp. 391–413.

COPPOCK, J. T., 1970, 'Land-use changes and the hedgerow,' in *Hedges and Hedgerow Trees*, Nature Conservancy (ed.) (Monks Wood Experimental Station Symposium Number 4).

— and SEWELL, W. R. D., 1975, 'Resource management and public policy – the changing role of geographical research', *Scottish Geographical Magazine*, 91, pp. 4–11.

CORBET, G. B., 1971, 'Provisional distribution maps of British mammals', *Mammal Review*, 1, pp. 95–142.

CORNWALLIS, R. K., 1969, 'Farming and wildlife conservation in England and Wales', *Biological Conservation*, 1, pp. 142–7.

COTGROVE, S., 1976, 'Environmentalism and Utopia', *Sociological Review*, 24, pp. 23–42.

COTTRELL, A., 1978, *Environmental Economics; an introduction for students of the resource and environmental sciences* (Arnold. London).

COUNCIL FOR EUROPE, 1977, 'List of rare, threatened and endemic plants in Europe', *Council of Europe, Nature Conservation Series Number* 14.

COUNCIL FOR THE PROTECTION OF RURAL ENGLAND, 1971, *Loss of Cover through removal of hedgerows and trees* (CPRE).

COUNTRYSIDE COMMISSION, 1971, *Changing Countryside Project – a Report* (The Countryside Commission).

COUSENS, J., 1974, *An Introduction to Woodland Ecology* (Oliver & Boyd. Edinburgh).

COX, C. B., HEDLEY, I. N., and MOORE, P. D., 1973, *Biogeography – an Ecological and Evolutionary Approach* (Blackwell. Oxford).

CRAMP, S., BOURNE, W. R. P., and SAUNDERS, D., 1974, *The Seabirds of Britain and Ireland* (Collins. London).

CREED, E. R., LEES, D. R. and DUCKETT, J. G., 1973, 'Biological method of estimating smoke and sulphur dioxide pollution', *Nature*, 244, pp. 278–80.

CRISP, D. T., 1966, 'Input and output of minerals for an area of Pennine Moorland – the importance of precipitation, drainage, peat erosion and animals', *Journal of Applied Ecology*, 3, pp. 327–48.

CROCKER, R. L., and MAJOR, J., 1955, 'Soil development in relation to vegetation and surface age at Glacier Bay, Alaska', *Journal of Ecology*, 43, pp. 427–48.

CROFTS, R. S., and COOKE, R. U., 1974, 'Landscape evaluation; a comparison of techniques', *University College London, Occasional Papers in Geography* 25.

CROIZAT, L., 1952, *Manual of Phytogeography* (Junk. The Hague).

CROSSLAND, J., 1978, 'Reporting pollution', *Environment*, 20, pp. 29–31.

CROWE, J., 1968, 'Toward a "definitional model" of public perceptions of air pollution', *Journal of the Air Pollution Control Association*, 18, pp. 154–8.

CUMMINS, K. W., COFFMAN, W. P., and ROFF, P. A., 1966, 'Trophic relationships in a small woodland stream', *Ver. Inst. Vèr. Limnol.*, 16, pp. 627–38.

CURTIS, J. T., and MCINTOSH, R. P., 1951, 'An upland forest continuum in the prairie-forest border region of Wisconsin', *Ecology*, 32, pp. 476–96.

DANSEREAU, P., 1951, 'Description and recording of vegetation upon a structural basis', *Ecology*, 32, pp. 172–229.

— 1957, *Biogeography; an Ecological Perspective* (Ronald. New York).

DARLINGTON, P. J., 1957, *Zoogeography – the Geographical Distribution of Animals* (Wiley. New York).

DARNELL, R. M., 1970, 'Evolution in the tropics', *American Zoologist*, 10, pp. 9–15.

DARWIN, C., 1859, *The Origin of Species* (Murray. London).

DASMANN, R. F., 1972, 'Towards a system for classifying natural regions of the world and their representation by national parks and reserves', *Biological Conservation*, 4, pp. 247–55.

— 1973, 'A rationale for preserving natural areas', *Journal of Soil and Water Conservation*, 28, pp. 114–17.

— 1974, 'Conservation, counter-culture and separate realities', *Environmental Conservation*, 1, pp. 133–7.

— 1975, *The Conservation Alternative* (Wiley. New York).

— 1976, *Environmental Conservation* (Wiley. New York).

DAVIDSON, J.. and WIBBERLEY, J., 1977, *Planning and the Rural Environment* (Pergammon. Oxford).

DAVIES, B. E., and PINSENT, R. J. F. H., 1975, 'Minerals and morbidity', *Cambria*, 2, pp. 85–93.

DAVIS, B. N. K., 1976, 'Wildlife, urbanization and industry', *Biological Conservation*, 10, pp. 249–91.

DAVIS, R. D., and BECKETT, P. H. T., 1978, 'The use of young plants to detect metal accumulation in soils', *Water Pollution Control*, 77, pp. 193–210.

DAVIS, W. M., 1906, 'An inductive study of the content of geography', *Bulletin of the American Geographical Society*, 38, pp. 67–84.

DAWSON, J. A., and THOMAS, D., 1975, *Man and his World* (Nelson. London).

DAY, T. J., 1972, 'Stand structure, succession and use of southern Alberta's Rocky Mountain Forest', *Ecology*, 53, pp. 472–8.

DEMPSTER, J. P., KING, M. L., and LAKHANI, K. H., 1976, 'The status of the swallowtail butterfly in Britain', *Ecological Entomology*, 1, pp. 71–84.

DIAMOND, J. M., 1975, 'The island dilemma; lessons of modern biogeographic studies for the design of nature reserves', *Biological Conservation*, 7, pp. 129–46.

DICE, L. R., 1952, *Natural Communities* (University of Michigan Press. Ann Arbor).

DICKSON, D., 1974, *Alternative Technology and the Politics of Technical Change* (Fontana/Collins. London).

DIMBLEBY, G. W., 1975, 'Archaeological evidence of environmental change', *Nature*, 256, p. 265.

DISNEY, R. H. L., 1968, 'The terms ecology and natural history', *Journal of Biological Education*, 2, pp. 235–7.

— 1970, 'On natural history and ecology', *Journal of Biological Education*, 4, pp. 183–6.

— 1975, 'Review of management policy for the Malham Tarn Estate', *Field Studies*, 4, pp. 223–42.

DOBRY, G., 1975, *Review of the Development Control System (Final Report)* (H.M.S.O. London).

DOBZHANSKY, T., 1950, 'Evolution in the tropics', *American Scientist*, 38, pp. 209–21.

— 1955, *Evolution, Genetics and Man* (Chapman & Hall. London).

DOKUCHAYEV, V. V., 1898, *Writing* (Akademia Nauk. Moscow).

DOWER, M., 1964, 'The function of open country', *Journal of the Town Planning Institute*, 50, pp. 132–40.

DOWNS, A., 1972, 'Up and Down with Ecology – the "Issue Attention Cycle"', *Public Interest*, 28, pp. 38–50.

DUFFEY, E., 1968, 'Ecological studies on the Large Copper Butterfly (Lycaena dispar Haw. batanus Obth) at Woodwalton Fen National Nature Reserve, Huntingdonshire', *Journal of Applied Ecology*, 5, pp. 69–96.

— 1970, *Conservation of Nature* (Collins. London).

— 1974, *Nature Reserves and Wildlife* (Heinemann. London).

— 1978, 'An ecological survey of National Trust properties', *Bulletin of the British Ecological Society*, 9, p. 11.

— MORRIS, M. G., SHEAIL, J., WARD, L. K., WELLS, D. A., and WELLS, T. C. E., 1974, *Grassland Ecology and Wildlife Management* (Chapman and Hall. London).

— and WATT, A. S. (eds), 1971, *The Scientific Management of Animal and Plant Communities for Conservation* (Blackwell. Oxford).

DUNBAR, M. J., 1972, 'The ecosystem as a unit of natural selection', *Transactions of the Connecticut Academy of Arts and Sciences*, 44, pp. 111–30.

— 1973, 'Stability and fragility in Arctic ecosystems', *Arctic*, 26, pp. 179–85.

DYER, K. F., 1968, 'Evolution observed – some examples of evolution occurring in historical times', *Journal of Biological Education*, 2, pp. 317–38.

EATON, J. S., LIKENS, G. E., and BORMANN, F. H., 1973, 'Throughfall and stemflow chemistry in a northern hardwood forest', *Journal of Ecology*, 61, pp. 495–508.

ECOLOGIST, 1972, *A Blueprint for Survival* (Penguin. Harmondsworth).

EDEN, M. J., 1974, 'The origin and status of savanna and grassland in Southern Papua', *Transactions of the Institute of British Geographers*, 63, pp. 97–110.

EDINGTON, J. M., and EDINGTON, M. A., 1977, *Ecology and Environmental Planning* (Chapman and Hall. London).

Edward, N., 1977, 'Scourges on the trees', *Country Life*, 162, pp. 1541–2.
Edwards, K. C., 1964, 'The importance of biogeography', *Geography*, 49, pp. 85–97.
Eggeling, W. J., 1964, 'A nature reserve management plan for the Island of Rhum, Inner Hebrides', *Journal of Applied Ecology*, 1, pp. 405–19.
Ehrenfeld, D. W., 1970, *Biological Conservation* (Holt, Rinehart & Winston. New York).
Ehrlich, P., 1968, *The Population Bomb* (Ballantine. London).
— and Harriman, R. L., 1971, *How to be a Survivor–a plan to save Spaceship Earth* (Pan/Ballantine. London).
Elkington, J., 1977, 'A Broads National Park?', *New Scientist*, 74, pp. 146–7.
Elton, C., 1927, *Animal Ecology* (University of Washington Press. Washington).
Elton, C. S., 1958, *The Ecology of Invasion by Plants and Animals* (Methuen. London).
Engelhardt, W. V., Goguel, J., Hubbart, M. K., Prentice, J. E., Price, R. A., and Trumpy, R., 1976, 'Earth resources, time and man – a geoscience perspective', *Environmental Geology*, 1, pp. 193–206.
Engleman, M. D., 1961, 'The role of soil arthropods in the energetics of an old-field community', *Ecological Monographs*, 21, pp. 221–38.
Ernst, W., 1977, 'Determination of the bioconcentration potential of marine organisms – a steady state approach. I. Bioconcentration data for seven chlorinated pesticides in mussels (*Mytilus edulis*) and their relation to solubility data', *Chemosphere*, 11, pp. 731–40.
Evans, F. C., 1956, 'Ecosystem as the basic unit in ecology', *Science*, 123, pp. 1127–8.
— and Lanham, U. N., 1960, 'Distortion of the pyramid of numbers in a grassland insect community', *Science*, 131, pp. 1531–2.
Everett, R. D., 1977, 'A method of investigating the importance of wildlife to countryside visitors', *Environmental Conservation*, 4, pp. 227–31.
Eyre, S. R., 1964a, 'The integration of geography through soil and vegetation studies', *Geography*, 49, p. 111.
— 1964b, 'Determinism and the ecological approach to geography', *Geography*, 49, pp. 369–76.
— 1968, *Vegetation and Soils* (Arnold. London).
— 1978, *The Real Wealth of Nations* (Arnold. London).
Fairchild, W. B., 1949, 'Renewable resources – a world dilemma', *Geographical Review*, 39, pp. 86–98.
Fenton, A. F., 1960, 'Lichens as indicators of atmospheric pollution', *The Irish Naturalists' Journal*, 13, pp. 153–9.
— 1964, 'Atmospheric pollution of Belfast, and its relationship to the lichen flora', *The Irish Naturalists' Journal*, 14, pp. 237–45.
Ferry, B. W., Baddeley, M. S., and Hawksworth, D. L. (eds.), 1973, *Air Pollution and Lichens* (Athlone Press. London).
Fischer, A. G., 1960, 'Latitudinal variations in organic diversity', *Evolution*, 14, pp. 64–81.
Fisher, S. G., and Likens, G. E., 1972, 'Stream ecosystem – organic energy budget', *Bioscience*, 22, pp. 33–5.
— 1973, 'Energy flow in Bear Brook, New Hampshire; an integrative approach to stream ecosystem metabolism', *Ecological Monographs*, 43, pp. 421–39.
Foley, G., 1976, *The Energy Question* (Penguin. Harmondsworth).
Forrester, M. J., 1974, 'Rebirth of a wilderness', *Sierra Club Bulletin*, 59, pp. 4–8.
Fosberg, F. R., 1976, 'Geography, ecology and biogeography', *Annals of the Association of American Geographers*, 66, pp. 117–28.
Fournier, F., 1960, *Climat et erosion; la relation entre l'erosion du sol par l'eau et les precipitations atmospheriques* (P.U.F. Paris).
Fraser Darling, F., 1963, 'The unity of ecology', *British Association for the Advancement of Science*, 20, pp. 297–306.
— 1967, 'A wider environment of ecology and conservation', *Daedalus*, 96, pp. 1003–19.
Fridriksson, S., 1968, 'Life arrives on Surtsey', *New Scientist*, 37, pp. 684–7.
— 1975, *Surtsey – Evolution of Life on a Volcanic Island* (Butterworths. London).

FRIEDERICHS, K., 1958, 'A definition of ecology and some thoughts about basic concepts', *Ecology*, 39, pp. 154–9.

FRIENDS OF THE EARTH, 1978, *The Whale Manual 1978* (Friends of the Earth. London).

FUTUYMA, D. J., 1973, 'Community structure and stability in constant environments', *The American Naturalist*, 107, pp. 443–6.

GANE, M., 1976, 'Nature conservation in relation to a national forest policy', *Forestry*, 49, pp. 91–8.

GARY, R., 1958, *The Roots of Heaven* (Simon & Schuster. New York).

GATES, D. M., 1965, 'Energy, plants and ecology', *Ecology*, 46, pp. 1–13.

— 1968, 'Towards understanding ecosystems', *Advances in Ecological Research*, 5, pp. 1–35.

— 1972, 'The flow of energy in the Biosphere', *Scientific American*, 225, pp. 88–100.

GAUCH, H. G., and WHITTAKER, R. H., 1972, 'Coencline simulation', *Ecology*, 53, pp. 446–51.

GAUSE, G. F., 1934, *The Struggle for Existence* (Baltimore).

GEHLBACH, F. R., 1975, 'Investigation, evaluation and priority ranking of natural areas', *Biological Conservation*, 8, pp. 79–88.

GEORGE, M., 1976, 'Land use and nature conservation in Broadland', *Geography*, 61, pp. 137–42.

GERSMEHL, P. J., 1976, 'An alternative biogeography', *Annals of the Association of American Geographers*, 66, pp. 223–41.

GILBERT, O. L., 1974, 'Air pollution survey by school children', *Environmental Pollution*, 6, pp. 175–80.

GIMINGHAM, C. H., 1972, *Ecology of Heathlands* (Chapman and Hall. London).

GLEASON, H. A., and CRONQUIST, A., 1964, *The Natural Geography of Plants* (Columbia University Press. New York).

GODWIN, H., 1929, 'The subclimax and deflected succession', *Journal of Ecology*, 17, pp. 144–7.

— 1956, *The History of the British Flora* (CUP. Cambridge).

GOLDSMITH, F. B., 1974a, 'An assessment of the Fosberg and Ellenberg methods of classifying vegetation for conservation purposes', *Biological Conservation*, 6, pp. 3–6.

— 1974b, 'An assessment of the nature conservation value of Majorca', *Biological Conservation*, 6, pp. 79–83.

— 1974c, 'Ecological effects of visitors in the countryside', pp. 217–31 in *Conservation in Practice*, Warren, A., and Goldsmith, F. B. (eds.) (Wiley. London).

— 1975, 'The evaluation of ecological resources in the countryside for conservation purposes', *Biological Conservation*, 8, pp. 89–96.

GOMEZ-POMPA, A., VASQUEZ-YANES, C., and GUEVARA, S., 1972, 'The tropical rain forest – a non-renewable resource', *Science*, 177, pp. 762–4.

GOOD, R., 1964, *The Geography of Flowering Plants* (Longmans. London).

GOODIER, R., and GRIMES, B. H., 1970, 'The interpretation and mapping of vegetation and other ground surface features from air photographs of mountainous areas in North Wales', *Photogrammetric Record*, 6, pp. 553–66.

GOODNIGHT, C. J., and WHITLEY, L. S., 1960, 'Oligochaetes as indicators of pollution', *Proceedings of the 15th Annual Waste Conference, Purdue University, Lafayette (Indianna)*, pp. 139–42.

GOSZ, J., LIKENS, G. E., and BORMANN, F. H., 1972, 'Nutrient content of litter fall on the Hubbard Brook Experimental Forest, New Hampshire', *Ecology*, 53, pp. 769–84.

GOULD, S. J., 1971, 'Speciation and punctuated equilibria; an alternative to phyletic gradualism', *Abstract of the 1971 Annual Meeting of the Geological Society of America*, pp. 584–5.

GRANT, S. A., 1968, 'Heather regeneration following burning – a survey', *Journal of the British Grasslands Society*, 23, pp. 26–33.

GREEN, B. H., 1972, 'Relevance of seral eutrophication and plant competition to the management of successional communities', *Biological Conservation*, 4, pp. 378–84.

GREEN, F. H. W., 1973, 'Aspects of the changing environment; some factors affecting the aquatic environment in recent years', *Journal of Environmental Management*, 1, pp. 377–91.

GREGORY, K. J., and WALLING, D. E., 1973, *Drainage Basin Form and Process* (Arnold. London).

GREIG-SMITH, P., 1964, *Quantitative Plant Ecology* (Butterworths. London).

GRIME, J. P., 1974, 'Vegetation classification by reference to strategies', *Nature*, 250, pp. 26–31.

GRINNELL, J., 1917, 'The niche relationships of the California thrasher', *Auk*, 34, pp. 427–33.
GRODZINSKA, K., 1977, 'Acidity of tree bark as a bioindicator of forest pollution in southern Poland', *Water, Air and Soil Pollution*, 8, pp. 3–7.
GROSSMAN, L., 1977, 'Man-environment relationships in anthropology and geography', *Annals of the Association of American Geographers*, 67, pp. 126–44.
HAGERSTRAND, T., 1976, 'Geography and the study of interaction between nature and society', *Geoforum*, 7, pp. 329–34.
HAGGETT, P., 1972, *Geography – A Modern Synthesis* (Harper & Row. New York).
HALFON, E., 1976, 'Relative stability of ecosystem linear models', *Ecological Modelling*, 2, pp. 279–86.
HALL, I. G., 1957, 'The ecology of disused pit heaps in England', *Journal of Ecology*, 45, pp. 689–720.
HAMILL, L., 1968, 'The process of making good decisions about the use of the environment by man', *Natural Resources Journal*, 8, pp. 279–301.
HAMMOND, A. L., 1972, 'Ecosystem analysis – biome-approach to environmental research', *Science*, 175, pp. 46–8.
HANSON, J. A., 1977, 'Towards an ecologically-based economic philosophy', *Environmental Conservation*, 4, pp. 3–10.
HARDING, G., 1968, 'Tragedy of the Commons', *Science*, 162, pp. 1243–8.
HARDING, P. T., 1975, 'Changes in the woodlands of West Cambridgeshire, with special reference to the period 1946–73', *Nature in Cambridgeshire*, 18, pp. 23–32.
HARDY, M., 1905, *Esquisse de la geographie de la vegetation des Highlands d'Ecosse* (Paris).
HARDY, M. E., 1920, *The Geography of Flowering Plants* (Clarendon. Oxford).
HARDY, P., and MATHEWS, R., 1977, 'Farmland tree survey in Norfolk', *Countryside Recreation Review*, 2, pp. 31–3.
HARPER, J. L., 1967, 'A Darwinian approach to plant ecology', *Journal of Applied Ecology*, 4, pp. 267–90.
HARRIES, J. E., 1973, 'Measurement of some hydrogen-oxygen-nitrogen compounds in the stratosphere from Concorde 002', *Nature*, 241, pp. 515–18.
HARRISON, C. M., 1969, 'The ecosystem and the community in biogeography', pp. 36–44 in *Trends in Geography – an Introductory Survey*, Cooke, R. U., and Johnson, J. H. (eds.) (Pergammon. Oxford).
— 1971, 'Recent approaches to the description and analysis of vegetation', *Transactions of the Institute of British Geographers*, 52, pp. 113–27.
— 1974, 'The ecology and conservation of British lowland heaths', pp. 117–30 in *Conservation in Practice*, Warren, A., and Goldsmith, F. B. (eds.) (Wiley. London).
HARRY, J., GALE, R., and HENDEE, J., 1969, 'Conservation – an upper-middle class movement', *Journal of Leisure Research*, 3, pp. 246–54.
HARVEY, B., and HALLETT, J. D., 1977, *Environment and Society – an introductory analysis* (Macmillan. London).
HARVEY, G., 1977, 'The Somerset Levels – a test case', *New Scientist*, 76, pp. 504–6.
HASSELROT, T. B., 1975, 'Bioassay methods of the National Swedish Environmental Protection Board', *Journal of the Water Pollution Control Federation*, 47, p. 851.
HAWES, R. A., and HUDSON, R. J., 1976, 'A method of regional landscape evaluation', *Journal of Soil and Water Conservation*, 31, pp. 209–11.
HAWKEY, P., 1976, 'The Farne Islands – a National Trust nature reserve', *Roebuck*, 14, pp. 27–32.
HAWKSWORTH, D. L., 1974a, 'Lichens and indicators of environmental change', *Environment and Change*, February, pp. 381–6.
— 1975b, *The Changing Flora and Fauna of Britain* (Academic Press. London).
— and ROSE, F., 1976, *Lichens as Pollution Monitors* (Arnold. London).
HELLIWELL, D. R., 1969, 'Valuation of wildlife resources', *Regional Studies*, 3, pp. 41–7.
— 1973, 'Priorities and values in nature conservation', *Journal of Environmental Management*, 1, pp. 85–127.

— 1974a, 'The value of vegetation for conservation. I. Four land areas in Britain', *Journal of Environmental Management*, 2, pp. 51–74.
— 1974b, 'The value of vegetation for conservation. II. M1 Motorway area', *Journal of Environmental Management*, 2, pp. 75–8.
— 1975a, 'The concept of waste and the conservation of nature', *Environmental Conservation*, 2, pp. 271–3.
— 1975b, 'The distribution of woodland plant species in some Shropshire hedgerows,' *Biological Conservation*, 7, pp. 61–72.
— 1976, 'The extent and location of nature conservation areas', *Environmental Conservation*, 3, pp. 255–8.
HERNANDEZ, H., 1973, 'Natural plant recolonization of surficial disturbances – Tuktoyaktuk Peninsula Region, Northwest Territories', *Canadian Journal of Botany*, 51, pp. 2177–96.
HESJEDAL, O., 1976, 'Remote sensing techniques in vegetation mapping', *Norsk Geografisk Tidsskrift*, 30, pp. 57–61.
HEWITT, K., and HARE, F. K., 1973, *Man and Environment; conceptual frameworks* (Commission on College Geography Resource Paper 20).
HILL, A. C., 1971, 'Vegetation – a sink for atmospheric pollutants', *Journal of Air Pollution Control Association*, 21, pp. 341–6.
HILL, A. R., 1975a, 'Biogeography as a sub-field of geography', *Area*, 7, pp. 156–60.
— 1975b, 'Ecosystem stability in relation to stresses caused by human activities', *Canadian Geographer*, 19, pp. 206–20.
HODSON, H. V., 1972, *The Diseconomies of Growth* (Pan/Ballantine. London).
HOGARTH, P. J., 1976, 'Ecological aspects of dragons', *Bulletin of the British Ecological Society*, 7, pp. 2–5.
HOLDGATE, M. W. (ed.), 1970a, *Antarctic Ecology*, Volume 1 (Academic Press. London).
— (ed.), 1970b, *Antarctic Ecology*, Volume 2 (Academic Press. London).
HOLDREN, J. P., and EHRLICH, P. R., 1972, 'One-dimensional ecology revisited', *Science and Public Affairs*, 27.
HOLLIMAN, J., 1974, *Consumer's Guide to the Protection of the Environment* (Pan/Ballantine. London).
HOLLING, C. S., and CHAMBERS, A. D., 1973, 'Resource Science – the nurture of an infant', *Bioscience*, 23, pp. 13–20.
HOOPER, M. D., 1971, 'The size and surroundings of nature reserves', in *The Scientific Management of Animal and Plant Communities for Conservation*, E. Duffey and A. S. Watt (eds.) (Blackwell. Oxford), pp. 552–62.
HOPKINS, B., 1955, 'The species-area relations of plant communities', *Journal of Ecology*, 43, pp. 409–26.
HORN, H. S., 1974, 'The ecology of secondary succession', *Annual Review of Ecology and Systematics*, 5, pp. 23–37.
HOWARD, J. A., 1970, *Aerial Photo-Ecology* (Faber & Faber. London).
HUGHES, M. K., 1974, 'The Urban Ecosystem', *Biologist*, 21, pp. 117–27.
HUNTINGDON, E., 1915, *Civilization and Climate* (Yale University Press. New Haven).
HURD, L. E., MELLINGER, M. V., WOLF, L. L. and MCNAUGHTON, S. J., 1971, 'Stability and diversity at three trophic levels in terrestrial successional ecosystems', *Science*, 173, pp. 1134–6.
HUTCHINS, L. W., 1947, 'The bases for temperature zonation in geographical distribution', *Ecological Monographs*, 17, pp. 325–35.
HUTCHINSON, C., 1970, 'People and Pollution – the challenge to planning', *Journal of the Society for Long Range Planning*, 2, pp. 2–7.
HUTCHINSON, G. E., 1970, 'The Biosphere', *Scientific American*, 223, pp. 45–53.
HUTCHINSON, J., 1974, 'Land restoration in Britain – by nature and by man', *Environmental Conservation*, 1, pp. 37–41.
HUTCHISON, S. B., 1969, 'Bringing resource conservation into the main stream of American thought', *Natural Resources Journal*, 9, pp. 518–36.
HUTTON, J., 1795, *Theory of the Earth* (Creech. Edinburgh).

ICHIKAWA, R., 1961, 'On the concentration factors of some important radionuclides in the marine food organisms', *Bulletin of the Japanese Society of Scientific Fisheries*, 27, pp. 66–74.

INTERNATIONAL ACADEMY AT SANTA BARBARA, 1976, *Environmental Periodicals Bibliography*, 5, 159 pp.

JARRETT, H. (ed.), 1966, *Environmental Quality in a Growing Economy* (Johns Hopkins Press. Baltimore).

JARVIS, P., 1973, 'North American plants and horticultural innovation in England, 1550–1700', *Geographical Review*, 63, pp. 477–99.

JEFFERS, J. N. R., 1973, 'Systems modelling and analysis in resource management', *Journal of Environmental Management*, 1, pp. 13–28.

— 1978, *An Introduction to Systems Analysis – with Ecological Applications* (Arnold. London).

JENNY, H., 1941, *Factors of soil formation* (McGraw Hill. New York).

— 1961, 'Derivation of state factor equations of soils and ecosystems', *Proceedings of the Soil Science Society of America*, 25, pp. 385–8.

JOHNSON, L. C., 1976, 'Factors affecting the establishment and distribution of Corsican Pine natural regeneration at Holkham National Nature Reserve', *Quarterly Journal of Forestry*, 70, pp. 95–102.

JOHNSON, R., 1975, 'Impact Assessment is on the way', *New Scientist*, 68, pp. 323–5.

JONSSON, E., DEANE, M., and SANDERS, G., 1975, 'Community reactions to odors from pulp mills – a pilot study in Eureka, California', *Environmental Research*, 10, pp. 249–70.

JORDAN, B., and ORMROD, S., 1978, 'The last great wild beast show', *New Scientist*, 77, pp. 736–8.

JUDSON, S., 1968, 'Erosion of the land, or what's happening to our continents?', *American Scientist*, 56, pp. 356–74.

KAMINSKI, H., 1976, 'Study for a European surveillance system for the determination of air and water pollution by means of environmental and earth research satellites', *International Journal of Ecology and Environmental Science*, 2, pp. 129–57.

KEAY, R. W. J., 1974, 'Changes in African vegetation', *Environment and Change*, 2, pp. 387–94.

KELCEY, J. G., 1975, 'Industrial development and wildlife conservation', *Environmental Conservation*, 2, pp. 99–108.

— 1976, 'Aspects of practical ecology', *Bulletin of the British Ecological Society*, 7, pp. 2–7.

KELLMAN, M. C., 1975, *Plant Geography* (Methuen. London).

KERCHER, J. R., and SHUGART, H. H., 1975, 'Trophic structure, effective trophic position and connectivity in food webs', *American Naturalist*, 109, pp. 191–206.

KILBURN, P. D., 1976, 'Environmental implications of oil shale development', *Environmental Conservation*, 3, pp. 101–16.

KLAGES, K. H. W., 1942, *Ecological Crop Geography* (Macmillan. New York).

KLOTZ, J. W., 1972, *Ecology Crisis – God's Creation and Man's Pollution* (Concordia Press. New York).

KNEESE, A. V., 1977, *Economics and the Environment* (Penguin. Harmondsworth).

KRUSE, H., 1974, 'Development and environment', *American Behavioural Science*, 17, pp. 676–89.

KUCHLER, A. W., 1967, *Vegetation Mapping* (Ronald Press. New York).

LAMB, H. H., 1972, *The Changing Climate; selected papers* (Methuen. London).

LAMBERT, J. M., and WILLIAMS, W. T., 1962, 'Multi-variate methods in plant ecology, IV. Nodal Analysis', *Journal of Ecology*, 50, pp. 775–802.

LANDNER, L., NILSSON, K., and ROSENBERG, R., 1977, 'Assessment of industrial pollution by means of benthic macrofauna surveys along the Swedish Baltic coast', *Vatten*, 3, pp. 324–79.

LANDSBERG, H. E., 1970, 'Man-made climatic changes', *Science*, 170, pp. 1265–74.

LANGBEIN, W. B., and SCHUMM, S. A., 1958, 'Yield of sediment in relation to mean annual precipitation', *Transactions of the American Geophysical Union*, 39, pp. 1076–84.

LAVE, L. B., and SESKIN, E. P., 1970, 'Air pollution and human health', *Science*, 169, pp. 723–33.

LEE, N., and WOOD, C., 1972, 'Planning and pollution', *The Planner*, 58, pp. 153–8.

LEOPOLD, L. B., CLARKE, F. E., HANSHAW, B. B., and BALSLEY, J. R., 1971, 'A procedure for evaluating environmental impact', *United States Geological Survey, Circular 645*.

LEWIS, G. F., 1904, 'A geographical distribution of vegetation in the basins of the Rivers Eden, Tees, Wear and Tyne', *Geographical Journal*, 23, pp. 313–31.

LEWIS, W. M., 1974, 'Primary production in the plankton community of a tropical lake', *Ecological Monographs*, 44, pp. 377–409.

LIDDLE, M. J., 1973, 'The effects of trampling and vehicles on natural vegetation' (Ph.D. thesis, University College of North Wales, Bangor).

— 1975, 'A selective review of the ecological effects of human trampling on natural ecosystems', *Biological Conservation*, 7, pp. 17–36.

LIETH, H., and WHITTAKER, R. H., 1975, *Primary Productivity of the Biosphere* (Springer-Verlag. New York).

LINDEMAN, R. L., 1942, 'Trophic-dynamic aspects of ecology', *Ecology*, 23, pp. 399–418.

LINTON, D. L., 1957, 'Geography and the social revolution', *Geography*, 42, pp. 13–24.

LOUCKS, O. L., 1970, 'Evolution of diversity, efficiency and community stability', *American Zoologist*, 10, pp. 17–25.

LOUSLEY, J. W. (ed.), 1953, *The Changing Flora of Britain* (Botanical Society of the British Isles).

LOVELOCK, J. E., MAGGS, R. J., and WADE, R. J., 1973, 'Halogenated hydrocarbons in and over the Atlantic', *Nature*, 241, pp. 194–6.

LOWE, P., and WORBOYS, M., 1976, 'The ecology of ecology', *Nature*, 262, pp.432–3.

LUCAS, G. L., and SYNGE, A. H. M., 1977, 'The I.U.C.N. Threatened Plants Committee and its work throughout the world', *Environmental Conservation*, 4, pp. 179–87.

LUNDE, G., and BJØRSETH, A., 1977, 'Human blood samples as indicators of occupational exposure to persistent chlorinated hydrocarbons', *Science of the Total Environment*, 8, pp. 241–6.

LVOVICH, M. I., 1969, *Water Resources of the Future* (Progreschenie. Moscow).

MABEY, R., 1974, *The Pollution Handbook; the ACE/Sunday Times Clean Air and Water Surveys* (Penguin. Harmondsworth).

MACARTHUR, R. H., 1955, 'Fluctuations in animal populations and a measure of community stability', *Ecology*, 36, pp. 533–6.

MCCARL, B., RAPHAEL, D., and STAFFORD, E., 1975, 'The impact of man on the world nitrogen cycle', *Journal of Environmental Management*, 3, pp. 7–19.

MCGEE, W. J., 1909, 'The conservation of natural resources', *Proceedings of the Mississippi Valley Historical Association*, 3, pp. 365–79.

MACHTA, L., 1971, *The role of the oceans and the biosphere in the Carbon Dioxide cycle*, Nobel Symposium 20 (Gothenberg. Sweden).

MACINKO, G., 1973, 'Man and the environment; a sampling of the literature', *Geographical Review*, 63, pp. 378–91.

MCNAUGHTON, S. J., and WOLF, L. L., 1970, 'Dominance and niche in ecological systems', *Science*, 167, pp. 131–9.

MCVEAN, D. N., and RATCLIFFE, D. A., 1962, 'Plant communities of the Scottish Highlands', *Nature Conservancy Monograph Number 1.*

MADDOX, J., 1972, *The Doomsday Syndrome* (McGraw Hill. New York).

MAITLAND, P. S., 1969, 'A preliminary account of the mapping of the distribution of freshwater fish in the British Isles', *Journal of Fish Biology*, 1, pp. 45–58.

MAJOR, J., 1958, 'Plant ecology as a branch of botany', *Ecology*, 39, pp. 352–63.

MALONE, T. F., 1976, 'The role of scientists in achieving a better environment', *Environmental Conservation*, 3, pp. 83–9.

MANN, K. H., 1969, 'The dynamics of aquatic ecosystems', *Advances in Ecological Research*, 6, pp. 1–81.

MANNERS, G., 1969, 'New resource evaluations', in *Trends in Geography*, Cooke, R. U., and Johnson, J. H. (eds.), pp. 153–63 (Pergammon. Oxford).

MANSHARD, W., 1975, 'Geography and environmental science', *Area*, 7, pp. 147–55.

MARGALEF, R., 1968a, 'On certain unifying principles in ecology', *American Naturalist*, 897, pp. 357–74.

— 1968b, *Perspectives in Ecological Theory* (University of Chicago Press. Chicago).

MARKS, P. L., and BORMANN, F. H., 1972, 'Revegetation following forest cutting – mechanisms for return to steady state nutrient cycling', *Science*, 176, pp. 914–15.

MARQUIS, S., 1968, 'Ecosystems, societies and cities', *American Behavioural Scientist*, 12, pp. 11–15.

MARSH, G. P., 1864, *Man and Nature; or Physical Geography as modified by Human Action* (New York).

MARTIN, P. S., 1967, 'Pleistocene overkill', *Natural History*, 76, pp. 32–8.

MASSACHUSETTS INSTITUTE OF TECHNOLOGY, 1970, *Study of Critical Environmental Problems* (MIT. Massachusetts).

— 1971, *Inadvertent Climate Modification* (MIT. Massachusetts).

MATHER, A. S., 1974, 'Areal variations in land-use productivities in the Northern Highlands', *Scottish Geographical Magazine*, 90, pp. 153–67.

MATTHEWS, J. R., 1939, 'The ecological approach to land utilization', *Scottish Forestry Journal*, 53, p. 28.

MATSON, E. A., HORNER, S. G., and BUCK, J. D., 1978, 'Pollution indicators and other microorganisms in river sediment', *Journal of the Water Pollution Control Federation*, 50, pp. 13–19.

MATTERN, H., 1966, 'The growth of landscape consciousness', *Landscape*, 16, pp. 14–20.

MEAD, W. R., 1969, 'The course of geographical knowledge', in *Trends in Geography*, Cooke, R. U., and Johnson, J. H. (eds.), pp. 3–10 (Pergammon. Oxford).

MEADOWS, D. H., MEADOWS, D. L., RANDERS, J., and BEHRENS, W. W., 1972, *The Limits to Growth* (Earth Island Press. London).

MEANS, R. L., 1969, 'The new conservation', *Natural History*, 78, pp. 16–25.

MEENTMEYER, V., and ELTON, W., 1977, 'The potential implementation of biogeochemical cycles in biogeography', *Professional Geographer*, 29, pp. 266–71.

MEETHAM, A. R., 1950, 'Natural removal of pollution from the atmosphere', *Quarterly Journal of the Royal Meteorological Society*, 76, pp. 359–71.

MELLANBY, K., 1967, *Pesticides and Pollution* (Collins. Glasgow).

MERRIAM, D. F., 1974, 'Resource and environmental data analysis', *United States Geological Survey Professional Paper* 921, pp. 37–45.

MESSENGER, K. G., 1968, 'A railway flora of Rutland', *Proceedings of the Royal Botanical Society of the British Isles*, 7, pp. 325–44.

MILLER, G. R., and WATSON, A., 1974, 'Heather moorland – a man-made ecosystem', in *Conservation in Practice*, Warren, A., and Goldsmith, F. B. (eds.), pp. 145–66 (Wiley. London).

MONCRIEF, L. M., 1970, 'The cultural basis for our environmental crisis', *Science*, 170, pp. 508–12.

MOORE, J. J., FITZSIMONS, P., LAMBE, E., and WHITE, J., 1970, 'A comparison and evaluation of some phytosociological techniques', *Vegetatio*, 20, pp. 1–20.

MOORE, N. W., 1962, 'The heaths of Dorset and their conservation', *Journal of Ecology*, 50, pp. 369–91.

— 1969, 'Experience with pesticides and the theory of conservation', *Biological Conservation*, 1, pp. 201–7.

— 1977a, 'Agriculture and nature conservation', *Bulletin of the British Ecological Society*, 8, pp. 2–4.

— 1977b, 'Conservation and agriculture', *Naturopa*, 27, pp. 19–23.

MOORE, P. D., and BELLAMY, D. J., 1974, *Peatlands* (Elek Science. London).

MORGAN, W. B., and MOSS, R. P., 1965, 'Geography and ecology; the concept of the community and its relationship to environment', *Annals of the Association of American Geographers*, 55, pp. 339–50.

MORRIS, M. G., and PERRING, F. H. (eds.), 1974, *The British Oak – its history and natural history* (Botanical Society of the British Isles. London).

MOSS, B., 1977, 'Conservation problems in the Norfolk Broads and rivers of East Anglia, England; phytoplankton, boats and the causes of turbidity', *Biological Conservation*, 12, pp. 95–114.

MOSS, C. E., RANKIN, W. M., and TANSLEY, A. G., 1910, 'The woodlands of England', *New Phytologist*, 9, pp. 113–49.

MOSS, M. R., 1974, 'The science of environment – towards a conceptual structure', *Pacific Viewpoint*, 15, pp. 3–18.

— 1975, 'Spatial patterns of precipitation reaction', *Environmental Pollution*, 8, pp. 301–15.
— 1976, 'Biogeochemical cycles as integrative and spatial models for the study of environmental pollution (the example of the sulphur cycle)', *International Journal of Environmental Studies*, 9, pp. 209–16.
MOSS, R. P., and MORGAN, W. B., 1967, 'The concept of community; some applications in geographical research', *Transactions of the Institute of British Geographers*, 41, pp. 21–34.
MOYAL, M., 1976, 'Endangered and threatened plant species', *Scottish Forestry*, 30, pp. 182–5.
MULLER, P., 1974, *Aspects of Zoogeography* (Junk (Publishers). The Hague).
MUNN, R. E., PHILLIPS, M. L., and SANDERSON, H. P., 1977, 'Environmental effects of air pollution – implications for air quality criteria, air quality standards and emission standards', *Science of the Total Environment*, 8, pp. 53–67.
MUNTON, D., and BRADY, L., 1970, *American Public Opinion and Environmental Pollution* (Behavioural Science Laboratory, Ohio State University. Columbus).
MYSLINSKI, E., and GINSBURG, W., 1977, 'Macroinvertebrates as indicators of pollution', *Journal of the American Water Works Association*, 69, pp. 538–44.
NAMKOONG, G., and ROBERDS, J. H., 1974, 'Extinction probabilities and the changing age structure of Redwood Forests', *American Naturalist*, 108, pp. 355–68.
NATURAL ENVIRONMENT RESEARCH COUNCIL, 1976, *Research on Pollution of the Natural Environment* (Natural Environment Research Council Publications Series B, number 15).
— 1977, *Amenity Grassland – the needs for research* (Natural Environment Research Council Publications Series C, number 19).
— 1978, *European Wetlands Campaign 1976–7; Report for the UK* (Nature Conservancy Council. London).
— /Society for the Promotion of Nature Conservation, 1977, *Otters 1977* (Nature Conservancy Council/Society for the Promotion of Nature Conservation. London).
NATURE CONSERVANCY COUNCIL, 1970, *The Nature Conservancy Research in Scotland. Report for 1968–70* (Nature Conservancy Council. Edinburgh).
— 1974, *Tree Planting and Wildlife Conservation* (Nature Conservancy Council. London).
— 1977a, *Nature Conservation and Agriculture* (Nature Conservancy Council. London).
— 1977b, *The Somerset Wetlands Project; a consultation paper* (Nature Conservancy Council, South West Region. Taunton).
NEWBIGIN, M., 1936, *Plant and Animal Geography* (Methuen. London).
NEWBOULD, P. J., 1960, 'The ecology of Cranesmoor, a New Forest valley bog. 1. The present vegetation', *Journal of Ecology*, 48, pp. 361–83.
— 1964, 'Production ecology and the International Biological Programme', *Geography*, 49, pp. 98–104.
NEWBY, H., BELL, C., SAUNDERS, P., and ROSE, D., 1977, 'Farmers' attitudes to conservation', *Countryside Recreation Review*, 2, pp. 23–30.
NICHOLSON, M., 1969, *The Environmental Revolution – a guide for the new masters of the world* (Penguin. Harmondsworth).
NOY-MEIR, A., 1973, 'Desert ecosystems; environment and producers', *Annual Review of Ecology and Systematics*, 4, pp. 25–52.
O'CONNOR, F. B., 1964, 'Energy flow and population metabolism', *Science Progress*, 52, pp. 406–14.
ODUM, E. P., 1959, *Fundamentals of Ecology* (Sanders. Philadelphia).
— 1962, 'Relationships between structure and function in the ecosystem', *Japanese Journal of Ecology*, 12, pp. 108–18.
— 1964, 'The new ecology', *Bioscience*, 14, pp. 14–16.
— 1968, 'Energy flow in ecosystems – a historical review', *American Zoology*, 8, pp. 11–18.
— 1969, 'The strategy of ecosystem development', *Science*, 164, pp. 262–70.
— 1974, 'Ecosystem theory in relation to man', in *Man's Natural Environment – a systems approach*, Russwurm, L., and Somerville, E. (eds.), pp. 31–42 (Duxbury Press. Massachusetts).
ODUM, H. T., 1957, 'Trophic structure and productivity of Silver Springs, Florida', *Ecological Monographs*, 27, pp. 55–112.

— 1960, 'Ecological potential and analogue circuits for the ecosystem', *American Scientist*, 48, pp. 1–8.

— and Odum, E. P., 1955, 'Trophic structure and productivity of a windward coral reef community on Eniwetok Atoll', *Ecological Monographs*, 25, pp. 291–320.

Olschowy, G., 1975, 'Ecological landscape inventories and evaluation', *Landscape Planning*, 2, pp. 37–44.

Olson, J. S., 1958, 'Rates of succession and soil changes on southern Lake Michigan sand dunes', *Botanical Gazette*, 119, pp. 125–70.

— 1963, 'Energy storage and the balance of producers and decomposers in ecological systems', *Ecology*, 44, pp. 322–31.

O'Riordan, T., 1971, *Perspectives on Resource Management* (Pion. London).

— 1976, *Environmentalism* (Pion. London).

Overman, M., 1968, *Water – solutions to a problem of supply and demand* (Aldus. London).

Ovington, J. D., 1961, 'Some aspects of energy flow in plantations of *Pinus sylvestris*', *Annals of Botany*, 25, pp. 12–20.

— 1964, 'The ecological basis of the management of woodland nature reserves in Great Britain', *Journal of Ecology*, 52, pp. 29–37.

Packard, V., 1960, *The Waste Makers* (Penguin. Harmondsworth).

Paine, R. T., 1966, 'Food web complexity and species diversity', *American Naturalist*, 100, pp. 65–76.

— 1969, 'A note on trophic complexity and community stability', *American Naturalist*, 103, pp. 91–3.

Palmer, C. J., Robinson, M. E., and Thomas, R. W., 1977, 'The countryside image – an investigation of structure and meaning', *Environment and Planning*, Series A, 9, pp. 739–49.

Paludan, C. T., 1976, 'Land use surveys based on remote sensing from high altitudes', *Geographica Helvetica*, 31, pp. 17–24.

Papadopoulou, C., and Kanias, G. D., 1977, 'Tunicate species as marine pollution indicators', *Marine Pollution Bulletin*, 8, pp. 229–31.

Park, C. C., 1977, *History of the Conservation Movement in Britain* (Conservation Trust. London).

Paterson, S. S., 1956, *The forest area of the world and its potential productivity* (Geography Department of the Royal University of Goteborg. Sweden).

Patrick, R., 1972, 'Benthic communities in streams', *Transactions of the Connecticut Academy of Arts and Sciences*, 44, pp. 269–84.

Patterson, W. C., 1976, *Nuclear Power* (Penguin. Harmondsworth).

Pears, N. V., 1968, 'Some recent trends in classification and description of vegetation', *Geografiska Annaler*, 50A, pp. 162–72.

— 1977, *Basic Biogeography* (Longmans. London).

Pearsall, W. H., 1964, 'The development of ecology in Britain', *Journal of Ecology*, 52, pp. 1–12.

Peirson, D. H., Cawse, P. A., Salmon, L., and Cambray, R. S., 1973, 'Trace elements in the atmospheric environment', *Nature*, 241, pp. 252–6.

Pennington, W., 1969, *The History of British Vegetation* (English University Press. London).

Perera, N., 1975, 'A physiognomic vegetation map of Sri Lanka', *Journal of Biogeography*, 2, pp. 185–203.

Perkins, R. J., 1975, 'The environment from a systems viewpoint', *International Journal of Environmental Studies*, 8, pp. 59–63.

Perring, F. H., 1967, 'Hedges are vital – a botanist looks at our roadsides', *Journal of the Institute of Highway Engineers*, 14, pp. 13–16.

— 1968, *Critical supplement to the Atlas of the British Flora* (Botanical Society of the British Isles. London).

— (ed.), 1970, *Flora of a Changing Britain* (Botanical Society of the British Isles. London).

— and Farrell, L., 1977, *British Red Data Books, 1. Vascular Plants* (Society for the Promotion of Nature Conservation. London).

— and Walters, S. M. (eds.), 1962, *Atlas of the British Flora* (Botanical Society of the British Isles. London).

PETERKEN, G. F., 1974, 'A method for assessing woodland flora for conservation value using indicator species', *Biological Conservation*, 6, pp. 239–45.

— 1977, 'Nature conservation and visual amenity in British woodlands', *Arboricultural Journal*, 3, pp. 96–9.

— and HARDING, P. T., 1974, 'Recent changes in the conservation value of woodlands in Rockingham Forest', *Forestry*, 47, pp. 109–28.

— and TUBBS, C. R., 1965, 'Woodland regeneration in the New Forest, Hampshire, since 1650', *Journal of Applied Ecology*, 2, pp. 159–70.

PHILLIPSON, J., 1966, *Ecological Energetics* (St Martins Press. London).

PIANKA, E. R., 1966, 'Latitudinal gradients in species diversity – a review of concepts', *American Naturalist*, 100, pp. 33–46.

PICKERING, M. E., 1977, 'A new environmental planning tool', *Surveyor*, 149, p. 13.

PICKETT, S. T. A., and THOMPSON, J. N., 1978, 'Patch dynamics and the design of nature reserves', *Biological Conservation*, 13, pp. 27–37.

PIELOU, E. C., 1975, *Ecological Diversity* (Wiley. New York).

PINCHOT, G., 1936, 'How conservation began in the United States', *Agricultural History*, 11, pp. 255–65.

POELMANS-KIRSCHEN, J., 1974, 'Consumption and the environment', *Naturopa*, 21, pp. 23–5.

POLICARPOV, G. G., 1966, *Radioecology of Aquatic Organisms* (North-Holland. Amsterdam).

POLUNIN, N., 1960, *Introduction to Plant Geography and some related sciences* (Longmans. London).

— 1969, 'Conservational significance of Botanical Gardens', *Biological Conservation*, 1, pp. 104–5.

POORE, M. E. D., 1955, 'The use of phytosociological methods in ecological investigations, 1. The Braun-Blanquet method', *Journal of Ecology*, 43, pp. 226–44.

— and MCVEAN, D. N., 1957, 'A new approach to Scottish mountain vegetation', *Journal of Ecology*, 45, pp. 401–39.

PORTEOUS, A., ATTENBOROUGH, K., and POLLITT, C., 1977, *Pollution – the professionals and the public* (Open University Press. Milton Keynes).

PRESTON, A., 1977, 'The study and control of environmental radioactivity and its relevance to the control of other environmental contaminants', *Atomic Energy Review*, 15, pp. 371–405.

PRICE, D. R. H., 1978, 'Fish as indicators of water quality', *Water Pollution Control*, 77, pp. 285–93.

RANDALL, R., 1978, *Theories and Techniques in Vegetation Analysis* (OUP. Oxford).

RANWELL, D. S., 1972, *Ecology of Salt Marshes and Sand Dunes* (Chapman and Hall. London).

RATCLIFFE, D. A., 1971, 'Criteria for the selection of nature reserves', *Advancement of Science*, 27, pp. 294–6.

— 1974, 'Ecological effects of mineral exploitation in the United Kingdom, and their significance to nature conservation', *Proceedings of the Royal Society of London*, 339A, pp. 355–72.

— 1976, 'Thoughts towards a philosophy of nature conservation', *Biological Conservation*, 9, pp. 45–53.

— 1977a, 'Nature Conservation – aims, methods and achievements', *Proceedings of the Royal Society of London*, 197B, pp. 11–29.

— 1977b, *A Nature Conservation Review* (Nature Conservancy Council. London).

— 1977c, 'The conservation of important wildlife areas in Great Britain', *Bulletin of the British Ecological Society*, 8, pp. 5–11.

RAUNKIER, C., 1934, *The Life Forms of Plants and Statistical Plant Geography* (Clarendon Press. Oxford).

RAWSON, D. S., 1956, 'Algal indicators of trophic lake types', *Limnology and Oceanography*, 1, pp. 18–25.

RAYMONT, J. E. G., 1966, 'The production of marine plankton', *Advances in Ecological Research*, 3, pp. 117–205.

REES, J. A., 1977, 'The economics of environmental management', *Geography*, 62, pp. 311–24.

REGIER, H. A., and COWELL, E. B., 1972, 'Applications of ecosystem theory, succession, diversity, stability, stress and conservation', *Biological Conservation*, 4, pp. 83–8.

REGNIER, A. P., and PARK, R. W. A., 1972, 'Faecal pollution of our beaches – how serious is the situation?', *Nature*, 239, pp. 408–10.

REICHLE, D. E. (ed.), 1973, *Analysis of Temperate Forest Ecosystems* (Springer-Verlag. Berlin).
REID, K., 1969, *Nature's Network* (Aldus Books. London).
REINERS, W. A., 1972, 'Structure and energetics of three Minnesota forests', *Ecological Monographs*, 42, pp. 71–94.
RENNIE, R. J., 1955, 'The uptake of nutrients by mature forest growth', *Plant and Soil*, 8, pp. 49–95.
RIBAULT, J. P., 1978, 'Alert – a convention for nature protection', *Forum (Council of Europe)*, 1, pp. 37–8.
RICHARDS, P. W., 1952, *The Tropical Rain Forest* (CUP. Cambridge).
RILEY, D., and YOUNG, A., 1966, *World Vegetation* (CUP. Cambridge).
RILEY, J. F., 1967, *Introducing Biology* (Penguin. Harmondsworth).
RIPLEY, S. D., and BUECHNER, H. K., 1967, 'Ecosystem science as a point of synthesis', *Daedalus*, 96, pp. 1192–9.
ROBINSON, H., 1972, *Biogeography* (MacDonald and Evans. London).
ROBINSON, J., 1972, 'Industrial and agricultural pollution and the environment', *Royal Society of Health Journal*, 93, pp. 62–8.
ROBSON, A. J., 1977, 'The effect of urban structure on the concentration of pollution', *Urban Studies*, 14, pp. 89–93.
ROSENZWEIG, M. L., 1968, 'Net primary productivity of terrestrial communities – prediction from climatological data', *American Naturalist*, 102, pp. 67–74.
ROWE, J. S., 1961, 'The level-of-integration concept and ecology', *Ecology*, 42, pp. 420–7.
RUMNEY, G. R., 1970, *The Geosystem – dynamic integration of land, sea and air* (Brown. Dubuque, Iowa).
RUSSWURM, L. H., 1974, 'A systems approach to the natural environment', in *Man's Natural Environment – a systems approach,* Russwurm, L. H., and Sommerville, E. (eds.) (Duxbury. Massachusetts).
RYTHER, J. H., 1969, 'Photosynthesis and fish production in the sea', *Science*, 166, pp. 72–6.
SALISBURY, E. J., 1922, 'The soils of Blakeney Point; a study of soil reaction and succession in relation to plant covering', *Annals of Botany*, 36, p. 391.
— 1925, 'Note on the edaphic succession in some dune soils with special reference to the time factor', *Journal of Ecology*, 13, p. 322.
SAUER, J. D., 1977, 'Biogeographical theory and cultural analogies', *World Archaeology*, 8, pp. 320–31.
SCHEFFER, V. B., 1976, 'The future of wildlife management', *Wildlife Society Bulletin*, 4, pp. 51–4.
SCHIMPER, A. W. F., 1903, *Plant Geography on a Physiological Basis* (Clarendon Press. Oxford).
SCHUMACHER, E. F., 1974, *Small is Beautiful* (Sphere Books. London).
SCLATER, P. L., 1858, 'On the general geographical distribution of the class *Aves*', *Proceedings of the Linnean Society (London), Zoology*, 2, pp. 130–45.
SCOTT, D., and BILLINGS, W. D., 1964, 'Effects of environmental factors on standing crop and productivity of an alpine tundra', *Ecological Monographs*, 34, pp. 243–70.
SCOTT, P., and RINES, R., 1975, 'Naming the Loch Ness Monster', *Nature*, 258, pp. 466–8.
SEARS, W. B., 1972, *Where there is life* (Dell. New York).
SEDDON, B., 1971, *Introduction to Biogeography* (Duckworth. London).
SELBY, M. J., 1969, 'Organic analogies and ecosystems in geographical studies', *New Zealand Journal of Geography*, 47, pp. 2–10.
SELLERS, W. D., 1965, *Physical Climatology* (University of Chicago Press. Chicago).
SELMAN, P. H., 1976, 'Wildlife conservation in structure plans', *Journal of Environmental Management*, 4, pp. 149–59.
SEMPLE, E. C., 1911, *Influences of Geographic Environment* (Holt. New York).
SHARMA, G. K., 1977, 'Cuticular features as indicators of environmental pollution', *Water, Air and Soil Pollution*, 8, pp. 15–19.
SHEAIL, J., 1971, 'The time factor in understanding the ecology of wildlife', *Journal of Biological Education*, 5, pp. 157–63.
— 1976, *Nature in Trust* (Blackie. Glasgow).

SHELFORD, V. E., 1963, *The Ecology of North America* (University of Illinois Press. Urbana).

SHERLOCK, R. L., 1922, *Man as a Geological Agent* (Witherby. London).

SHIMWELL, D. W., 1971, *Description and Classification of Vegetation* (Sidgewick and Jackson. London).

SILVERBERG, R., 1973, *The Dodo, the Auk and the Oryx – Vanished and Vanishing Creatures* (Penguin. Harmondsworth).

SIMBERLOFF, D. S., and WILSON, E. O., 1969, 'Experimental zoogeography of islands; the colonization of empty islands', *Ecology*, 50, pp. 278–96.

— and WILSON, E. O., 1970, 'Experimental zoogeography of islands; a two year record of colonization', *Ecology*, 51, pp. 934–7.

SIMMONS, I. G., 1966, 'Ecology and land use', *Transactions of the Institute of British Geographers*, 38, pp. 59–72.

— 1974, *The Ecology of Natural Resources* (Arnold. London).

— and SIMMONS, C. M., 1973, 'Environmentalism and education; a context for "conservation"', *School Science Review*, 54, pp. 574–9.

SIMON, N., and GERONDET, P., 1970, *Last Survivors – the Natural History of Animals in Danger of Extinction* (Stephens. London).

SINGER, S. F. (ed.), 1975, *The Changing Global Environment* (Reidol. Dordrecht).

SLATER, F. M. and AGNEW, A. D. Q., 1977, 'Observations on a peat bog's ability to withstand increasing public pressure', *Biological Conservation*, 11, pp. 21–7.

SLOBODKIN, L. B., 1962, 'Energy in animal ecology', *Advances in Ecological Research*, 1, pp. 69–101.

SMITH, D. D., and WISCHMEIER, W. H., 1962, 'Rainfall erosion', *Advances in Agronomy*, 14, pp. 109–48.

SMITH, F. E., 1972, 'Spatial heterogeneity, stability and diversity in ecosystems', *Transactions of the Connecticut Academy of Arts and Sciences*, 4, pp. 309–35.

SMITH, R., 1900, 'Botanical survey of Scotland. 1. Edinburgh district', *Scottish Geographical Magazine*, 16, pp. 385–416.

SMITH, R. O., 1976, 'Planning and ecology in a county council', *The Planner*, 62, pp. 42–4.

SMITH, V. R., 1977, 'A qualitative description of energy flow and nutrient cycling in the Marion Island terrestrial ecosystem', *Polar Record*, 18, pp. 361–70.

SOUTHERN, H. N., and LOWE, V. P. W., 1968, 'The pattern of distribution of prey and predation in tawny owl territories', *Journal of Animal Ecology*, 37, pp. 75–97.

SPEIGHT, M. C. D., 1973, 'Ecological change and outdoor recreation', *Discussion Papers in Conservation, Number 4* (University College, London).

STAMP, L. D., 1962, *The Land of Britain – its use and misuse* (Longmans. London).

STODDART, D. R., 1965, 'Geography and the ecological approach – the ecosystem as a geographical principle and method', *Geography*, 50, pp. 242–51.

— 1967, 'Growth and structure of geography', *Transactions of the Institute of British Geographers*, 41, pp. 1–19.

— 1969, 'Climatic geomorphology – review and re-assessment', *Progress in Geography*, 1, pp. 159–222.

STONE, E. C., 1965, 'Preserving vegetation in Parks and Wilderness', *Science*, 150, pp. 1262–7.

STOTT, P. A., 1978, 'Tropical Rain Forest in recent ecological thought: the reassessment of a non-renewable resource', *Progress in Physical Geography*, 2, pp. 80–98.

STREETER, D. T., 1974, 'Ecological aspects of oak woodland conservation', in *The British Oak – its History and Natural History*, Morris, M. G., and Perring, F. H. (eds.), (Botanical Society of the British Isles. London).

SUTTON, P., 1975, *The Protection Handbook of Pollution Control* (Alan Osborne and Associates. London).

SWAN, F. R., 1970, 'Post-fire response of four plant communities in south-central New York state', *Ecology*, 51, pp. 1074–82.

SWINNERTON, G. S., 1976, 'Land quality and land use in England and Wales', *Landscape Research News*, 1, pp. 10–12.

TANSLEY, A. G., 1904, 'Problems in ecology', *New Phytologist*, 13, pp. 191–200.

— 1935, 'The use and abuse of vegetational concepts and terms', *Ecology*, 16, pp. 284–307.

— 1947, 'The early history of modern plant ecology in Britain', *Journal of Ecology*, 35, pp. 130–7.

— 1949, *The British Isles and their Vegetation* (CUP. Cambridge).

TAYLOR, B. W., 1957, 'Plant succession on recent volcanoes in Papua', *Journal of Ecology*, 45, 233–43.

TAYLOR, J. A., 1968, 'Reconnaissance vegetation surveys and maps (including a preliminary report of the Vegetation Survey of Wales, 1961–71)', in *Geography at Aberystwyth*, Bowen, E. G., Carter, H., and Taylor, J. A. (eds.), pp. 87–110 (University of Wales Press. Cardiff).

— 1974, 'The ecological basis of resource management', *Area*, 6, pp. 101–6.

TAYLOR, L. R. (ed.), 1970, *The Optimum Population for Britain* (Academic Press. London).

TEAL, J. M., 1957, 'Community metabolism in a temperate cold spring', *Ecological Monographs*, 27, pp. 283–302.

TERBORGH, J., 1970, 'Distribution on environmental gradients – theory and a preliminary interpretation of distributional pattern in the avifauna of the Cordillera Vilcabamba, Peru', *Ecology*, 52, pp. 22–40.

— 1974, 'Preservation of natural diversity – the problem of extinction prone species', *Bioscience*, 24, pp. 715–22.

THAMES WATER AUTHORITY, 1977, *Report of the Thames Migratory Fish Committee* (Thames Water Authority. London).

THOMAS, A., 1975, *The Follies of Conservation* (Stockwell. Ilfracombe).

THOMAS, A. S., 1960, 'Changes in vegetation since the advent of myxomatosis', *Journal of Ecology*, 48, pp. 287–306.

THOMAS, W. L. (ed.), 1956, *Man's Role in Changing the Face of the Earth* (University of Chicago Press. Chicago).

THOREAU, H. D. (ed.), 1960, *Walden, or Life in the Woods* (New American Library. New York).

TIVY, J., 1954, 'Reconnaissance vegetation survey of certain hill grazings in the Southern Uplands', *Scottish Geographical Magazine*, 70, pp. 21–35.

— 1971, *Biogeography – a Study of Plants in the Ecosphere* (Oliver and Boyd. Edinburgh).

TOMLINSON, R. W., 1972, 'An approach to biogeography', *South African Geographer*, 4, pp. 85–8.

TOYNBEE, A., 1972, 'The religious background of the present environmental crisis', *International Journal of Environmental Studies*, 3, pp. 141–6.

TUBBS, C. R., 1968, *The New Forest – an Ecological History* (David and Charles. Newton Abbot).

— 1974, 'Woodlands – their history and conservation', in *Conservation in Practice*, Warren, A., and Goldsmith, F. B. (eds.), pp. 131–44 (Wiley. London).

— and BLACKWOOD, J. W., 1971, 'Ecological evaluation of land for planning purposes', *Biological Conservation*, 3, pp. 169–72.

TWIGG, H. M., 1959, 'Freshwater studies in the Shropshire Union Canal', *Field Studies*, 1, pp. 116–42.

UNITED NATIONS, 1972, *Statistical Yearbook* (United Nations. New York).

UNESCO, 1970, *Use and Conservation of the Biosphere* (Natural Resources Research X, UNESCO. Paris).

UNWIN, D. J., and HOLTBY, F. E., 1974, 'Public perception of cleaner air in the Manchester area, 1970', *Cambria*, 1, pp. 43–51.

URSIC, S. J., and DENDEE, F. E., 1965, 'Sediment yields from small watersheds under various land uses and forest cover', *Proceedings of the Federal Inter-Agency Sedimentation Conference, United States Department of Agriculture Miscellaneous Publication 970*, pp. 47–52.

USHER, M. B., 1973, *Biological Management and Conservation* (Chapman and Hall. London).

— and MILLER, A. K., 1975, 'The development of a nature reserve as an area of conservational and recreational interest', *Environmental Conservation*, 2, pp. 202–4.

— TAYLOR, A. E., and DARLINGTON, D., 1970, 'A survey of visitors' reactions on two Naturalists' Trust nature reserves in Yorkshire', *Biological Conservation*, 2, pp. 285–91.

VAN DYNE, G. M. (ed.), 1969, *The Ecosystem Concept in Resource Management* (Academic Press. London).

VENRICK, E. L., BACKMAN, T. W., BARTRAM, W. C., PLATT, C. J., THORNHILL, M. S., and YATES, M. S., 1973, 'Man-made objects on the surface of the Central Pacific Ocean', *Nature*, 241, p. 271.

VISHER, S. S., 1955, 'Comparative agricultural potentials of the world's regions', *Economic Geography*, 31, pp. 82–6.

VUILLEUMIER, B. S., 1971, 'Pleistocene changes in the fauna and flora of South America', *Science*, 173, pp. 771–80.

WALLACE, A. R., 1876, *The Geographical Distribution of Animals* (Macmillan. London).

WALLWORK, K. L., 1976, *The extent and development of derelict land in England and Wales* (Royal Town Planning Institute. London).

WALTON, K., 1968, 'The unity of the physical environment', *Scottish Geographical Magazine*, 84, pp. 5–15.

WALTON, W. C., 1970, *The World of Water* (Weidenfeld and Nicolson. London).

WARD, B., and DUBOS, R., 1972, *Only One Earth – the care and maintenance of a small planet* (Penguin. Harmondsworth).

WARD, R. C., 1975, *Principles of Hydrology* (second edition). (McGraw Hill. New York).

WARD, S. D., and EVANS, D. F., 1976, 'Conservation assessment of British limestone pavements, based on floristic criteria', *Biological Conservation*, 9, pp. 217–33.

— JONES, A. D., and MANTON, M., 1972, 'The vegetation of Dartmoor', *Field Studies*, 3, pp. 505–33.

WARMING, E., 1909, *Oecology of Plants* (OUP. Oxford).

WARREN, A., and GOLDSMITH, F. B. (eds.), 1974, *Conservation in Practice* (Wiley. London).

WASSINK, E. C., 1959, 'Efficiency of light energy conversion in plant growth', *Plant Physiology*, 34, p. 356.

WATSON, L. J., and HANHAM, R. Q., 1977, 'Flower colour and environment – the case of butterfly weed in Oklahoma', *Professional Geographer*, 29, pp. 374–7.

WATTS D., 1974, *Principles of Biogeography* (McGraw Hill. New York).

WATTS, G. D., HORNBY, R., LAMBLEY, P. W., and ISMAY, J., 1975, 'An ecological review of the Yare Valley near Norwich', *Transactions of the Norfolk and Norwich Naturalists' Society*, 23, pp. 231–48.

WAY, J. M., 1970, 'Roads and the conservation of wildlife', *Journal of the Institute of Highway Engineers*, 17, pp. 5–11.

WEAVER, J. E., and CLEMENTS, F. E., 1938, *Plant Ecology* (McGraw Hill. New York).

WEEDEN, R. B., and KLEIN, D. R., 1971, 'Wildlife and oil; a survey of critical issues in Canada', *Polar Record*, 15, pp. 479–94.

WELLS, T. C. E., 1968, 'Land-use changes affecting *Pulsatilla vulgaris* in England', *Biological Conservation*, 1, pp. 37–44.

WEST, G. C., 1976, 'Environmental problems associated with Arctic development, especially in Alaska', *Environmental Conservation*, 3, pp. 218–24.

WESTERN, D., and VAN PRAET, C., 1973, 'Cyclical changes in the habitat and climate of an East African ecosystem', *Nature*, 241, pp. 104–6.

WESTHOFF, V., 1970, 'New Criteria for nature reserves', *New Scientist*, 46, pp. 108–13.

WESTMAN, W. E., 1977, 'How much are nature's services worth?', *Science*, 197, pp. 960–3.

WHITE, G. F., 1949, 'Towards an appraisal of world resources', *Geographical Review*, 39, pp. 625–39.

WHITE, L., 1967, 'The historical roots of our ecological crisis', *Science*, 155, pp. 1203–7.

WHITMORE, T. C., 1975, *Tropical Rain Forests of the Far East* (Clarendon Press. Oxford).

WHITTAKER, G. A., and MCCUEN, R. H., 1976, 'A proposed methodology for assessing the quality of wildlife habitat', *Ecological Modelling*, 2, pp. 251–72.

WHITTAKER, R. H., 1953, 'A consideration of climax theory – the climax as a population and pattern', *Ecological Monographs*, 23, pp. 41–78.

— 1967, 'Gradient analysis of vegetation', *Biological Review*, 49, pp. 207–64.

— LEVIN, S. A., and ROOT, R. B., 1973, 'Niche, habitat and ecotype', *American Naturalist*, 107, pp. 321–38.

— and LIKENS, G. E., 1975, 'The biosphere and man', in *Primary Productivity in the Biosphere*, Leith, H., and Whittaker, R. H. (eds.), pp. 305–28 (Springer-Verlag. Berlin).

WIELGOLASKI, F. E., 1975a, 'Biological indicators on pollution', *Urban Ecology*, 1, pp. 63–79.

— (ed.), 1975b, *Fennoscandian Tundra Ecosystems*, two volumes (Springer-Verlag. Berlin).

WILHM, J. H., and DORIS, T. C., 1968, 'Biological parameters for water quality criteria', *Bioscience*, 18, pp. 477–81.

WILLIAMS, N. V., and DUSSART, G. B. J., 1976, 'A field course survey of three English river systems', *Journal of Biological Education*, 10, pp. 4–14.

WILLIAMS, W. T., and LAMBERT, J. M., 1959, 'Multi-variate methods in plant ecology. i. Association analysis in plant communities', *Journal of Ecology*, 47, pp. 83–101.

— 1960, 'Multi-variate methods in plant ecology. ii. The use of an electronic digital computer for association-analysis', *Journal of Ecology*, 48, pp. 689–710.

— and LANCE, G. N., 1966, 'Multi-variate methods in plant ecology. iv. Similarity analysis and information analysis', *Journal of Ecology*, 54, pp. 427–45.

WILSON, D. S., 1976, 'Evolution on the level of communities', *Science*, 192, pp. 1358–60.

WILSON, E. O., and SIMBERLOFF, D. S., 1969, 'Experimental zoogeography of islands – defaunation and monitoring techniques', *Ecology*, 50, pp. 267–78.

WOLMAN, M. G., 1967, 'A cycle of sedimentation and erosion in urban river channels', *Geografiska Annaler*, 49A, pp. 385–95.

WOODWELL, G. M., 1970, 'Effects of pollution on the structure and physiology of ecosystems', *Science*, 168, pp. 429–33.

WRIGHT, L. W., and WANSTALL, P. J., 1977, *The vegetation of Mediterranean France – a review* (University of London, Queen Mary College, Department of Geography Occasional Papers Number 9).

WRIGHT, S., and TIDD, W. M., 1933, 'Summary of limnological investigation in Western Lake Erie in 1929 and 1930', *Transactions of the American Fisheries Society*, pp. 271–85.

WULFF, E. V., 1950, *Introduction to Historical Plant Geography* (Waltham. Massachusetts).

WYNNE-EDWARDS, V. C., 1962, *Animal Dispersion in relation to Social Behaviour* (Oliver and Boyd. Edinburgh).

— 1965, 'Self-regulating systems in populations of animals', *Science*, 147, pp. 1543–8.

ZELINSKY, W., 1966, *A Prologue to Population Geography* (Prentice Hall. Englewood Cliffs, New Jersey).

ZIMMERMAN, E. S., 1951, *World Resources and Industries* (Harper. New York).

ZOBLER, L., 1962, 'An economic-historical view of natural resource use and conservation', *Economic Geography*, 38, pp. 189–94.

Index

Items which are referred to in either Figures or Tables in the text, are included in this index *in italics*.